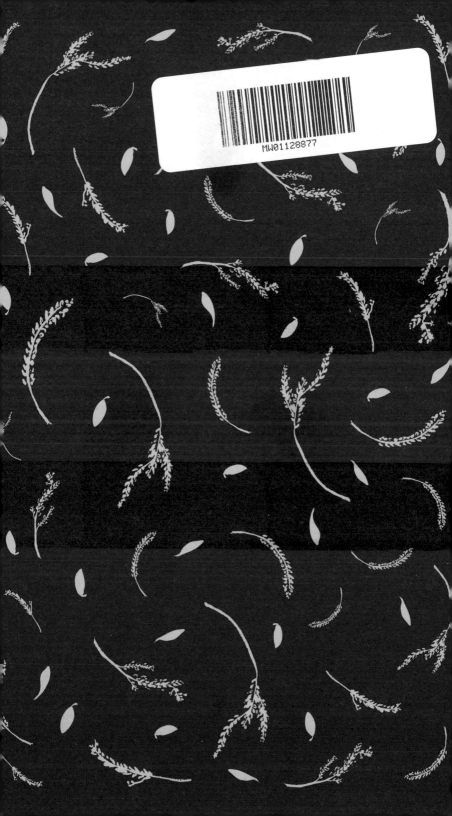

# LOST WONDERS

Tom Lathan

# LOST WONDERS

*10 tales of extinction from
the 21st century*

PICADOR

First published 2024 by Picador
an imprint of Pan Macmillan
The Smithson, 6 Briset Street, London EC1M 5NR
*EU representative:* Macmillan Publishers Ireland Ltd, 1st Floor,
The Liffey Trust Centre, 117–126 Sheriff Street Upper,
Dublin 1, D01 YC43
Associated companies throughout the world
www.panmacmillan.com

ISBN 978-1-5290-4792-9

Copyright © Tom Lathan 2024
Illustrations copyright © Claire Kohda 2024, unless stated otherwise

The right of Tom Lathan to be identified as the
author of this work has been asserted by him in accordance
with the Copyright, Designs and Patents Act 1988.

All rights reserved. No part of this publication may be reproduced,
stored in a retrieval system, or transmitted, in any form, or by any means
(electronic, mechanical, photocopying, recording or otherwise)
without the prior written permission of the publisher.

Pan Macmillan does not have any control over, or any responsibility for,
any author or third-party websites referred to in or on this book.

1 3 5 7 9 8 6 4 2

A CIP catalogue record for this book is available from the British Library.

Typeset in Janson Text by Jouve (UK), Milton Keynes
Printed and bound by CPI Group (UK) Ltd, Croydon, CR0 4YY

This book is sold subject to the condition that it shall not, by way of
trade or otherwise, be lent, hired out, or otherwise circulated without
the publisher's prior consent in any form of binding or cover other than
that in which it is published and without a similar condition including
this condition being imposed on the subsequent purchaser.

Visit **www.picador.com** to read more about all our books
and to buy them. You will also find features, author interviews and
news of any author events, and you can sign up for e-newsletters
so that you're always first to hear about our new releases.

For Twoo

# Contents

*Introduction* 1

*Note on the Illustrations by Claire Kohda* 5

1 THE BUILDERS' STORY: A MALAYSIAN MICROSNAIL
(*Plectostoma sciaphilum*) 7

2 PARADISE, LOST: ST HELENA OLIVE
(*Nesiota elliptica*) 43

3 A TALE OF THREE SNAILS: A POLYNESIAN TREE SNAIL
(*Partula labrusca*) 81

4 THE LITTLE MASKED BIRD: PO'OULI
(*Melamprosops phaeosoma*) 123

5 DRIFTWOOD STOWAWAY: BRAMBLE CAY MELOMYS
(*Melomys rubicola*) 165

6 BIRD OF THE EVENING: CHRISTMAS ISLAND PIPISTRELLE
(*Pipistrellus murrayi*) 191

7 RUN, FOREST, RUN!: CHRISTMAS ISLAND FOREST SKINK
(*Emoia nativitatis*) 229

8 The Gritador: Alagoas Foliage-Gleaner
(*Philydor novaesi*) and Cryptic Treehunter
(*Cichlocolaptes mazarbarnetti*) 255

9 Ghost of the Galápagos: Pinta Island Tortoise
(*Chelonoidis abingdonii*) 281

10 The Eye of Potosí: Catarina Pupfish
(*Megupsilon aporus*) 319

Epilogue 351

*Acknowledgements* 355

*Picture Acknowledgements* 361

*Notes and References* 363

# *Introduction*

SIXTY-SIX MILLION YEARS ago, an asteroid struck the surface of a shallow sea. On impact, it tore an eighteen-mile-deep crater into the seabed. Fires engulfed 70 per cent of the Earth's forests, giant tsunamis tore up coastlines, and a thick blanket of dust and soot clogged the air, making photosynthesis impossible for almost all plant life. Food chains collapsed and 99.9999 per cent of all living organisms died. This event was so significant to the course of life on Earth that we now use it to define the boundary between the current geological era, the Cenozoic, and the last, the Mesozoic. Since our planet's creation, 4.5 billion years ago, there have only been four other extinction events of this magnitude, each of which left the Earth, and life on it, radically altered.

Today, we stand with both feet firmly planted inside what many scientists believe to be the Earth's sixth mass-extinction event. The boundaries are hazy, with disagreements about when exactly this event began. However, the theories all identify our species as the cause, with extinctions now occurring up to a thousand times more frequently than in the sixty million years before humans. Accounts from the front lines of this crisis speak of forests that, having once teemed with insect and bird life, now lie deathly silent, and of three-thousand-mile oceanic dead zones – stretches of oxygen-depleted coastal waters in which next to nothing can survive. According to the United Nations, around a million species are now threatened with extinction, many within the next few decades. This bleak picture is one that has become familiar to many of us in recent years. Less familiar, however, are the stories of the species that make up these increasingly foreboding statistics.

In the seventeenth century, European naturalists often studied new species entirely apart from their contexts. This meant that when birds of paradise arrived in Amsterdam without feet – an alteration made by hunters to fit the specimens into boxes – they were presumed to be naturally footless. Similarly, Victorian taxidermists encountering the skin of a walrus for the first time had no idea that folds of excess skin were a part of that species' form, and so stuffed it almost to bursting. Stripped of any place within a habitat or ecosystem, these animals became unreal and otherworldly. Species now arrive with us in extinction statistics and headlines similarly hewn of context – disconnected from their ecologies and histories, without us knowing what they ate and what ate them, what moved into their shells after they died, and which plants now grow out of control in their absence. Like footless birds and over-inflated walruses, they too become unreal and otherworldly, and we are left none the wiser as to what has truly been lost.

I wanted to understand what was happening to the natural world, beyond the headlines. I wanted to know what was being lost, and what the sudden absence of a species meant for the ecosystem it had vanished from. In 2019, I asked the International Union for Conservation of Nature (IUCN) – the global body that assesses the conservation status of the world's wildlife – for a list of species that had recently gone extinct. In the spreadsheet I received from them were hundreds of entries, detailing species from all over the world. But eleven entries stood out to me. Each was a species that had been declared extinct within the twenty-first century; however, each had also been seen alive at some point during the period of time between 1 January 2000 and the date of its extinction. These were species which, having lived and died in the last two decades, were undeniably of our time – twenty-first-century extinctions.

I immediately began researching the stories of these species. Soon, the spreadsheet the IUCN had sent me – whose spartan columns listed for each species only the common name, scientific name, date of last sighting, and extinction date – came alive with detail. There was the little brown bat, so small it could easily fit in an egg cup, and the rat that lived on a tiny coral cay in the middle of the ocean. The slow plodding footsteps of giant tortoises echoed through my mind as I stared for hours at photographs of beautiful but impossibly minute snails. I listened to the songs of long-lost birds over and over again and, in museum collections, I gingerly held the first collected specimens of species in my hands. Speaking with the biologists, ecologists, and conservationists who had studied and tried to save these species, I learned of the extraordinary passion and ingenuity of the people attempting to hold back the rising tide of extinction. And gradually, over the next few years, this research became the book you are holding today.

By necessity, this account of extinction in the twenty-first century is far from complete. For every species that has been declared extinct by the IUCN, there are many more waiting to be assessed. This is partly because, when a species vanishes, the IUCN allows a grace period of several years before it is formally declared extinct to ensure it really has gone. It's also because, like almost all conservation bodies, the IUCN is severely under-resourced, especially considering its mammoth task. As a result, there is a long queue of vanished species waiting to be assessed. And those are just the species that we know about. There are an estimated 8.7 million plant, fungi, and animal species living on Earth. Of these, only around 2.1 million have been identified and described, according to the IUCN. This gaping blind spot in our scientific knowledge means that alongside the extinctions we do know of, there are likely many so-called 'dark extinctions' occurring every year – when species vanish before we even know of their existence. For these reasons, the stories in this book can only offer us a snapshot of extinction; in fact, by the time I had finished writing, more extinctions had already been declared. The species described here are best viewed as representatives whose stories offer us a glimpse of a far broader process, each a stand-in for other extinctions that we will probably never know occurred.

When I first received the list of recent extinctions from the IUCN, a detail about one species, the Christmas Island pipistrelle, drew my attention. The last individual of this species had vanished on my twenty-third birthday. Tucked away in a box of things from my past, I had kept a memento from that date: a matchbox-sized cardboard music box given to me by a friend that, when wound, struck the melody of 'Happy Birthday'. Memories from that day, and that time in my life, came flooding back to me; and extinction, a phenomenon that can feel so distant and divorced from our personal lives, suddenly felt much closer. The species featured in this book have lived and died within the lifetimes of most of the people who will read it; some will recognize in these pages the dates of birthdays, weddings or personal tragedies. For others, none of the dates will hold any immediate significance. However, almost all of us can look back over social media and emails and see what we were doing on the days these species left our world: a meeting, a photo, a mood, a like. Either way, it is my hope that these stories will fill all who read them with a new sense of connectedness to these lost wonders, to the broader natural world and the unfolding extinction crisis.

Tom Lathan
March 2024

# Note on the Illustrations by Claire Kohda

THE ILLUSTRATIONS IN this book figuratively take the species back to their homes. *Plectostoma schiaphilum*'s hill habitat is restored, Lonesome George the giant tortoise is returned to Pinta, and the Christmas Island pipistrelle is back in the night sky over Christmas Island. Their homes and elements of their stories are drawn in pencil, while the species themselves are outlined in ink to remind us that a true return is impossible. There is a permanent separation between these extinct species and our world.

# 1 The Builders' Story

A Malaysian Microsnail
*Plectostoma sciaphilum*

IMAGINE YOU HAVE shrunk to the size of a flea. Now, this book towers over you. Each line of text stretches ahead of you like a road. Welcome to the world of microsnails – any species of snail with shells measuring 5mm or less in length. The shell of the smallest, *Angustopila psammion*, is just 0.46–0.57mm long. It's a scale that's difficult to visualize, which is why microsnails are often photographed next to everyday household objects for scale – dwarfed by the red tip of a matchstick, or perfectly perched inside the eye of a needle. 'Even if you were to put your nose to the ground and one of the smaller microsnails crept past, you could still miss it entirely,' explains Dutch evolutionary biologist Menno Schilthuizen. It is in this minuscule world that the story of extinction in the twenty-first century begins.

After the Second World War, Michael Wilmer Forbes Tweedie wasted no time getting back to what he loved. A British zoologist and curator at the Raffles Museum (now Raffles Museum of Biodiversity Research) in Singapore, Tweedie had been imprisoned by the Japanese during the war. On his release, he'd been appointed as the museum's director, a promotion that came with a plethora of new responsibilities, but also gave him latitude to return to a project that he'd started in 1938. Tweedie had long been fascinated by the microsnails that lived on and around the limestone hills that were dotted all over Malaysia. He believed that because many were isolated, they might harbour unique species of microsnail found nowhere else. All that remained was for him to find them.

The word 'hill' doesn't quite do justice to the often-magnificent sculptural forms that Southeast Asia's limestone hills take. Far from the gentle, rolling contours that might come to mind when you read the word, they look like sudden interruptions to the landscape – monolithic lumps of rock that rise near-vertically out of flat land, etched with crevices and topped with bare rock or caps of vegetation. They take this form thanks to the limestone they are made up of, which is easily eroded by the elements. Rain has worked on these hills over millions of years, creating spectacular folds and protuberances.

In 1947, Tweedie visited a limestone hill called Bukit Panching in what is now the eastern Malaysian state of Pahang. It was relatively small: in area no larger than London's Russell Square, and roughly the height of the London Eye. From here, miles of rainforest spread out in all directions, buzzing with the calls of insects and birds, the occasional trumpeting of elephants and the whooping chatter of monkeys. It was hard, in fact, to tell where the rainforest ended and Bukit Panching began; almost every inch of the hill was densely populated with mosses, trees, climbing plants, and other vegetation, as if the rainforest was swallowing this rebellious nubbin of rock back into itself.

Most people would have entirely missed the shell that Tweedie found at Bukit Panching that day. Spotting it with the naked eye would have been like picking out a breadcrumb on a sandy beach. Tweedie, however, had devised a special microsnail-collecting method. He gathered debris from the base of Bukit Panching, then dried it in the sun and threw it into a bucket of water, which he 'stirred vigorously', as he later described. Anything that contained air – plant matter, dead insects, and of course the buoyant shells of snails – floated to the surface in one tangled mass, which Tweedie skimmed off and dried for a second time. To any passer-by, the scene that played out at the base of Bukit Panching that day would have seemed odd; Tweedie looked more like he was making mud pies than searching for scientific specimens. However, his method was precise. Once the debris dried, he sifted it through pans with increasingly small meshes – in a sense, panning for snails, much as you would gold – until all that remained were tiny and fragile swirls of red, yellow and pink. These were the microsnails he was hunting for: an entirely new species, he hoped, undocumented and nameless.

Although Tweedie had perfected the art of panning for microsnails, determining exactly what he'd found at Bukit Panching was beyond his expertise. For this, he turned to Woutera Sophie Suzanna van Benthem Jutting, a renowned malacologist and curator at the Zoological Museum, Amsterdam, who had agreed to analyse and identify any interesting snails he found in the field. So it was that the shells he collected at the base of Bukit Panching were packed into vials, loaded into a crate, and shipped halfway around the world to Amsterdam, along with numerous snail specimens collected from other Malaysian hills. And this happened not a moment too soon; in 1948, Malaysia plunged into war as the Malayan National Liberation Army fought to end centuries of colonial subjugation and replace it with communism. With much of the countryside embroiled in conflict, it was no longer safe to visit limestone hills, and Tweedie had to halt his snail-collecting project yet again.

Van Benthem Jutting was a long-term collaborator of Tweedie's. Yet they were far from equals, van Benthem Jutting being both better known and busier. Letters sent over the years they worked together reveal the differences in their status. While Tweedie often sheepishly asks about her progress with the project, van Benthem Jutting makes a point of reminding him of her busy schedule and fobs him off with vague promises. 'I am so behind-hand in my work, and this will remain so till my death,' is one of her many replies. Van Benthem Jutting's tardiness was no reflection of her commitment to or passion for her work, however. During the Nazi occupation of Holland, she had returned to the Zoological Museum, Amsterdam, carefully packed up the most important specimens in the collection – including a crate of snails that Tweedie had sent before war had broken out – and hidden them in an underground bomb shelter. There, these specimens had waited out the war, protected first from German bombs, then those of the Allies, before emerging safely. Now, in 1949, she was unloading carefully wrapped specimens of snails sent to her by Tweedie from another war zone.

Van Benthem Jutting was the first person to analyse Tweedie's specimens. The minuscule shells, however, defied analysis: even when holding them close to her face, she struggled to render in them any clear form. Some were different shapes – sharply pointed and cone-like, or round and squat; that much she could see and vaguely feel, but the finer details were lost. Under her microscope though, the detail came alive. Translucent and cast in deep yellow and red hues, from mustard to plum to maroon, they formed striking shapes. Some looked as though they had

*Woutera van Benthem Jutting working in her laboratory at the Zoological Museum, Amsterdam, in 1955. Three years earlier she published the paper that described* Plectostoma sciaphilum *to science. (Courtesy of Barbara van Benthem Jutting.)*

been tied messily around themselves in knots; others had the appearance of having grown around an object that had been removed, leaving a conspicuous empty space. All the shells had thin ribs protruding from their coils. They jutted out prominently on some of the shells, like the frills of a lionfish, while on others they gave the impression of pleats. Despite these differences, there was something about these shells, in their colours and their lines, that gave them the appearance of being variations on a single idea – as though they were tiny glass sculptures blown by the same artist.

All told, van Benthem Jutting identified and named twenty-three new species of microsnail from the samples Tweedie had collected from hills in Malaysia. These species were grouped within a single genus, *Opisthostoma* (some were later reclassified as *Plectostoma*). Among them was a minuscule snail from Bukit Panching, which van Benthem Jutting named *sciaphilum*. She noted that its shell was 'a delicate spiral sculpture . . . ornamented with elegant, oblique white ribs, distantly placed and standing wing-like.' At its opening, the shell had a thin lip of lemon yellow; from there its coil gradated to orange, then red, burgundy, and finally a deep copper. Emerging from inside, the snail's body would have been a pale shade of amber, and translucent, like a thin streak of honey, with a tiny black dot of an eye on either side of its head. This species is the subject of this chapter: a snail the size of a sesame seed, with a perfect, helter-skelter-shaped shell coloured in the shades of a sunset, measuring just under 3mm long, and found nowhere else in the world but on Bukit Panching.

*Sciaphilum*, the name that van Benthem Jutting chose for the species, means 'lover of shadows' (from the Greek 'skia' meaning shadow, and 'philea' meaning love). This name most likely refers to Tweedie's belief that it lived in dark cracks and crevices in its limestone-hill home. However, it is also evocative of a local legend. According to one Malay folk tale from the

*The 3mm-long shell of* Plectostoma sciaphilum *next to the tip of a pencil. (© The Trustees of the Natural History Museum, London.)*

nearby region of Perak, cattle left to graze too close to limestone hills like Bukit Panching became weak and emaciated, the frailest amongst them even dying. Their blood was said to have been magically drained from their shadows by tiny vampiric snails, which burst like berries when crushed, and gushed not juice but blood.

In 1952, van Benthem Jutting's findings were published in *The Bulletin of the Raffles Museum,* and *Plectostoma sciaphilum,* lover of shadows, was inducted into the pantheon of formalized human knowledge. This was the beginning of our shared history with *Plectostoma sciaphilum*, a story that came to a violent end just half a century later. Last seen alive in 2001, the snail was entirely wiped out soon after, becoming the first recorded extinction of the twenty-first century. However, despite our acquaintance with this species beginning only very recently, the story of this tiny snail started much earlier, when the world was a very different place.

DURING THE CARBONIFEROUS period (359–299 million years ago), vast swampy forests covered much of the land. *Lepidodendron*, an extinct genus of huge tree-like vascular plants that could reach fifty metres in height, dominated the landscape.* Insect life flourished in this oxygen-rich environment; two-metre-long centipedes and metre-long cockroaches scuttled across the ground, and dragonflies the size of seagulls zipped through the air above. The place where Tweedie would one day collect *Plectostoma sciaphilum* was at this point still under the sea, a slightly raised but otherwise featureless part of the seabed. However, around 330 million years ago a massive building project began here.

Organisms such as brachiopods (clam-like creatures that creep along the seafloor on long tongue-like feet) that secrete calcium carbonate as a waste product inadvertently laid down the foundations of a hill. On their deaths, their bodies were added to the mound, which the next generation would dutifully build upon. The ice-cream-cone-shaped formations of rugose corals, the tiny teeth-like feeding apparatus of conodonts, the rigid skeletons of bryozoans, and the shells of single-celled organisms called foraminifera: all formed the bricks that, over millions of years, eventually became Bukit Panching. Having grown larger and larger beneath the waves, tectonic movements then shifted the hill, along with the entire surrounding area, up above sea level. Bukit Panching, for millions of years a part of

---

* The organic deposits laid down in these dense, heavily vegetated environments would one day become the coal that fuelled the Industrial Revolution.

the seascape, was now a part of a sprawling landscape, and the seabed that once surrounded it became bedrock beneath what soon became a thriving rainforest.

The building of Bukit Panching didn't end there, however. Now exposed to the open air, the soft sediments covering the hill began to erode approximately 140 million years ago, when dinosaurs such as *Hylaeosaurus* and *Iguanodon* still walked the Earth, until the limestone rock was exposed. Erosion and dissolution went to work on its surface, shaping and sculpting deep grooves in the rock that gave it the appearance of having been folded. Rainwater bore into the limestone, carving labyrinthine cave systems that snaked through the rock, like arteries.

Bukit Panching was not alone. All around it, for miles and miles, other similar hills rose with it. By the time Bukit Panching welcomed its first species – trees, mosses, lichens, then invertebrates – there were thousands of these hills, standing above rainforest canopies across Southeast Asia. Limestone hills are a signature of the Malaysian countryside. In some places they crowd together; in others there are just a few scattered here and there. Bukit Panching had three close neighbours, each one pointed upwards, shaped curiously like the tip of a finger, together resembling the buried hand of a god. Over millions of years, a separate ecosystem that supported different, unique species developed on each hill.

*Plectostoma* snails colonized many of Malaysia's limestone hills, along with those further afield. It is thought that, tens of millions of years ago, an unidentified ancestor species of all *Plectostoma* snails roamed across Southeast Asia, which was then a single connected landmass. They travelled via paths of exposed limestone bedrock, stumbling across hill after hill and establishing populations on each of them.

A snail builds its shell using an electric current, produced by an organ called the mantle, to arrange calcium ions into strips. Strip after strip is slowly placed in an ever-expanding spiral that hardens into a protective exoskeleton. Invaluable to the snail, this shell shields its body from both predators and hot weather; by withdrawing deep inside, a snail can avoid overheating, which can kill it. Limestone hills were like evolutionary jackpots for the mystery ancestor of *Plectostoma* snails. The hills offered not only shelter from predators in the form of cracks and crevices in the rock, but also what essentially amounted to an unlimited supply of calcium, absorbed directly from the limestone through the sole of the snail's foot, for shell building.

Over time, very gradually, the shape of the landscape changed once again. The limestone bedrock that had connected the hills and provided a means of travel between them for *Plectostoma*'s ancestor became almost completely buried deep under sediments and a blanket of rainforest. Unlike most species of microsnail, those of the *Plectostoma* genus are entirely dependent on limestone and are unable to survive away from it. The rainforest, therefore, was likely a boundary their ancestors could not cross. And the hills, which they were now dependent on, became like islands that the snails could not leave.

There was, however, still one means of travel; but it depended entirely on luck. This is what biologists call 'passive colonization' whereby, rather than transporting itself to a new location, a species is spread by something else. Some *Plectostoma* species occur on more than one hill. It is possible that they were once picked up by storms and carried through the air or swept from one hill to another by floodwater. 'My personal favourite theory is that bearded pigs are responsible,' Schilthuizen tells me. These large foraging herbivores cover vast distances through the rainforest in search of fruiting trees, often pausing along the way to scratch their coarse fur on rocks at the base of limestone hills. 'You can see gouges in these rocks where generation after generation of bearded pigs have all rubbed against them,' he says. It might have been during these scratching sessions, so the theory goes, that *Plectostoma* snails were sometimes brushed up into the thick wiry coats of pigs, before later disembarking at other limestone hills the next time their bearded travel companions stopped for a scratch.

Pigs and hurricanes aside, most of the snails lived only on a single hill. Marooned, each population became one of the eighty-two species that make up the *Plectostoma* genus.

JUNN KITT FOON, a Malaysian biologist and research associate in malacology at the Australian Museum in Sydney, describes the shells of all the different *Plectostoma* species as 'telling stories'. Each has what he describes as 'quirks' in its shape: 'You have *Plectostoma* snails that have fascinating rib shapes that come out from the shell and fan out like the petals of flowers, while others coil and twist in convoluted ways,' he tells me. 'All of the hills have been evolutionary battlegrounds between predator and prey, and the differences between the individual snail species' shells tell stories about how the specific predators on each hill prey on the snails, and how the snails have found ways to defend themselves.'

On Bukit Panching, *Plectostoma sciaphilum* responded to the unique threats and opportunities that arose around it. The spectacular *Atopos* slug, whose pale grey body is bejewelled with thousands of glistening bright blue bead-like studs, was one such threat. This nocturnal predator spent daylight hours hidden in deep crevices in the limestone rock, emerging at night to hunt. Twenty times larger than *Plectostoma sciaphilum*, the *Atopos* wrapped its body around its prey, then either threaded its sharp proboscis into the shell's opening or used it to drill directly through the shell and extract the jelly-like flesh within. The *Atopos* is believed to have played a part in shaping the evolutionary development of the shells of all *Plectostoma* species, including *Plectostoma sciaphilum*; over time the species grew long and closely clustered ribs along the coil of its shell, which made it harder for the slug to get its proboscis close enough to drill. Foon calls these sharp protrusions 'armour'. Fireflies are also thought to have played a part in shaping the shells of all *Plectostoma* snail species. The larvae of *Pteroptyx*, a genus of firefly found all over Southeast Asia, enter *Plectostoma* shells through their apertures, their heads disappearing inside while they feast. Possibly in response to this threat *Plectostoma sciaphilum* evolved its spiralling, helter-skelter-like shell structure. At the top, in the tightly whorled peak, the snail could tuck its body away.

Variations of this story played out on limestone hills across Southeast Asia. While *Plectostoma sciaphilum* was developing its unique shape, colours, and frill-like ribs, other *Plectostoma* snails and microsnails of related genera were similarly responding to the ecologies of their own particular limestone-hill homes, developing what Foon describes as 'unique patterns and shapes'. He tells me about a microsnail endemic to Gunung Rapat, a limestone hill in the Kinta Valley in Perak, called *Whittenia vermiculum*, which evolved a remarkable shell that coils on four axes, essentially tying itself in knots. 'Maybe it developed in this way to keep out predators, or maybe it did so to help it float during floods,' says Foon. 'In water, these snails are very buoyant, like air bubbles, because of the shapes of their shells. They are incredible.' *Plectostoma concinnum*, meanwhile, evolved on Bukit Gomantong with a tightly wound spire, much like *Plectostoma sciaphilum*'s, but with a sharp turning just before its aperture that points almost directly upwards. This feature, which makes the snail's body even harder for predators to reach, is called a 'tuba' – named after the musical instrument whose mouthpiece twists towards the player.

While we can make educated guesses as to what has

influenced the shapes of microsnail shells, the precise beats of each particular story – including *Plectostoma sciaphilum*'s – are a mystery. Part of the problem is how difficult they are to study. *Plectostoma* snails are so highly specialized towards their specific habitats that, in most instances, they are unable to survive in a laboratory setting for more than two or three weeks. Thor-Seng Liew, a biologist and one of the foremost authorities on *Plectostoma* snails, has attempted to breed the tiny molluscs of this genus in captivity for years. 'In the laboratory we replicate everything from the snails' natural habitat,' Liew tells me when we speak. 'We control the temperature and humidity precisely, and we even bring blocks of limestone rock from their habitat into the lab. But, sooner or later, they always die.' This likely happens because of the difference in UV exposure the snails receive in their natural and captive habitats, Liew explains. In the wild, sunlight has a cleansing role in the lives of these snails, in that it controls the bacteria and fungi that grow on their bodies. Without it, Liew's lab snails are likely defenceless, and quickly succumb to infections. Whatever the precise causes of their fragility, it's clear that *Plectostoma* snails are inseparable from the hills they live on, many of which are themselves vanishingly small.

WHEN TWEEDIE DISCOVERED *Plectostoma sciaphilum* in 1947, Bukit Panching supported a diverse and vibrant array of wildlife. In the caves that snaked inside the hill – where, in the walls, the fossilized remains of the ancient creatures which built Bukit Panching hung suspended in the rock – wrinkle-lipped free-tailed bats roosted. These bats preyed upon insects found in and around the hill, but they were also themselves givers of life; their excrement formed the nutritional basis for an entire ecosystem. Flies, cockroaches, snails, millipedes, beetles, and crickets feasted on this guano, and they in turn fed larger animals, such as centipedes, tailless whip scorpions, blind snakes and crabs.

At dusk, the bats left to hunt insects in the surrounding forest, billowing like plumes of smoke in long, pulsing murmurations. At the same time, swiftlets, returning to their roosts from these same hunting grounds, filed into the caves, as though the species had together devised a day shift and a night shift for their activities. Once inside, the returning swiftlets added their own chirruping calls to the cacophony of clicks and squeaks made by the bats' pups. The birds' nests – strikingly white, shaped like petals, and sculpted from strands of thick, quick-drying mucus, which the birds secreted from specially adapted salivary glands – may have drawn humans into the caves. Shimmying up long wooden poles to reach the cave ceilings, humans visited limestone hills to collect the nests of swiftlets, a culinary delicacy in China since at least the fourteenth century. Farmers and horticulturalists may also have frequented Bukit Panching to collect the guano that carpeted its cave floors to be used as fertilizer, and to harvest the orchids that grew on its cliff faces and summit.

The hill also served humans in another crucial way. Its porous structure meant that when rain came, Bukit Panching became a natural reservoir that emptied slowly into the surrounding soil, which kept the land around the hill moist and arable and provided a source of fresh water in times of drought. 'The stone is constantly dripping with water,' explains Foon, who has been familiar with the hills and their caves since he was a child. 'The hills are like natural water towers.' Millions of years of rainwater is stored in each hill, and it seeps out, mixed with the lime from the rock.* The hills continue to be built in this way. When the

---

* A soft clay called 'moonmilk' also naturally occurs on the walls of many limestone caves. Often marked with the imprints of the feet of the animals that live in that cave, in Bukit Panching, this substance may have

water in the lime–water mix evaporates, lime is left behind in striking shapes that hang from cave ceilings as stalactites, or stream down the hillsides as stone 'waterfalls', as Foon describes them.

It is thought that millions of *Plectostoma sciaphilum* were living on Bukit Panching at any one time. As evening fell each day, they emerged from the cracks and crevices in the rock. While individuals of the species were easy to miss, when they congregated together, concentrated in huge numbers on the moss from which they ate microscopic lichen and algae, the autumnal hues of their shells lent the limestone rock a mottled appearance, as though it was somehow rusting away. On rainy nights, and especially during Malaysia's October–December monsoon season, the snails came together to mate. Most land snails are hermaphrodites, but *Plectostoma* snails are either male or female – the key morphological difference between them, other than their genitalia, being a dash of bright red coloration on the tip of the male's shell. *Plectostoma sciaphilum* was never observed courting or mating. However, based on the behaviour of other *Plectostoma* snails, the male likely would have mounted the shell of the female to court her. Then, if the invitation was accepted, the two snails mated face to face, temporarily joining to form a single symmetrical shape. After parting ways, the female deposited her tiny 0.4mm-long eggs, from which hatched minuscule juveniles, each wearing a little white speck of a shell on its back.

The larvae of fireflies sought out these dense gatherings of the snail. Entering through the shell's aperture, one could swiftly consume a snail's body in its entirety – if it was successful in navigating the snail's helter-skelter shell to its top – leaving nothing behind but an empty husk. As if to belie the brutality of

---

read like a register of attendance. In it might have been the footprints of bats or scorpions, or the tracks of animals that had visited the caves; perhaps even the thin lines traced by snails.

this act, the captivating light displays of the adult fireflies pulsed through the nearby treetops each night, as though miles upon miles of flashing green fairy lights had been threaded around the branches of every tree and then triggered in unison. It is a spectacular light display which Tweedie might well have seen as he left Bukit Panching with a handful of shells in his backpack. As he went, he would have passed huge limestone boulders with curiously smooth edges that had been worn down by elephants and bearded pigs that visited to scrape dead skin and dirt from their bodies. Everywhere on Bukit Panching, there would have been signs of life.

TODAY, IT IS impossible to retrace Tweedie's footsteps, because the ancient hill of Bukit Panching no longer exists. It has been completely removed from the landscape and, in the spot where the hill once rose up over the surrounding rainforest, there is instead a mysterious lake. Viewed from above in satellite images, it looks as though someone has taken a pair of scissors to the terrain and carefully cut along Bukit Panching's perimeter, leaving a perfect stencil of the hill in its place. On Google Maps, the words 'Bukit Pancing' float eerily in space over this dark body of water, next to a single five-star review of the hill's 'mountain peak'.*

Like this hill that is now a lake, the rainforest around Bukit Panching has also been replaced. Instead of the many thousands of plant species that used to live here, row after row of equidistant oil-palm trees are marshalled into perfect formation. The resulting landscape – with its straight lines, uniformly spaced trunks, and vegetative homogeneity – is an eerie sight, and viewed from above this is even more pronounced. The land takes

---

* 'Bukit Pancing' is an alternative spelling of 'Bukit Panching'.

on the appearance of being covered not in trees but in green-tinted bubble-wrap. It is as though the rainforest is still beneath, but is slowly suffocating under this sprawling sheet of plastic. And, in a sense, there is some truth to this.

'Slash and burn' is the practice of clearing and burning vegetation – whether wild forest, existing farmland or plantations – to invigorate land for agriculture. Across Southeast Asia, limestone hills frequently find themselves surrounded by fires set for this purpose, as rainforest and oil-palm plantations are habitually burned to ash.* During the fire season flames lap at the bases of the hills; the air around them fills with smoke and the caves with a cloying smog. The heat and smoke kill vegetation, insects, bats, and all manner of other organisms – including, of course, snails.

When the fires die down, plantation owners spray pesticides to suppress the native plants and trees that continue to sprout on their land. If left to their own devices, these charred fields would once again become rainforest. The only places left for the forest to reclaim are the limestone hills, and so this is where life returns first after the fires.† Life cannot return to Bukit

---

* These fires cause hazes that pollute entire countries. In 2015, when 100,000 fires had been detected in Indonesia alone, creating emissions greater than the daily average from all economic activity in the US, the orange haze covered parts of Indonesia, Malaysia, and Singapore, grounding flights and shutting down schools and businesses, and causing half a million cases of respiratory infection. Mass prayers for rain took place in Ulema, Indonesia, and warships were kept on standby to evacuate residents. A study by Harvard and Columbia Universities found that 100,300 premature deaths had been caused by the 2015 haze in Indonesia, Malaysia, and Singapore. A spokesperson for the Meteorology, Climatology and Geophysics Agency, Sutopo Puro Nugroho, called it 'a crime against humanity of extraordinary proportions.'

† Some, however, have been exposed so frequently to the high temperatures from slash and burns that they remain barren.

Panching, however. The transformation it has undergone is permanent. So, where is Bukit Panching? Bizarrely, at least a part of it might be closer than you'd think.

CONCRETE IS THE second most used substance in the world, after only water. The runways, roads, and pavements that transport us each day, as well as the foundations and sewage pipes that lie underneath our homes and offices; the dams, canals, and sea walls that we build to contain and guide bodies of water, and the bridges that carry us over them; swimming pools, bank vaults, and headstones – all are, more often than not, made of concrete.

This most ubiquitous of materials is created by mixing cement with water and an aggregate such as sand. The binding quality of the cement is activated by the water, forming a paste that hardens into a strong, stone-like material when mixed with the aggregate. Adjusting the ratios of cement, water, and aggregate produces concrete that is stronger and denser, or weaker but more malleable. Concrete has been used by humans for millennia; the largest unsupported concrete dome, for instance, which measures 43.3m in diameter, was created for the Pantheon in Rome two thousand years ago. With the fall of the Roman Empire, the secrets of concrete production were lost. This was until they were rediscovered by British bricklayer and stonemason Joseph Aspdin in 1824. By the turn of the twentieth century, the material had not only made a comeback but had completely revolutionized construction due to its low cost and versatility. The world now uses 30 billion tonnes each year, as of 2021. That is almost four tonnes per person and, as Jonathan Watts wrote in *The Guardian* in 2019, 'enough to patio over every hill, dale, nook and cranny in England'. Cement, the magical binding agent that makes all of this construction possible, is itself created by mixing a few ingredients together and heating them to 1400 °C. The key ingredient is calcium carbonate, the chemical compound that is the primary constituent of limestone.

It's easy to predict what happened next in the story of *Plectostoma sciaphilum*. As Malaysia's economy expanded throughout the mid-twentieth century, while the country was under British colonial rule, financiers and investors looked for new prospects, particularly in the heavy industries. Malaysia's many limestone hills, including Bukit Panching, presented an especially promising opportunity and, in the 1980s, cement manufacturing companies sprang up across the country.

Quarrying began at Bukit Panching in the early 1980s and within twenty-five years the hill was completely gone. In 2008, Menno Schilthuizen and conservation scientist Gopalasamy Reuben Clements, who had been monitoring the destruction of the hill, suggested that *Plectostoma sciaphilum* had gone extinct as a result. They made this judgement on the basis of satellite imagery alone. In the authors' words, they had identified an extinction here on Earth 'from space'.

When I speak with Schilthuizen, I ask him about what he and Clements describe in their paper as the 'emotional impact' of these satellite images. 'I think the emotion comes from the fact that I had experienced this place on the ground,' he says. Dramatic, towering cliffs coated in moss and climbers, surrounded by 'fairytale-like' tropical rainforest – the scene comes alive as Schilthuizen describes it to me. And then, of course, there were the cracks and crevices in the rock, in which the lives of countless *Plectostoma sciaphilum* played out. 'As biologists, we experience these snails and their habitats at the scale of centimetres,' he tells me. 'Whereas the scale at which you see what's really happening to them is planetary.' What moves

Schilthuizen is this stark contrast, between the intimate perspective of the malacologist and 'the sterile view from space.'

Later, in-person surveys of the site confirmed that the species was indeed gone and in 2015, *Plectostoma sciaphilum* was declared extinct. On Google Earth Pro, you can view the destruction of Bukit Panching in a timelapse. In these satellite images, it is as though the hill is the cake at a party where, with each advancing frame, a fresh slice is carved off. As ridiculous a comparison as this might seem, it is entirely appropriate. With one company reporting a profit of $150 million from a single year of quarrying at one single site in Malaysia, this was clearly quite a cake.

Since the shells of *Plectostoma sciaphilum* were themselves made of calcium carbonate, they would have emerged from the cement kiln fused with the limestone rock on which the species'

*This lake is all that remains of Bukit Panching,* Plectostoma sciaphilum's *limestone hill home. In the background is the nearby hill, Bukit Charas. (Photo by Thor-Seng Liew.)*

entire existence had taken place. Broken down into powder and packaged up, this cement was sold, then processed into concrete or used to join bricks or pipes. No records survive that can tell us where this cement ended up (the company responsible for the hill's destruction was dissolved not long after the quarry site was exhausted), but Malaysian cement is traded internationally, and with the huge amount of cement being produced from the quarrying of Bukit Panching, it really could be anywhere: within the concrete foundations under your feet, or filling the potholes on your street.

So it was that *Plectostoma sciaphilum*, a species whose lineage stretched back tens of millions of years, was consigned to history, the final irony being that the shells of these snails, which they had built to shield themselves from danger, now lived on in another incarnation: as part of the bricks and mortar of walls that protect from harm the very species that had destroyed them.

IT WOULD BE absurd to suggest that things do not exist until humans have discovered them. The dinosaurs would have existed whether or not Richard Owen had coined the term in 1842. However, since the very concept of 'existence' is a product of language and the particular understanding of life that our species possesses, is there perhaps a kernel of truth in the idea that the discovery and naming of something is itself a life-giving act?

We are the only species on Earth that keeps a record of things such as the number of whorls on a snail's shell, or the angle at which this shell is carried by its owner. If van Benthem Jutting had not witnessed and made note of these details, *Plectostoma sciaphilum* would now be just one of the hundreds, or thousands, or tens of thousands of species that become extinct each year without us ever having discovered them. These animals slip quietly away without us having the opportunity to witness and record their existence. Are they, in their anonymity, less real than *Plectostoma sciaphilum*, which, for as long as our species remains literate, will be remembered? Perhaps not, but van Benthem Jutting's act of writing the word '*sciaphilum*' above her measurements and observations of a then-nameless animal is the reason that this species now lives on, not least in the pages of this book.

In practical terms, there is one very real sense in which taxonomists can grant life to the organisms that they name. Once a species becomes known to us, and its place in an ecosystem is revealed, environmental protections and conservation strategies can be devised to protect it. For species that we don't know about, however, this is obviously not possible. Instead, their survival is left to chance in a world where every habitat and ecosystem is, in some way or another, shaped by human actions. In this way, the act of naming a species can be the difference between life and death.

In 2014, the Dutch biologist Jaap Vermeulen and Malaysian biologist Mohammad Effendi Marzuki provided the perfect demonstration of this when they weaponized the taxonomical process against the largest cement company in the world. Lafarge Cement was scheduled to begin quarrying the preserved southern part of a limestone hill called Gunung Kanthan in Perak when Vermeulen and Marzuki discovered a number of new species at the site – one of which was a tiny land snail endemic to the hill. In their research paper announcing the discovery of this new snail species, they wrote:

> We name this species '*Charopa*' *lafargei* after Lafarge whose declared goals for biodiversity include minimising

and avoiding damage to important habitats, minimising and avoiding species mortality and stress, and minimising and reversing habitat defragmentation (Lafarge/WWF 2012), and whose biodiversity 'aspiration' is to have a Net Positive Impact on biodiversity (Lafarge, 2014). The decisions the company makes regarding Gunung Kanthan will determine the future existence of this snail.

Coming just a few months after Liew recommended that *Plectostoma sciaphilum* be declared extinct, this move made an impact. The story was reported by the international press, prompting Lafarge to launch an official investigation and promise to preserve a small part of the site for conservation. Three years later, the company signed a biodiversity conservation agreement with Fauna & Flora International, giving the conservation charity access to review its quarry sites in Malaysia and further afield, and to suggest ways to improve biodiversity management. In this instance, the usually benign act of naming a species could well be considered life-giving; it bought *Charopa lafargei* time. However, despite this small victory for this particular species, the broader picture for snails on limestone hills does not give much cause for optimism.

THE MOST RECENTLY available records, from 2021, show that of the 1,393 limestone hills in Malaysia, 116 were being or had already been quarried. Perak, in the west of the country, could be thought of as the epicentre of the quarrying industry. The province's capital city, Ipoh, is surrounded by a staggering forty-two active quarries, which collectively produce 48.1 per cent of the country's limestone. However, a similar fate is being planned for far less exploited regions, such as in Malaysian Borneo. This does not bode well for Malaysian wildlife, a huge proportion of which is found on limestone hills.

Biologists have referred to limestone hills as 'arks of biodiversity', on account of the many different types of habitat they contain – including caves, forests, fissured cliffs, and streams – and the crucial role they play as safe havens when surrounding ecosystems are damaged or destroyed. Of all Malaysian species of flora, 14 per cent can be found on limestone hills. The Bau karsts in the Malaysian province of Sarawak are inhabited by up to 40 per cent of all butterfly, macro-moth, and stick insect species found in the region. Up to 22 per cent of all Bornean species of fish, amphibian, snake, and mammal can be found on limestone hills. In addition to this, new species are being found all the time. A 2005 survey of an already well-scrutinized hill yielded new species of millipede, scorpion, and dipluran (tiny earwig-shaped insects also commonly known as two-pronged bristletails) all in the space of three hours. Snails are particularly dependent on limestone hills. Foon and his colleagues surveyed twelve hills for a 2017 paper and found thirty-nine to sixty-three snail species on each, including thirty species potentially new to science. The number of snails that are endemic to individual limestone hills can vary widely – from one or two on smaller hills, to as many as fifty on large hills such as Gunung Subis in Sarawak province.

In January 2022, the journalist and conservationist Natasha Zulaikha attempted to quantify the non-extractive value of limestone hills in Perak and compare it to the extractive value from quarrying.* Alongside biodiversity, she cites palaeontology and archaeology as adding value to the hills – in 2020, Malaysia's first ever *Stegodon* (an extinct genus of Proboscidea, related to the elephant) was found in a Perak limestone cave. The hills are

---

* In 2020, limestone quarrying, tin mining and rock crushing in Perak contributed £88 million to the GDP of the state.

*Gunung Kanthan has been quarried continuously for nearly fifty years. With more than 80 per cent of the northern side of the hill gone (pictured), its endemic species including* Charopa lafargei *cling to life on its southern side, along with the Sakyamuni Cave Monastery. (Photo by Toby Smith.)*

also important for geotourism, culture, and religion. In Perak alone, there are thirty limestone cave temples serving Buddhist, Taoist, and Hindu devotees, which also draw in tourists. Gunung Kanthan, the hill-home of *Charopa lafargei*, houses the one-hundred-year-old Sakyamuni Cave Monastery, Perak's oldest Buddhist cave monastery. Associated Pan Malaysia Cement Sdn Bhd (APMC), a subsidiary of YTL Corporation Berhad – the Malaysian mega-corporation which owns, amongst other overseas assets, a 100 per cent stake in Wessex Water – is currently in the process of litigating the right to evict the monks so that the hill, along with their monastery, can be quarried. As of April 2024, a petition to halt this action and declare Gunung Kanthan a national heritage site has 32,000 signatures.

Zulaikha also cites a 2014 study in Thailand that revealed the economic value of bats dwelling in limestone caves. The wrinkle-lipped free-tailed bat (*Mops plicatus*) has 'prevented losses of up to RM5 million [£1 million] annually from the rice production industry' by predating on the white-backed planthopper (*Sogatella furcifera*) – a pest notorious for damaging or destroying rice crops across Asia and Oceania. Project Pteropus, a Malaysian conservation group, similarly discovered that flying foxes living in the caves of limestone hills are crucial for the pollination of durian trees.

For decades, many of the hills in Malaysia have been quarried without ecological impact assessments – and quite unnecessarily, it turns out. According to Ramli Mohd Osman from Malaysia's Mineral Research Centre, up to six times more limestone can be extracted from disused tin mines than from all the remaining limestone hills of Malaysia combined. These underground deposits host none of the unique biodiversity of the limestone hills and so provide an opportunity to balance both economic and environmental interests simultaneously. So why don't cement companies quarry these deposits instead? Surveys to assess the size and depth of underground limestone deposits are a small upfront cost in such a venture; limestone hills, however, can be easily sized up with no need for surveys. For most companies operating in Malaysia – where leases to quarry limestone were established in British colonial times, before the conservation of limestone hills was considered important – that also have a pack of expectant investors snapping at their heels, it is an easy decision to make.

However, during the Covid-19 pandemic, the cement industry – like many other industries – halted work. The land was, in a sense, allowed to take a breath, after decades of non-stop quarrying. In her article, Zulaikha calls this 'a rare opportunity' – a small moment in which the government, states, and companies could choose to 'make a fresh start and balance economic gains and the conservation of nature.' This largely hasn't come to pass, unfortunately. During the early months of the pandemic, production dropped to the lowest levels since the late 1980s, but the Malaysian cement industry is once again firing on all cylinders, setting a new production record of 2.7 million tonnes in March 2023 alone.

On the face of it, the future of Malaysia's limestone hills and the species that depend on them seems grim. Unless taxonomists start renaming species after corporations left, right, and centre, it's unlikely that the fate of a snail – or even an entire ark's worth of animals – will do much to dent the appetite of cement companies. But there is still much that can be done to steer that appetite so that the overall damage to the natural world is minimized, according to Liew. It was Liew who wrote the Red List assessment that convinced the International Union for Conservation of Nature (IUCN) to declare *Plectostoma sciaphilum* extinct, and much of what we know about this and other *Plectostoma* species is thanks to his work. These days, however, his focus is on what he sees as the 'bigger picture' for snails like *Plectostoma sciaphilum* – the protection of their limestone-hill homes.

'There are enough hills to be quarried to produce cement for development,' Liew says, 'but so far the problem has been that cement companies haven't known which hills are more ecologically important than others.' All limestone hills, it turns out, are not made equal – at least not in terms of their biodiversity value. Some hills, Liew explains, are connected to the same bedrock and clustered close enough together that species have been able to spread between them. Quarrying on these hills, where there are few or no endemic species, can have a comparatively small ecological impact. At the opposite end of the spectrum, 'lenticular' limestone hills are isolated and not connected to others through bedrock, and these are the most likely to house endemic species. According to Liew, providing cement companies with this knowledge is key to preventing future extinctions.

In 2021, Liew, Foon, and Clements published a seven-part report entitled 'Conservation of limestone ecosystems of Malaysia'. The culmination of a five-year project, it is the most comprehensive accounting of the country's limestone hills to date, compiling geological and biodiversity reports with the authors' first-hand observations on the condition of each hill. The idea behind the report was to create a kind of handbook for quarry companies, Liew tells me, to help fill the enormous knowledge gap many of them had about the hills they were quarrying. With this resource, Liew and his colleagues hope to steer companies away from vulnerable, isolated hills and towards hills that share bedrock with many others. 'Also,' says Liew, 'now that we know which hills are already degraded, because of past quarrying or fires for example, we can recommend that these hills be quarried instead of others.' However, as of 2024, the resource has not yet been used by companies or the government authorities that control mining policy. 'I hope one day it will,' says Liew.

Some people believe that working with cement companies is a mistake. According to Liew, these people worry that companies will only collaborate with them in order to greenwash their environmental reputations. Although he is of the opinion that it is better to collaborate with cement companies than not, Liew understands this other school of thought. 'I'm not saying I'm doing the right thing,' he says, 'I just think that this is a way forward.'

This 'way forward' is not only about having a seat at the table when cement companies are deciding where to quarry. It also means getting these companies to finance biodiversity surveys and other scientific work – research that, in such a cash-strapped sector, might otherwise not be possible. Since 2016, Liew, Foon, and Clements have been collaborating with YTL Cement; the company provides them with access to Gunung Kanthan and funding to study the hill's snails. This partnership has already yielded positive results for one species. Populations of *Pollicaria elephas*, a snail with striking red antennae and a lemon-yellow shell, were located dangerously close to a quarry site on Gunung Kanthan. After studying the species' ecology, the biologists translocated the snails from their original location to a safer site on the same hill, far away from the quarry.

Perhaps, with further collaborations between cement companies and biologists, more species can be spared the fate of *Plectostoma sciaphilum*. With over 700 of the 900 limestone hills in Peninsular Malaysia lacking any kind of biodiversity record, there will be many more out there waiting to be saved.

FROM THE SHORE of the dark lake that now lies in Bukit Panching's place, another hill looms in the distance, jutting out from the seemingly never-ending canopy of oil-palm trees. This is Bukit Charas, a hill that has escaped the unfortunate fate of many limestone hills by virtue of a cave inside it. From the base of the hill, a track leads to a metal staircase and a ticketing booth, where a small fee is charged for entry. The path continues upwards, before splitting in two: one branch leads to the summit of Bukit Charas, the other to 'Gua Charas' ('gua' is the Malay word for 'cave'). The entrance to the cave is foreboding and enormous, and the darkness inside engulfing; but, then, your eyes adjust to the dim light.

This place is sacred to both Hindus and Buddhists and is the site of religious festivals and pilgrimages throughout the year. Statues and shrines of both faiths line the path that leads deeper into the cave, which, today, carries in its air the sweet, unmistakable smell of milk. People surround one statue, a Shiva lingam – a two-and-a-half-metre-tall cylindrical votive object that represents the deity Shiva and is a site of devotion for members of the Hindu sect of Shaivism. On the floor are offerings they have made: flowers, grass, dried rice, fruit, and leaves. The lingam has been ceremonially bathed in water and now milk is poured over it while Shaivist priests chant. We leave this scene behind, however, and pass more shrines and statues – including one dedicated to the unnamed Thai Buddhist monk who is said to have designated the cave a place of worship in the 1960s – before arriving at the main attraction: the Sleeping Buddha.

This eight-metre-long statue depicts the Buddha reclining on a platform. Propping his head up with one arm, he is not quite sleeping and instead looks out, hazy-eyed, towards the entrance of the cave. For two hours each day, positioned as he is beneath a hole that has been bored through the cave's ceiling, the Sleeping Buddha is bathed in sunlight. But for the rest of the day, he is perhaps kept awake by the three fluorescent lights that hang over him. Or maybe his sleep is disturbed by something else.

Set to one side of the path to the Sleeping Buddha, a smaller Buddhist shrine commemorates the story of the Wednesday Night Buddha. Here, the Buddha is depicted with a serene expression on his face, watching while an elephant bows reverently at his feet and a kneeling monkey presents him with a gift. The story goes as follows: one day, the Buddha decided to leave his monastery and enter the forest so that the monks, who argued about everything, would be forced to learn how to resolve their differences without him. Setting up camp in a cave inside a tall, round hill, he meditated. Although he had moved quietly, the Buddha's arrival had not escaped the attention of the animals of the forest, and soon he was visited by an elephant and a monkey. Eager to honour him, the elephant devised a way to heat the pond in front of the cave so that the Buddha could bathe in comfort. The monkey, meanwhile, presented him with a honeycomb, which he had stolen from a nearby beehive.

'I know you mean well,' the Buddha said, rejecting the honeycomb. 'But to squeeze honey from the comb will kill the bees inside. We shouldn't harm them.'

In response, the monkey carefully returned the bees to their hive and re-presented the now-empty honeycomb to the Buddha, who finally accepted it and gave the monkey the praise he had so desired.

This story, which encourages us to consider the consequences of our actions on even the smallest of beings, is commemorated at Gua Charas, which sits a mere 3km from the dark lake marking the place where Bukit Panching once stood. As a metaphor for the story of Bukit Panching's destruction, this fable is almost too perfect; the limestone rock, cavernous and fluted as a honeycomb, was home to many tiny animals. But there is one crucial difference. Here the monkey ignored the Buddha's advice and squeezed. Perhaps this is what troubles the Sleeping Buddha at Gua Charas, who is himself – along with the reinforced ceiling of the cave and the long, winding path that leads from its entrance – made of concrete.

Today, the specimens of *Plectostoma sciaphilum* that van Benthem Jutting used to describe the species for the first time can be found at the Naturalis Biodiversity Center in Leiden, the Netherlands, and at London's Natural History Museum. These collections are the only physical remains of the species – but *Plectostoma sciaphilum* exists in one other form, too. For a 2014 study led by Liew, one of van Benthem Jutting's *Plectostoma sciaphilum* shells was placed inside a high-resolution micro-CT scanner, and a 3D model was knitted together from the X-ray images produced.

Shortly after I spoke with Liew, he emailed me this 3D model. I loaded the file in an animation app on my phone and stared at it in awe. This tiny speck of a shell, which seventy years earlier had been shipped halfway around the world so that its secrets might be teased out under a microscope, now laid itself bare in the palm of my hand. Though some of the shell's edges had become rounded and chipped with age – wear that was now immortalized in this 3D model – I could see in it all of the qualities van Benthem Jutting had described so vividly. More than this, though, I soon found myself venturing beyond even what van Benthem Jutting could have seen through her microscope. First, I rolled it over and over under my fingertip, taking in its helter-skelter shape from every angle; pinching outwards, the snail's cave-like aperture filled my screen. And then, I went inside. Here, columns propped up the shell from its centre, as though it were the vaulted ceiling of a great cathedral. As I travelled up the spiral, the aperture disappeared from view behind me, obscured by a sharp bend, at which point I spotted a hole in the wall of the shell that was probably the work of an *Atopos* slug drilling with its proboscis for its dinner. Eventually, the vaults and columns of the shell converged into a single point, as I reached the end of the spiral – the top of the shell, where in life the snail it belonged to would have sought refuge from predators.

Although not quite walking a mile in its shell, this experience brought me closer to *Plectostoma sciaphilum* and its minuscule world. But for all the intricate structural detail Liew's model revealed, none of the shell's colour had been captured in this new simulacrum, which was rendered instead in shades of grey and white, almost like a ghost.

IN DECEMBER 2022, about a year after I'd started interviewing people to find out all I could about *Plectostoma sciaphilum*, Junn Kitt Foon told me a story. Close to the lake where Bukit Panching used to stand there is a long winding road. Following the bends of a small river, this road carves through oil-palm plantations and pockets of rainforest until a turn-off leads to a house. From the outside, the house looks much like many others in the area: timber-framed and topped with a corrugated metal roof, from which paper lanterns hang. The inside, however, is like nowhere else. Crystals of all shapes, sizes, and colours line every table, shelf, and cabinet, and giant lumps of rock speckled with glinting minerals stand in the corners of the rooms like sculptures. This is the Crystal House, a local museum created and maintained by a retired miner in his own home.

Foon visited the Crystal House in 2006, when he was still in high school. At one point during his visit, the old miner (who was only too happy to recount the story of his enormous collection) pointed him in the direction of one particular cabinet. Behind the glass, Foon saw spiky balls of aragonite crystal clustered together in slabs that looked, from afar, like crown-of-thorns starfish. Blocks of creamy-pink quartz, dull white calcite, and other crystals crowded every inch of remaining shelf space. The contents of this cabinet, the miner told him, had come from a cave in a nearby hill that had recently been quarried to dust. Just before the dynamite was laid, the miner had visited the cave and salvaged what he could, so that a part of it might live on in his museum.

'He couldn't stand the idea of losing the entire cave,' Foon told me.

The crystals in this cabinet are all that remains of Bukit Panching. They exist now as a kind of disembodied monument to a hill that took hundreds of millions of years to build, but only a few dozen to destroy.

After we spoke, Foon sent me the four photographs he had taken of the Bukit Panching crystals in 2006. He told me that a decade later he had tried to return to the Crystal House, but couldn't find his way back to it through the hive of winding roads. It had been there and then it had vanished, just like Bukit Panching and the tiny snail that had for millions of years made the hill its home.

# 2 Paradise, Lost

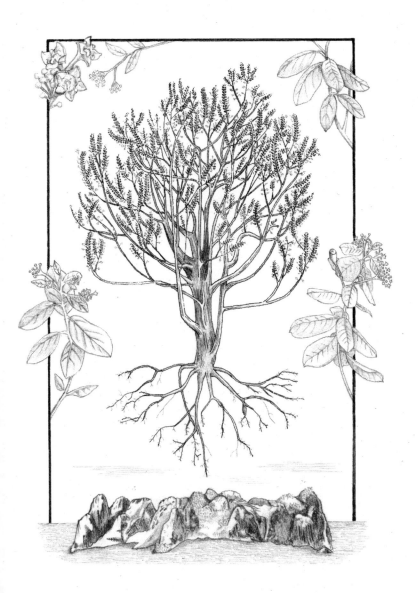

St Helena Olive
*Nesiota elliptica*

I VISIT THE Royal Botanic Gardens, Kew in February 2022, just after Storm Eunice has swept across the UK. Outside, the tail end of the storm rattles the branches of the trees, but inside the Temperate House the air is still. This Victorian glasshouse, which opened in 1863, is the largest of its kind in the world and contains 1,500 plant species from all over the planet – including many rare and threatened species. Today it also shelters a steady stream of waterlogged tourists, who huddle around the various entrances, waiting for the rain to stop.

'Travel the world in this glittering cathedral' – this is how the Kew website advertises the Temperate House, and it's a surprisingly accurate description. Towering trees, shrubs, bushes, spectacular flowers, and succulents normally dispersed across six continents line up here in rows. Little signs, much like you'd see at a garden centre, tell you the names of the plants and their countries of origin. Larger signs inform you when you're about to cross from one continent to another. From the Chilean wine palm (*Jubaea chilensis*) to the Australian tree fern (*Dicksonia antarctica*) to the Himalayan balsam (*Impatiens glandulifera*): in here, every step you take transports you over mountains, deserts, and oceans. But in amongst this veritable pick 'n' mix of botanical treats there is something that is out of place. A large, round terracotta pot stands just off the path at one end of the main room. Behind it there is a sign with a species name, just as there is for every other plant here, but nothing grows from the soil inside this pot. The dark, rich loam looks poised and ready to nurture something; however, that something isn't here. It's as if someone left the doors to the Temperate House open over the weekend and Eunice swept through and carried off whatever had been growing here.

I watch the empty pot stop a passing couple dead in their tracks. They look back and forth between the sign, the pot, and then at the empty space above it, lingering on the latter, as if there is something there that only they can see. After what feels like several minutes, the couple drift away to another part of the glasshouse, without saying a word to each other. 'St Helena olive' reads the sign behind the empty pot. There's a close-up photograph of deep-pink flowers next to the name. There are other photos, too: an elderly man wearing a bucket hat and glasses, standing amongst the foliage of a young tree; and a smiling, friendly looking woman, shot in black and white. Look at the sign a little closer and the mystery of the empty pot is explained: 'The St Helena olive, *Nesiota elliptica*, is extinct.'

This species takes its common name partly from St Helena, the tiny island in the middle of the South Atlantic Ocean where it once grew. The 'olive' part of its name, however, is a mysterious misnomer. An evergreen tree from the Rhamnaceae family (commonly called buckthorns), it wasn't an olive tree at all. Its winding dark-brown branches were clad in oblong leaves with pale furry undersides. Perhaps it was the grey-green coloration of these leaves that led to the name, or its woody seed pods, which were roughly the size of small olives – nobody knows. Fully grown, it could reach 10m in height, though no known photographs show the entire mature tree. From the limited observations made, its tiny deep-pink five-petalled flowers seemed to bloom from spring to autumn, when it also produced seeds. It grew nowhere else in the world but on St Helena.

The story of the St Helena olive ends in a dramatic and desperate attempt to save it, and a part of that story took place here at Kew. Now, an empty pot stands in the Temperate House, the round, dark surface of the soil inside marking this ending definitively, like a full stop. Though the story ends here, it began long, long ago, with the very thing that's missing from this lonely pot of earth: a seed.

ONE OF THE chief evolutionary problems that all plants have had to solve is how to spread – if you cannot move, what can you do to make sure your seeds are dispersed far and wide? The answer often lies in the design of those seeds themselves and their capacity to use the momentum of something else to move. Some seeds ride the winds on sail-like fins, while others use buoyant casings to float on water. Others dress themselves up in fruity assemblages and hail rides from passing birds, whose digestive systems double up as the perfect gestation chambers. In the world of seeds, if you're light enough, or delicious enough, life can take you anywhere. There are some places, however, that are so difficult to get to that even these tried-and-true methods might not be enough.

'One vast rock, perpendicular on every side, like a castle, in the middle of the Ocean' was how an anonymous historian described St Helena in 1759. Once rumoured to be the tip of the mythical sunken continent of Atlantis, this South Atlantic island is incredibly remote – 'Further away from anywhere than anywhere else in all the world,' writes the author Julia Blackburn. If you were to jump in a boat at Jamestown, St Helena's capital, and travel east, you'd see nothing but ocean for 1,900km, after which you'd finally reach Angola. In the opposite direction, 3,500km would get you to Brazil. To the north, you'd hit Côte d'Ivoire after about 2,250km. Sailing south, you'd just go on and on until the Antarctic ice sheet stopped you. On Google Earth, this 8km by 16km island is swallowed up in blue long before you've zoomed out far enough to see any of these reference points. In fact, for the people on St Helena, their closest neighbours sometimes aren't even on Earth, but are the astronauts orbiting our planet aboard the ISS.

For any seed, getting to this mid-ocean outpost would be challenging. Southern Africa is the only sensible departure point for windsurfers trying their luck riding the south-eastern trade winds to St Helena. However, the distance involved is immense, and the likelihood of being dumped in the sea before arriving at your destination is sky-high. Similarly, for those hoping to travel by bird gut, the odds of being excreted somewhere over the open ocean are frankly miserable. All of this is assuming that the seed in question has adapted to utilize these transport methods in the first place.

'We know lots of reasons why the St Helena olive shouldn't have been on St Helena,' says Mike Fay, a plant geneticist and the leader of Kew's Conservation Genetics team. 'The seed capsules are dry, so they're of no interest to fruit-eating birds, and in fact there are no fruit-eating birds that migrate between southern Africa and St Helena. The seeds are too heavy to be blown by wind, and we know that the seeds die if you soak them in seawater.' So, how did a tree whose seeds were seemingly ill-adapted for long-distance travel ever make it to St Helena? No one really knows. But the most likely hypothesis, it turns out, has little to do with clever seed design and everything to do with luck.

Without a single flap of its wings, an albatross can travel 1,000km in a day. These gigantic birds marshal the power of the wind so effectively that flying can require less energy than sitting. As a result, they can spend months at sea without touching land, sometimes even circumnavigating the globe. They spend most of their lives in the air, but they return to land once a year to breed. And it is likely, after one such occasion, millions of years ago, that an albatross or similar bird took off from somewhere in southern Africa with an unintended passenger – a seed – nestled tightly between its feathers, or stuck by mud to the bottom of its foot.

Over days, weeks, perhaps even months, and over thousands of miles of open ocean, this seed remained in place. Through the fiercest of winds and the thickest of downpours, it clung on. It survived even the calm, windless days, when the albatross would land on the surface of the sea and bob about like a cork, waiting for the wind to return. Throughout all this, the seed lay dormant, waiting – until, quite suddenly, there was a dramatic change in its surroundings. The stowaway seed dropped out from the soft bed between the feathers of the albatross. By a stroke of extraordinary luck, this happened not over the thousands of miles of ocean that was all around, but over the rich volcanic soil of St Helena.

This origin story might sound ludicrously far-fetched – a long chain of improbable events that, taken together, border on the miraculous. However, view it from the right perspective and it becomes all too plausible. St Helena emerged from the ocean 14.3 million years ago, squeezed as molten rock from an undersea volcano, which settled and cooled repeatedly to build the island. 14.3 million years is a timescale that is difficult for us humans to comprehend (the entire story of our own species could play out forty-seven times in 14.3 million years), but this is the amount of time the St Helena olive had to play with in making the jump to St Helena. 'The message from this is that rare events happen over geological time,' Fay tells me. 'And, for the St Helena olive, it only had to happen once.'

From the work of Fay and others, sequencing and studying the genome of the St Helena olive, we know that the species was about 12 million years old at the time of its extinction. Whether it speciated on St Helena, or did so in southern Africa and then later travelled to the island, isn't known. Nonetheless, that first seed, dropped by an albatross, would have found itself

*The tiny deep-pink flowers of the St Helena olive in bloom. (Photo by Andrew Jackson.)*

in an alien world. For a while, it would have been alone, the only one of its kind on the island. First just a tiny, fragile shoot, centimetres high, growing up into a delicate sapling and then – its supple stems thickening and dividing, and stiffening into branches – a majestic tree, adding its distinct grey-green colour to the island's cloud-forest canopy. From its branches sprang white blossoms and pink flowers. Insects living on the island arrived to pollinate it and, eventually, seeds were produced from which the next generation grew, and then the next and the next and so on. And in this way the tree became not an alien in St Helena's landscape but a part of it.

'IF ANYONE SAID to me, you've got one chance and you can go anywhere in the world, at any time in history, for one day,' says Phil Lambdon, 'I would go back to St Helena in probably about the year 1500.' Lambdon, a plant ecologist who has studied the flora of St Helena extensively, talks to me animatedly about what someone in this situation might witness. The place he evokes is a curious one, populated by very large bees and flies, and strange trees that have evolved from daisies, with extremely soft wood and leaves that cluster right at the tips of their branches. From Lambdon's experience studying the ecosystem of St Helena, he says it feels as though things are missing: types of species that scientists would expect to find there are absent, and the species that are present have evolved in unusual ways to fill niches that they don't quite know how to fill. 'It's almost like an ecosystem that's been put together by someone who's read about an ecosystem in a book, but has never actually seen one before,' he says, before rephrasing this more rhetorically as: 'If you just wiped out evolution and gave it another run, how would it turn out?'

Species took strange turns on isolated St Helena, like the now-extinct St Helena giant earwig (*Labidura herculeana*), which grew to over 8cm in length, or the spiky yellow woodlouse (*Pseudolaureola atlantica*), which is covered in long spines and brightly fluoresces under ultraviolet light. The St Helena olive, while not an outright oddball among the other plants in its tribe (Phyliceae), was nevertheless genetically unique, being the sole occupant of the genus *Nesiota* (from the Greek, meaning 'island dweller'). Today, over 500 endemic species call St Helena home, including 420 terrestrial invertebrate species and 45 vascular plant species. That means that 30 per cent of the total endemic species found in all UK territories combined are squeezed into an area only a little larger than Manchester.

Despite St Helena's extraordinary level of biodiversity, it represents a tiny fraction of what a time-travelling plant ecologist would find in the year 1500. Today's St Helena is, ecologically speaking, a shadow of its former self, with only an estimated 1 per cent of the native vegetation intact, dotted about in patches.

And there's a simple reason for this. In 1502, humans discovered St Helena, and this event fundamentally undermined the entire bedrock of the island's unique ecosystem. Once isolated from the world, St Helena was now connected, via shipping routes and navigators' charts, to far-flung places. It was as if a drawbridge had been lowered over the vast moat that had, for millions of years, protected the island.

DURING THE AGE of Exploration (1400–1600 CE), when Europeans travelled the world in search of goods, trade, and new lands, many believed that the Garden of Eden was a physical place. Explorers like Christopher Columbus searched for Eden during their travels; it was thought that it could be in some uncharted corner of the world, cut off by impenetrable mountains or vast seas. Remote islands were often seen as good candidates, so their discovery was always a cause for celebration.

In 1502, St Helena was discovered by the Portuguese admiral João da Nova, who was travelling from India to Portugal. Though not explicitly described as Eden, accounts of the discovery frame the island as a gift from God. Just as Columbus had in letters described the Orinoco River as streaming 'from paradise', the Portuguese historian João de Barros wrote in 1552 that da Nova 'was fortunate, because God revealed to him a small island . . . where he took in water . . . God appears to have created this island in that very location in order to nourish all those who come from India . . . it offers the best water on the whole journey.' Da Nova christened the island with the name of a saint whose story is emblematic of the meeting of the holy and mythical, and of discovery and exploration. Flavia Julia Helena, who was canonized St Helena, patron saint of discoveries, was a Roman empress and the mother of

Constantine the Great. In around 320 CE she travelled to Jerusalem and, under a temple commemorating the goddess Venus, found what was believed to be the cross on which Jesus had been crucified.

One person who undoubtedly bolstered the reputation of St Helena as a gift from God was the island's first resident, Fernão Lopes. Once a knight of the Portuguese empire, Lopes had rebelled against his country's colonizing mission in Goa, joined a local resistance movement, and married a Muslim woman. After he was captured, he was publicly tortured for three days, during which his nose, ears, right hand, and left thumb were cut off, and children were invited to smear excrement in his wounds. In 1516, as the ship transporting him from Goa back to Lisbon anchored at St Helena, the now disfigured and disgraced Lopes took himself ashore and hid in the forest until the ship's crew gave up searching for him and left.

To Lopes, who wanted nothing more than to hide away from the world, St Helena really did become something of a paradise. Over time he cultivated vast fields of crops using seeds left for him by visiting sailors. Barley, mulberries, red peppers, turnips, radishes, pomegranates, dates, mangoes, lemons, and more all grew by Lopes's hand. He also kept animals. When he'd jumped overboard, the crew had left food for him before sailing away, in the form of seven goats and a cockerel (the latter becoming his pet-cum-companion, who followed him around and slept on a shelf above his bed).* There were already goats on

---

* The story of Lopes and his cockerel is quite beautiful. Lopes saved the bird from the tide, dried it off by the fire in his cave, and fed it rice. In a 1906 book, *Heroes of Exile*, the author Hugh Charles Clifford writes of this relationship: 'thus love, of a sort, came into the life of this lonely man, the love of and for a creature which owed him everything, to which he, the outcast, played the part of an omnipotent and beneficent Providence.' Clifford imagines the pair, wandering the forest, foraging, 'an outlandish couple, the man crippled and gnarled, the bird dismally

the island, released over the years by sailors hoping to establish a living meat larder for future visits – a common practice at the time. And, from these and Lopes's goats, a large population formed.

Now, when sailors visited St Helena they really did find a land of plenty. Not only fresh water, but also an all-you-can-eat buffet of goat meat, along with vegetables, fruits, and herbs. And when the sailors returned home with stories of this verdant place and its attendant hermit, comparisons with biblical stories of John the Baptist and even Adam rippled through a devoutly Christian Europe. Concurrently, however, the ecosystem, which previously included no mammals, amphibians, or reptiles, absorbed the influx of new animals and plants, including rats that scurried ashore from visiting ships.*

It was the English who eventually settled St Helena in 1659. Perhaps drawn to the island by tales of an Edenic paradise that had rippled out from Lopes's time there, the English had wrested control of St Helena from the Dutch, Spanish, and Portuguese. Eden, of course, would have been the crown jewel of British conquests; but a useful stopover for ships that helped stave off scurvy, thirst, and hunger in sailors was almost as valuable. Rather than govern it directly, however, the island was placed in the hands of the English East India Company (EEIC), which had been formed in 1600 to trade in the Indian Ocean,

---

bedraggled, as befits a fowl which has been long cooped on board a ship, and has thereafter suffered grievously in an encounter with the waves. Yet I fancy that each was happy after his fashion.'

* In 1515, the year before Lopes arrived, St Helena hosted an even more unusual guest: a rhinoceros. It was a gift for the Portuguese king from Alfonso d'Albuquerque, the same general Lopes had rebelled against in Goa and who had ordered the knight's torture and disfigurement. A sketch of this rhinoceros was sent to the German artist Albrecht Durer, who created a woodblock print – now widely known as *Durer's Rhinoceros*.

and which brought with it its own ideas about St Helena's identity and purpose.

Islands like St Helena were referred to by the EEIC as 'factories'. The ambition for such places was as simple and single-minded as the name suggests. The colony on St Helena, headquartered at the newly constructed capital of Jamestown, would be guided by the company's belief that the island's Edenic promise should be harnessed and channelled into profit. Every factory needs workers, and this is where St Helena took on another dimension – it became a prison. White English farmers were offered land, homes, livestock, and wages to come to St Helena; Black Africans, and Asians, were promised death if they didn't. All told, tens of thousands of enslaved people lived out their lives on the island between 1659 and 1839.* During this time, life was eked out of the island as it was from the people enslaved upon it; resources were used up, land farmed, all space utilized. In the words of Julia Blackburn, '[the island's] rocks, its trees, its soil, the birds that flew above it and the sea creatures that swam around it, as well as the people and animals who had recently adopted it as their new home – [were] seen in terms of . . . money.' If, previously, St Helena had been worshipped and revered as an idyll, over this period it was consumed.

Natural vegetation was cleared en masse, so that every imaginable crop could be grown instead. Forests were felled to build and then heat new homes, to fuel lime kilns, to smelt, and to create fencing. Endemic St Helena redwood (*Trochetiopsis erythroxylon*) and St Helena ebony trees (*Trochetiopsis melanoxylon*) were beholden to the tanning industry, whose workers

---

* The horrors endured by enslaved people on St Helena are remembered by the island's trees. 'Under the trees', the official location that was printed on eighteenth-century notices advertising slave auctions on St Helena, was a reference to a pair of trees still growing at the bottom of Napoleon Street in Jamestown today, where a fresh nail was hammered into one of their trunks for each person sold.

used acid derived from their barks to tan hides. Feral cats, introduced to control legions of crop-destroying rats, turned their attention almost exclusively to young starlings. And new plants invaded ecosystems, choking out native species. Feral goats, whose numbers had ballooned astronomically by the time the English had arrived, and were never effectively controlled thereafter, tore through vegetation like lawnmowers, leading to widespread soil erosion. During heavy rain, so much soil was washed from the land that the sea around St Helena reportedly turned black.

In 1771, HMS *Endeavour*, commanded by James Cook, and with the naturalist and botanist Joseph Banks on board, arrived at St Helena. Banks immediately remarked on St Helena's barrenness. 'Even that valley,' he wrote in his diary, when the ship moored close to Jamestown after having travelled around the steep, bare cliffs of the island, 'resembles a large trench, in the bottom of which a few plants are to be seen.' The mountains, he added, 'are as bare as the cliff next [to] the Sea. Such is the apparent barrenness of the Island in its present cultivated state.' All around, Banks could not 'see any signs of fertility'. Almost all the habitats on the island had been destroyed. He ended his diary entry by unfavourably comparing St Helena with the Cape of Good Hope, 'by nature a mere desert', which had been made abundant by the Dutch who owned it at the time: 'Was the Cape now in the Hands of the English it would be a desert,' he says, 'as St Helena in the hands of the Dutch would as infallibly become a paradise.'

Banks urged the East India Company to start reforestation on the island, a project which eventually commenced in the early nineteenth century. With little understanding of the importance of preserving the endemic species of the island, however, the company brought in tree and plant specimens from around the world. The Governor of St Helena, Colonel Robert Patton, ordered the planting of thousands of a Mediterranean tree, *Pinus pinaster*. And in 1805, William Burchell, a botanist from London, brought herbarium specimens from Madeira to St Helena, and wrote: 'I carefully sowed them in a garden up the Valley above the Town where three plants grew up and plentifully produced seed.' He noted that the new seedlings thrived 'luxuriantly' and, by the time he returned to England in 1815, the tree species was 'flourishing in the greatest abundancy being a situation similar to that in which I had originally found it at Madeira, and now therefore become a native wild plant belonging to the Flora Heleniana.' Burchell also created a botanic garden on the island, whose plants (inevitably) escaped into the habitats around it. Patton's Mediterranean *Pinus pinaster* trees continued to thrive too, with 11,000 saplings planted by his successor, Alexander Beatson, in just one year between 1811 and 1812.

As more and more non-native trees took root, the St Helena olive made its first appearance in the annals of science. William Roxburgh, a Scottish surgeon and botanist sailing from India to Edinburgh, stopped on St Helena between 1813 and 1814. While there to recuperate from illness before the next leg of his journey, Roxburgh couldn't help but explore the strange island he'd wound up on. Hiking the steep ridges and valleys of St Helena, he took stock of the flora in extensive notes that became the first printed account of the island's plant life. And in the highest parts of the island he came across a tree that no one had yet recorded. 'A native of the most elevated parts of Diana's Peak, and of the Sandy Bay range, where it grows to be a pretty large, but low spreading tree, there called the wild Olive,' reads Roxburgh's description of the St Helena olive, published in the appendices of Alexander Beatson's 1816 book, *Tracts Relative to the Island of St. Helena*. Perhaps in reference to its elliptic – or oval-shaped – leaves, he gave it the species name '*elliptica*'.

Roxburgh's description of the St Helena olive gives us no indication of how abundant the species was when he discovered it. However, it ends in what now reads as a foreboding and premonitory statement: 'the wood is dark-coloured . . . hard, and very useful.' It's possible that some St Helena olive trees found their way into Jamestown homes around this time, their 'hard, and very useful' wood fashioned into pot handles, broomsticks, and fruit bowls. Perhaps, by the time Roxburgh arrived, the species had already been whittled away from the more accessible parts of the island and persisted now only in the remotest corners of its range.

NON-NATIVE PLANTS ARRIVED on St Helena by a variety of means, transforming the island's landscapes. First, there were the seeds planted by Lopes and other early visitors. Then, the EEIC's cash crops like coffee and tobacco, as well as staple food crops like yams to feed the island's enslaved and indentured workers. Then there were the trees brought to fill in the gaps in the island's forests left by ravenous goats and deforestation. Beyond these purely functional additions to St Helena's flora, many plants were also brought to the island for no other reason than that they were particularly desirable amongst the European elite. By the time Roxburgh discovered the St Helena olive, a plant craze was sweeping European society, with lavish gardens populated by plants gathered from the far corners of the Earth springing up across the continent. St Helena – which had been a stopover for sailors since its discovery – became a stopover for plants to rest and recover between the weeks-long legs of their globe-spanning voyages, and the St Helena Botanic Garden became a 'convalescent home for sickly plants'. Like a departure lounge at a busy international airport, the island was a place where plants native to Asia, Africa, and the Americas rubbed stems. By 1820, up to 6,000 exotic plants were entering Britain each year by way of St Helena. Willows, apricots, pears, poplars, sycamores, aspens, filberts, horse chestnuts, nectarines – all arrived on the island, to be offloaded from ships at Jamestown port and taken to the garden to recuperate. And, inevitably, some of these escaped and established themselves in the wild.

*Xerochrysum bracteatum*, a species commonly known as the everlasting daisy, ended up on St Helena for a very different reason. In 1815, the British empire exiled Napoleon Bonaparte to St Helena after his defeat at the Battle of Waterloo. Furniture and wallpaper at the height of Parisian fashion were shipped to the island for the benefit of its famous exile. And so was the everlasting daisy. A native of Australia, this vibrant, metre-tall flower had first been described in 1803 by the French botanist Étienne Pierre Ventenat, based on a specimen he had observed growing in Paris in the garden of Josephine de Beauharnais, Napoleon's first wife. Reportedly, seeds of the everlasting daisy were gathered from this garden and sent to St Helena with a shipment of Napoleon's books. From Longwood House – the converted farmhouse where, under the watchful eye of British officials, Napoleon turned to gardening in order to soothe the wounds of his defeat – the everlasting daisy spread across St Helena. However, by far the most pernicious plant to join the island's ecosystem was yet to arrive.

Not long after Napoleon's death on the island in 1821, St Helena reached something of an existential impasse. After 182 years of EEIC rule, the island hadn't materialized into the successful 'factory' the company had intended it to be; in fact, the EEIC spent as much as twenty-five times more on the island than it got back in annual profit. In 1834, control of St Helena was transferred back to the British crown in exchange for £100,000, and, for most St Helenians, this spelled disaster. The immediate closure of the over 700-strong garrison of soldiers the EEIC had maintained on the island, and the termination of civil servants' contracts, plunged many into poverty.

Having recently abolished slavery, in 1840 the British made St Helena a staging post for intercepting transatlantic slave ships crossing to the Americas; over 25,000 formerly enslaved people were brought ashore from those ships between 1840 and 1872. During this time, the island became a quarantine site from which some formerly enslaved people were shipped off to the British West Indies and forced into indentured labour working for the crown, and where others, including 3,000 children, lived in dire conditions in specially built 'accommodation huts', while they worked in the lowest-paid jobs on the island.* Other than the captured vessels these people arrived on, fewer and fewer ships stopped at St Helena as the nineteenth century wore on. The invention of the steamship cut the time it took to travel by sea from the Cape of Good Hope to Britain by three-quarters. Meanwhile, the Suez Canal, which opened in 1869, provided European ships with a direct route to Asia. There was no longer much need for a mid-Atlantic stepping stone like St Helena, and the island's local economy, already on the ropes, flatlined.

In the search for a solution to turn the island's fortunes around, everything from whaling to ostrich-farming was considered. Then in 1874, a gap in the market was spotted. For some time, the amount of post being sent by the British public had been increasing. Between 1850 and 1870, the number of letters sent annually rose from 327 million to 877 million (or 13.2 letters per person to 47.5). The British Post Office

---

* In 2008, during the construction of St Helena Airport, a burial ground was discovered containing the remains of over 10,000 formerly enslaved people. A plan was put in place for a 'peaceful and respectful final resting place' and for a memorial at the site, which is described as 'the most significant physical remaining trace of the Trans-Atlantic Slave Trade on Earth.' The plan was endorsed by the St Helena government, led by an appointee of the UK Foreign Office, but in 2024 has still not been enacted. At the time of writing, the St Helena government may face legal action over this failure.

required a huge quantity of mailbags and string to tie all of this post together. St Helena, which had provided sanctuary for masses of plants destined for English gardens, could perhaps now be rescued itself by a plant. New Zealand flax (*Phormium tenax*), an evergreen perennial named after the country of its origin where it is an integral part of traditional Maori culture, has long, strap-shaped leaves which can grow to four metres in height and be processed to make twine, rope, paper, and other products. Crucially, this hardy plant can grow in just about any soil conditions – even, it turned out, within the degraded, goat-ravaged soil of St Helena. And so it was decided: St Helena would become a flax factory.

In 1875, the first shipment of New Zealand flax fibre left the island and the St Helena olive made its first appearance in literature since its discovery. John Charles Melliss, a British engineer and naturalist born on St Helena in 1835, embarked on an exhaustive mission to catalogue the status of the island's flora. In his resulting 1875 book, Melliss describes the St Helena olive as a 'handsome indigenous plant known as the Wild Olive of Diana's Peak, growing amongst ferns and other native vegetation'. He either failed to locate the trees Roxburgh had described in the Sandy Bay range in 1813, or – perhaps more likely – the St Helena olive no longer inhabited this area. Either way, roughly fifty years after Roxburgh's visit, all Melliss could find of the species was a handful of trees, dotted about the northern side of the island's central ridge. 'Very few trees now remain, probably not more than twelve or fifteen at the most . . .' he writes.

For nearly a century, the fortunes of St Helena's flax industry waxed and waned. During the two world wars, demand for flax soared (it was used to make rope, coats, parachute harnesses, tarpaulins, and linen to surface the wings of warplanes), but at other times the industry had to be subsidized. At its peak, flax covered 10 per cent of the entire island and 30 per cent of cultivated land, employed 300–400 people out of a general population of 5,246, and provided 99 per cent of the island's exports. All of this culminated in the island turning a profit for the first time in its history in 1951. Images from the time show swathes of the island that were previously green turned the signature gold of sun-drying flax. On a 1960s St Helena postage stamp, the landscape depicted beside a profile portrait of Queen Elizabeth II is dominated not by trees but by flax, planted in neat rows. The buzz of machinery used to harvest the flax became a feature of the island's soundscape: 'Wherever you went you could hear the distant mewling of the flax strippers like Scottish bagpipes in the hills,' explained the documentary film-maker Charles Frater, who visited the island in 1962.

This boom didn't last long, however. Synthetic fibres soon replaced natural fibres, and in 1965 the British Post Office abandoned flax string for nylon. In the annual review of 'the colony' – a report made about the island's people and economy to be sent back to London – the end of St Helena's flax industry is described: 'The flax industry . . . which has provided employment where otherwise there would have been none, virtually came to an end in December 1965.' In the aftermath of the crash, businesses folded and flax mills were shuttered. But even when scores of people lost their jobs, the flax didn't stop working. From abandoned fields, the hardy flax plants spread uphill. In the cloud forest, the St Helena olive's home, the plants thrived, bedding their roots into the soil and smothering native vegetation. The St Helena olive had survived centuries of ecological destruction. But now, the wild and fast-spreading flax was tightening like a noose around the tree's last refuge.

IT WAS JUST the slightest interruption in the dark-green canopy of St Helena's cloud forest that drew George Benjamin's attention in 1977. Benjamin, a former flax-mill worker now employed by the St Helena government as a forest guard, spotted a speck of grey-green foliage from a mile away as he scanned Diana's Peak from an adjacent hillside. Even at this distance, he could tell this was a plant he hadn't seen before. But as much as this dot of colour piqued his interest, the tree was completely inaccessible. A thick blanket of head-high New Zealand flax surrounded Diana's Peak, and it would take a team of machete-wielding workers days to cut a path through it. Knowing that this was an operation that his bosses would never approve, Benjamin filed this mysterious tree away in the back of his mind.

Born in 1935 on St Helena, Benjamin had harboured a fascination with the endemic flora of the island since childhood, sparked by an enthusiastic schoolteacher. He had a gift – a remarkable ability to identify individual plants from great distances. Benjamin would scan the sides of valleys and cliffs by eye, searching for endemic plants. Many of the species he was looking for, which had not been seen since the nineteenth century, had taken on an almost mythical status on the island. But Benjamin kept searching. It wasn't until 1976 that his persistence was rewarded, when he spotted a she cabbage (*Lachanodes arborea*), an endemic tree that was presumed extinct. Then, a year later, he spotted the dot of grey-green on Diana's Peak, and wondered if it might be the St Helena olive, which hadn't been seen since Melliss had recorded 'twelve or fifteen, at most' 102 years earlier.

This brief brush with rediscovery could have been it for the St Helena olive. The flax might have simply swallowed it up, without anyone ever confirming that the species persisted in this solitary tree. But this was not to be, thanks to yet another of St Helena's lost endemics. In 1980, Quentin Cronk, a botany student from Cambridge University, travelled to the island to search for endemic plants – in particular the St Helena dwarf ebony (*Melhania ebenus*), a flowering plant that hadn't been seen since 1890. St Helena's governor introduced Cronk to Benjamin to be his guide on the island. 'We got on extremely well. We had the same kind of passion for plants, and so we started plotting how we would find the ebony,' Cronk, now Professor of Botany at the University of British Columbia, Vancouver, tells me. At the end of an exhausting week hiking the island's steep terrain, he and Benjamin found what they were looking for – the last two St Helena ebonies, clinging to the side of a vertiginous cliff. Benjamin's brother, Charlie, whose work as a fisherman often saw him scaling cliffs to reach otherwise inaccessible fishing spots, was enlisted to help. He managed to take cuttings and a flowering shoot, the latter of which he held between his teeth so his hands would be free to climb back up the cliff.

'The rediscovery of the ebony kick-started a big conservation project on St Helena,' Cronk says. 'Suddenly, all eyes were focused on the island's endemic flora.' The UK and St Helena governments, the IUCN, the World Wildlife Fund (WWF), the Royal Botanic Gardens, Kew, and others immediately embarked on a collaborative initiative to rescue the island's endemic plants. Appointed as the island's first Conservation Officer in 1982, Benjamin could at long last arrange for a path to be cut through the flax, so that he and Cronk could examine the mystery tree he'd spotted in 1977. Sure enough, it was the St Helena olive, growing from the side of Diana's Peak at an awkward angle, the only representative of its species, and an anomaly amongst cabbage trees, ferns, and flax. An adult tree, its thick branches wound this way and that, draped in moss and lichen. Benjamin had now rediscovered three of the island's long-lost endemic plants in the span of six years. Steadson Stroud, a botanist and another of St Helena's native sons, then added two more rediscoveries to the tally: the bastard gumwood (*Commidendrum rotundifolium*) and the St Helena boxwood (*Mellissia begoniifolia*).

Soon Benjamin's propagation facility burst into life, as cuttings and seeds from plants thought lost for decades or even centuries took root. It was a new dawn for the island's flora, albeit with one notable exception. Whatever was tried, cuttings and seeds taken from the last St Helena olive invariably failed to take root or germinate. Then Benjamin noticed a change in the last wild St Helena olive on Diana's Peak. Wilting leaves and increasingly bare branches – the telltale signs of dieback.

A CONSERVATION INITIATIVE called Project Popeye was launched in 1984. Named after the hero of the cartoon *Popeye* and his unending struggle to defend his girlfriend Olive, the project saw scientists and staff from Kew join Benjamin, Cronk, and others in trying to save the St Helena olive. Over the next few years, the atmosphere surrounding the last wild tree changed drastically. This remote and inaccessible spot, which had been left alone by humans for so long, was suddenly awash with them. Boots trampled through the flax leading up to the tree, and conservationists lumbered along its branches – counting insects, collecting seeds, and cutting off thin sprigs to take away with them. A platform was built around the tree to support both the work of the conservationists and the tree's ailing branches, and soil samples were dug out of the ground around its base. This solitary tree, once completely anonymous and forgotten, became famous across St Helena as conservationists took to the airwaves to talk about its plight on local radio, and in the island's schools Benjamin spoke to the next generation of islanders about this rarest of trees.

With the last tree in decline, much of the work of Project Popeye was aimed at getting it to reproduce. Andrew Jackson, a twenty-six-year-old horticulturalist from Kew, was sent to the island. 'The first challenge was getting pollen to study,' he says when we speak. When Jackson arrived on St Helena, the weather was wet and misty. As a result, pollen was being washed from the flowers before he could collect it. Jackson had to get creative. 'Next time you go into a supermarket and buy a freshly baked loaf of bread, you'll see these transparent bags that have tiny pinpricks in them so they're breathable... That's what we used to protect the flowers,' he explains. Slipped over branch tips like miniature raincoats, these bread bags kept the flowers' pollen safely in place, ready for collection each morning. Jackson sent the pollen he collected from the last wild St Helena olive for testing. 'If the pollen is completely infertile then, to be honest, you haven't got a hope,' says Jackson. On this front at least, there was good news: the pollen was fertile and germinating.

The next mystery to solve was how the pollen of the St Helena olive was spread. Cronk had earlier hypothesized that a mystery pollinator – possibly a fly – was out there somewhere on the island, waiting to be revealed. 'But nobody had actually seen anything actively pollinating the tree,' says Jackson. To remedy this, he began conducting nightly stake-outs of the tree, during which he meticulously recorded the comings and goings of moths, beetles, snails, mites, spiders, flies, and other invertebrates. Night after night, he watched as visitors crawled or slithered along the tree's branches, while others whizzed through the air seemingly oblivious to the lonely flowers beneath them. Then, finally, an unfamiliar hoverfly appeared, propped its forelegs up against the petals of a flower and plunged its head inside. This was *Sphaerophoria beattiei*, another species endemic to St Helena, which had been discovered in 1977, the same year that Benjamin had rediscovered the St Helena olive. Repeated observations soon confirmed that it was the pollinator of the St Helena olive. However, given that the last wild tree had not yet reproduced, the hoverflies clearly weren't doing enough. 'I think the St Helena olive would have preferred to live a few hundred metres down and, there, its pollinators probably would have been more effective,' Jackson explains. To supplement their efforts, Jackson, Benjamin, and others began pollinating the tree by hand. Simulating the passage of hoverflies between flowers, they gently dabbed the anthers to collect pollen, then transferred it to the stigmas, using fine-bristled artists' paintbrushes.

Despite hundreds of hand-pollinations, however, the tree only very rarely produced seeds, and those seeds almost always failed to germinate. This turned out to be the result of something called 'self-incompatibility'. Designed to minimize inbreeding, self-incompatibility is an adaptive trait that prevents a plant from pollinating itself – a sound idea on paper, but less helpful when the plant in question is the last of its species. In the case of the St Helena olive, its rate of self-incompatibility

*The hoverfly* Sphaerophoria beattiei *pollinating the last wild St Helena olive in 1990. Endemic to St Helena, this species has its roots in southern Africa, just like the olive. (Photo by Andrew Jackson.)*

(99 per cent) was extraordinarily high. Usually, there is a back-up method that botanists can rely on in situations like this. By inducing cuttings to take root, new plants can be propagated from their parent, bypassing the need for seed entirely. But in the case of the St Helena olive, it took over 400 failed attempts to produce even one successful propagule from a cutting taken by Benjamin in 1986.

With the tree rapidly declining, the amount of healthy tissue suitable for propagation was decreasing. On 31 December 1988, it was arranged for the healthiest cuttings to be transported to Kew for micropropagation – a more advanced propagation technique whereby thousands of plants can be produced in a laboratory from a small piece of plant tissue. This procedure, which was impossible on St Helena where the right equipment wasn't available, was thought to be the best hope for the species, but it had to be done before the cuttings degraded. Transported first by boat to Ascension Island, 1,300km away, the cuttings were loaded onto a military jet, which flew them to RAF Brize Norton in Oxfordshire. Meeting them on touchdown was Mike Fay, then leader of Kew's micropropagation unit at the Jodrell Laboratory. Fay and his team would be performing the micropropagation attempt on the cuttings, in hope of saving the species from extinction.

'Like an intensive care unit for baby plants,' is how Fay describes the micropropagation unit at Kew to me when we speak. Here, on New Year's Eve 1988, he carefully unwrapped and laid out the St Helena olive's cuttings. The shoots were blackened, possibly due to exposure to sub-zero temperatures during the flight, and degraded, but Fay and his team perse-vered. First, to get rid of any bacteria or fungi, the cuttings were gently bathed in a soapy solution. Then they were soaked in chlorine, which kills bacterial and fungal spores on the out-side of plant material. After this, they were placed in a sterile laminar air cabinet, a piece of equipment most commonly seen in hospitals in which the most delicate of tissue and blood samples are worked on. At Kew, the cabinet was used for the most critical work on vulnerable plants. From the cuttings, Fay collected what he thought might be the most viable buds and then planted them in a nutrient jelly, a mixture of minerals, plant growth hormones, and a small amount of sugar. Then, with the propagation attempt underway, Fay and his team cleaned up, hurriedly changed, and left Kew to join their friends and families to see in the New Year.

It was hoped that, left alone in the dark of the laboratory, the propagated buds would soon begin to stir; that thin, fragile roots would sprout and weave themselves into the surrounding jelly. Then, as people around the world made resolutions for the year ahead and returned to work or school, a new hope for the species would take root. However, none of the buds survived, a fact that was only evident after several days of watching and waiting. When it was clear that the process had failed, Fay and his team analysed the cuttings and discovered that they were riddled with fungi and bacteria, both inside and out. A team of mycologists Fay asked to investigate the cuttings 'gave up counting when they got to seven or eight different species of fungi,' he tells me.

Back on St Helena, pollination and propagation attempts continued, even as the condition of the last wild tree worsened. In October 1990, one of its main branches, which formed one third of its canopy, was found to have died; inside were rings of fungal infection. Over the next three months, three more branches died, including one with an immature seed capsule growing on it. A part of the tree's trunk was also discovered to be embedded in the soil and decayed, suggesting that the tree had fallen over decades earlier, before its rediscovery, and then had continued to grow sideways. When this had occurred, it is thought that as much as half of the tree's original canopy had broken away, and its root system had been damaged, after which fungi and bacteria had set to work attacking the tree. Perhaps, even as Benjamin had stood eyeing the tree from across the valley in 1977, wondering what on Earth it was, this process had already been underway.

On 13 October 1994, it was discovered that the last wild St Helena olive tree had died. A local radio station announced the news to the island. When we speak, Jackson tells me about a day during which he spent time with the tree alone, before its death. 'I remember standing on top of Diana's Peak, just me,' he says. 'George wasn't there, and it was a clear day. For three hundred and sixty degrees all the way around was just the ocean. And it dawned on me just how remote the chances of that original colonization were. That, and the subsequent divergence and evolution of *Nesiota*, is remarkable. And here was the last wild one, on its last legs, in front of me. It moved me somehow.'

The only hope now lay in the sapling that Benjamin had managed to propagate from a cutting, along with three younger plants grown from seed. Thin, spindly, and vulnerable, two were planted in George Benjamin's back garden so that, no matter the time of day, he would be on hand to tend to them. From his kitchen window, he could keep an eye on them.

There was a chance that the four saplings might provide a turning point for the species. If all grew into adult trees, pollen could be circulated between them; then as these trees produced seeds and saplings of their own, perhaps the genetic knot caused by generations of inbreeding could slowly be untangled. One day far into the future, when the responsibility of St Helena's conservation had passed into the hands of others – perhaps some of the children that Benjamin had spoken to during his school visits – the St Helena olive would return again to the island's cloud forest.

At first, there were positive signs – the four saplings seemed to grow healthily. But then, as if taking a cue from their parent tree from beyond the grave, they started to decline, one after another. 'They all went in exactly the same way . . . after a prolonged wet period, possibly preceded by drought,' says Rebecca Cairns-Wicks, a botanist and conservationist who has worked with the endemics of St Helena since 1999, and is now the coordinator of the St Helena Research Institute. Each time a tree showed signs of decline, staff hurriedly dug a metre-deep trench around it to provide better drainage. All this did was delay the inevitable, however. 'Once the dieback started there was nothing that could be done to stop it,' says Cairns-Wicks.

In the middle of all this, Benjamin wrote to Quentin Cronk. A 'cri de coeur' (a cry of the heart) is how Cronk describes this letter to me – 'Things are really bad! Can you help?' wrote Benjamin. Cronk, having recently been appointed Director of the University of British Columbia Botanical Garden and Centre for Plant Research, could not help directly. However, via a mutual acquaintance he contacted the Scottish Labour MP Tam Dalyell, who submitted a written question to Parliament voicing Cronk's concerns that St Helena didn't have sufficient support from the UK government to prevent the olive's extinction. 'This caused a little bit of panic in the UK government of the day,' says Cronk. However, when questioned, officials on St Helena downplayed the issue. 'It was a wonderful opportunity for them to say, "This is really important, we need more funds for this",' Cronk tells me. 'But the government of St Helena were so worried about being seen to be in charge that they wrote back and said the olive was safe and would not succumb to extinction.' And, with that, the opportunity was lost.

Eventually, all that remained of the St Helena olive was a single sapling. This young tree fared better than the other saplings. It had been pollinated by hand and had even started to produce seeds, but then, just like the others, it deteriorated. 'It was a really sad, slow demise,' says Cairns-Wicks. In December 2003, the last St Helena olive died in George Benjamin's back garden and the species became extinct.

'Continuity' is a word I've heard time and time again when speaking with conservationists. Whether protecting an endangered species or restoring a degraded ecosystem, conservation is a field where sustained, long-term commitment is needed, and where the slightest lapse in funding, staffing, or attention can see years of progress surrendered in an instant. On St Helena, continuity is particularly difficult to maintain. For one thing, there is the sheer scale of the task at hand: around fifty endemic plant species, each with its own specific husbandry needs. 'You basically need to have one person working on each [critically endangered] species, taking care of it and propagating it, and not stopping until you're at a population in the thousands,' says Cairns-Wicks. This might sound simple enough, on paper. However, with average salaries on St Helena far lower than in places like the UK mainland, many St Helenians opt to leave the island for work. Getting specialist staff to St Helena is also a costly endeavour, and convincing them to stay on a small island thousands of miles from anywhere is another challenge.

For decades, Benjamin had provided much of the necessary continuity for St Helena's plant conservation movement. Along with Steadson Stroud, he worked tirelessly, not only to rediscover species and locate new populations of endemic plants – providing crucial genetic diversity for these species – but to inspire the rest of the island to care for these plants, too. So it was that, as species like the bastard gumwood slowly tiptoed back from the brink of extinction in Benjamin's shade house, other endemics sprang up in the gardens of ordinary

St Helena residents who had been moved to action by his passionate evangelizing for the unique plants of their island. 'One thing that struck me immediately, when I first arrived on St Helena, was how the word "endemic" just tripped off many people's tongues,' says Cairns-Wicks. 'Even kids knew what it meant, and that was thanks to George planting these endemics in primary schools and speaking about them.' So crucial was Benjamin's role in conservation on the island that in a 1991 report on the progress of Project Popeye, Andrew Jackson specifically cited 'the identification of a successor to G. Benjamin' as 'an urgent priority'.

Benjamin retired in 1995 and although he remained involved in conservation, responsibility for the island's endemic plants passed to others. In 2002, a year before the St Helena olive's extinction, the UK government finally granted British passports to the residents of St Helena (a right they'd sought for decades), making it much easier to travel abroad. This coincided with the decision to shrink the footprint of the local government – by far the largest employer – which instigated an enormous exodus. Continuity and momentum in conservation on the island were lost and the St Helena olive 'slipped between the gaps,' says Cairns-Wicks.

Unlike *Plectostoma sciaphilum* – the microsnail from Chapter One, whose extinction was observed after the fact, in satellite images – the demise of the St Helena olive played out in real time, right in front of the people who were striving to save it. 'It was a very humbling experience to be there at the point of the death of a species,' Cairns-Wicks tells me, when I ask her what it was like to witness the extinction. 'It was like burying—' she begins, but interrupts herself mid-sentence. 'The tree was . . . sort of anthropomorphic by the end.' When I ask Cronk the same question, he tells me, 'I found the extinction so depressing that I didn't even want to talk about it.' Benjamin, who had been at the heart of plant conservation on the island for three decades, was particularly affected. 'I think it broke his heart that the St Helena olive was the plant that they didn't manage to save,' says Mike Fay. Benjamin passed away on 30 April 2012 at the age of seventy-six, just under a decade after the olive's extinction.

While the effort to save the St Helena olive failed, the work of Benjamin, Cronk, Jackson, Fay, and others on the species facilitated important scientific discoveries. In 2001, Fay co-authored a paper that was published in the scientific journal *Nature*. James Richardson, a former Ph.D. student of Fay's, had used genetic material extracted from Jackson's cuttings of the last wild St Helena olive – which Fay had attempted to propagate at Kew in 1988 – to conduct a phylogenetic study of the Rhamnaceae plant family. Building on this work, Richardson, Fay, and their colleagues created a 'molecular clock', which enabled them to precisely date the radiation of taxa within Rhamnaceae, including a group consisting of *Phylica*, *Nesiota*, and *Noltea* that radiated from southern Africa. A fascinating and detailed portrait emerged of the five genera of this group diverging and escaping southern Africa in all directions, like sparks from an exploding firework. Some of these genera travelled north, through the African continent, while others dispersed west and east, into the Atlantic and Indian Oceans. Some species embarked on extraordinary and surprising journeys, like *Phylica arborea*, a tree that had dispersed to the Tristan da Cunha archipelago, 2,450km south of St Helena, before somehow finding its way to New Amsterdam, an island over 8,000km south-east in the Indian Ocean. All of this could be read in the DNA of these plants, like a travel itinerary, with each stop on their routes timestamped in the molecular clock.

It was thanks to the St Helena olive that all of this was possible. Knowing that St Helena itself had formed 14.3 million years ago, before which the olive couldn't possibly have dispersed there, the researchers had a fixed point in time by which to 'calibrate' their clock. La Réunion – at 2 million years old, a comparatively young volcanic island – in the Indian Ocean, which is home to a species of *Phylica*, provided another, more recent anchor point. 'If you take those two dates and calibrate the [phylogenetic] tree from both directions, when you get down into the middle of the tree, you end up with basically the same answer,' says Fay. This meant that Richardson and Fay had an independent method of dating the radiation of *Phylica* species from southern Africa. 'If the St Helena olive hadn't been rediscovered and we hadn't been able to get intact DNA from it, we wouldn't have ever been able to ask these bigger scientific questions,' says Fay.

However, this work also raised questions about the St Helena olive itself. Comparisons between the 'genetic fingerprints' of the last wild tree and the saplings that briefly survived it showed that these plants were all essentially identical to one another. 'For a tree not to have any genetic variability at all in the progeny tells you that it'd probably been on the verge of extinction for a long time,' says Fay. Even without fungal and bacterial infections, root damage, discontinuity, and government blunders, it's entirely plausible that the single, severely inbred St Helena olive tree was beyond saving. 'I think there is a real question hanging over the last tree,' says Cairns-Wicks on this topic, 'which is, from the moment it was rediscovered by George Benjamin in 1977, did it have a future? We'll never know.'

SOME PEOPLE HAVE suggested that the St Helena olive might still exist, somewhere on St Helena. *The species had been presumed extinct for over a century before George Benjamin rediscovered it,* the argument goes – *so who's to say it isn't lurking out there somewhere, so well hidden that it has not only avoided our attention, but that*

*of the myriad threats that caused the species' decline?* In 2008, Phil Lambdon and Shayla Ellick, a local conservationist, surveyed the entirety of St Helena, documenting every plant species they found and compiling the most exhaustive record of the island's flora to date. Although not exclusively aimed at locating the St Helena olive, the survey provided a conclusive verdict on the species' survival: there was not a trace of it on the island. There is, however, somewhere else we can look for it.

A short walk from the Temperate House, in Kew's Jodrell Laboratory, Mike Fay hands me a small translucent envelope, apologizing as he does so for the temperature in his office. The heating is on the fritz, but in truth I've hardly noticed; in the envelope, resting in the palm of my hand, are leaves cut from the last wild St Helena olive. This was the tree George Benjamin spotted in 1977 to resurrect the species, the tree that had been studied and puzzled over for decades before its demise. These leaves had been set aside from the cuttings that Fay had spent New Year's Eve 1988 attempting to propagate, he tells me. I look down at the envelope. Through the thin white paper, I can see how much the leaves look like olive leaves. 'We've got a couple more stops to make,' says Fay. I return the envelope to his desk and follow him out into the corridor.

Fog pours from the monolithic freezer as the door is opened. Inside, colour-coded test tubes are crammed into drawers, each coated in a thick layer of frost. In this quiet corner of the Jodrell Laboratory, three such gigantic freezers house the 60,000 plant genomic DNA specimens of the Kew DNA and Tissue Collection in a state of permanent deep freeze. '−80 °C is somewhat overkill,' Fay tells me. 'The specimens would probably be fine stored at room temperature, but the freezers give an extra level of insurance.' The frozen DNA in this bank could remain viable for hundreds of years, meaning that as technology advances, even more can be gleaned from it.

In a fridge nearby, a small vial of clear liquid is waiting to be re-analysed. This is a portion of the genetic material from the last wild St Helena olive. I peer into the receptacle at what looks like tap water, as Fay explains how modern techniques will reveal the story of this species in far greater detail than ever before.

We next find the olive in the basement of Kew's herbarium collection. A slim Manila folder contains the St Helena olive specimens, marked with red lines indicating the presence of a type specimen (an evacuation priority in the event of a fire). Collected by Roxburgh in 1824, this specimen is sewn onto the first page within this thin dossier and comprises a forked sprig with a few dozen leaves, each fork ending in a pressed flower. The leaves lack the grey-green colour Benjamin spotted in the canopy of the forest; instead, they are a deep, matt black, and the paper they are on is smeared with dark smudges – remnants of the mercury in which all the samples here were dipped in the nineteenth century to protect them from insects. The remaining pages in this folder contain more cuttings, bundles of broken flowers and seeds folded into makeshift envelopes, and sketches of the flowers and seeds, roughly drawn between cuttings. On one page there is a handwritten note, penned by Melliss in 1867. Fay unfolds it and we slowly decipher his handwriting. Soon, it becomes clear: this is Melliss's record of the last sighting of the St Helena olive before Benjamin's rediscovery of it, in which he states that only a few specimens survive. With its carefully sewn-in and taped-in cuttings, handwritten letters, and sketches, this folder feels like a scrap book – lovingly assembled to chronicle, as best it can, something precious and fleeting. Soft tissue paper protects each page, through which I spot a photograph: the ghostly image of George Benjamin climbing the last wild tree.

BACK ON ST HELENA, a successor to Benjamin has been found. A native St Helenian, Vanessa Thomas-Williams is recognizable as the friendly looking woman in one of the photographs next to the empty pot at the Temperate House, and today she speaks to me over Zoom. She began working with Benjamin at the age of sixteen, straight after finishing secondary school. Paintbrush in hand, she had visited the last wild tree with him and dabbed pollen between its flowers, and the two worked side by side to germinate seeds and pollinate cuttings. When the wild tree died in 1994, Thomas-Williams watched alongside Benjamin as its trunk was cut up into chunks. And eighteen years later in 2012, she was at his side again, as he lay on his deathbed. 'His last words to me were: "Make sure you look after the endemics,"' she tells me. 'And the next day he died.'

*George Benjamin and his team, including Vanessa Thomas-Williams, look out to sea from Man-and-Horse cliffs on St Helena, in 1995. (Photo by Quentin Cronk.)*

Thomas-Williams has worked hard to fulfil this promise. Along with Cairns-Wicks and many others, much of the continuity in conservation that lapsed after Benjamin's retirement has been restored. Now the island's Nursery Officer, Thomas-Williams leads the team keeping the island's most vulnerable plant species alive, as well as managing a mini Millennium Seed Bank on St Helena, in which seeds from the island's plants are safeguarded. Under her care, the bastard gumwood has recovered from a population of just one to over 6,000. These trees have now been planted in the Millennium Forest, a bold conservation project aimed at restoring the Great Wood – a large forest destroyed in the seventeenth century. Thomas-Williams was appointed MBE for services to conservation in 2022.

At Thomas-Williams's plant nursery a banner hangs on the wall, bearing the words 'extinction is forever'. Working underneath this warning, the nursery team carefully tend to the next generation of endemics, so that these plants might one day return to the soil. Sitting in a corner is a part of the trunk of the last wild St Helena olive. Split up into sections, the last wild tree has spread across the island, logs and branches venturing into nurseries and St Helenians' homes, where they are a constant reminder of what has been lost, and what stands to be saved.

# 3 A Tale of Three Snails

A Polynesian Tree Snail
*Partula labrusca*

When Justin Gerlach got back to his room on 27 August 1992, he was ravenous. He'd spent the day hiking across the Temehani Plateau, on the French Polynesian island of Raiatea. One flank of a now-extinct volcano that, millions of years ago, had spewed Raiatea up from the depths of the South Pacific Ocean, today the plateau had seemed to Gerlach eerily reminiscent of British moorland – covered as it was in low vegetation and cloaked in a chilly fog. Now back at the guest house, the 22-year-old malacologist from Cambridge University was looking forward to one of the delicious home-cooked meals Marie-Isabelle, his host, had been making for him throughout his trip. Would he be dining alone tonight, or with company? The morning before, Marie-Isabelle had brought a snail in from her garden to show her guest while he ate breakfast. Before he could think about any of this, however, Gerlach had important guests of his own to attend to.

The vast majority of people who visit the Temehani Plateau do so to look at flowers. Clustered at the feet of thistles, white five-petalled flowers spring up in the little pockets of relative shelter that these bushes provide. These are Tiare Apetahi (*Sclerotheca raiateensis*) flowers and their name, which means 'only on one side', describes their striking characteristic of only producing petals on one side, making them appear curiously like ghostly hands, waving. They are one of the rarest flowers in the world, growing only here on the Temehani Plateau, and are known to have a delicate, sweet smell. The nineteenth-century British–Tahitian scholar and ethnologist Teuira Henry described them as 'open[ing] simultaneously with slight, exploding sounds when touched by the first rays of the rising sun'. In the oral traditions of the Polynesian people, the Tiare Apetahi is said to be the embodiment of a queen called Tiai Tau. One version of the myth says that she watched from the Temehani Plateau as her doomed lover, the king Tamatoa, left for war; then Tiai Tau – whose first name translates as 'waiting' – is said to have planted her hand into the ground, transforming it into the Tiare Apetahi, a flower forever 'waving' seawards to her lover.

When Gerlach visited the Temehani Plateau, he had been looking for a different but equally rare treasure. Like the flower, this treasure could also be thought of as waiting – not for a doomed lover, however, but for someone or something that could rescue it.

Stomach rumbling, Gerlach now slowly tipped the lunchbox he had used for collecting specimens onto its side, rolling six snails with dusky-red shells the size of acorns out onto his bed. He always took stock like this during fieldwork, examining and cataloguing what he'd collected at the end of each day. The bed was ideal for this purpose, providing both a soft landing and a broad workspace that live specimens struggled to escape from.

The snails Gerlach tipped out onto his bed that day belonged to a species called *Partula labrusca*, named after the winemaking grape *Vitis labrusca* on account of their burgundy colour by the evolutionary biologist Henry E. Crampton, who collected the first specimens in 1934. The genus *Partula* was named in 1821 after the fact that all snails of this genus give birth to live young (most snails lay eggs). Its namesake is the Roman goddess of childbirth, who was believed to be responsible for deciding the moment of partum in human lives – the separation of mother from child. There were once seventy-seven *Partula* species, with some living as far west as New Guinea but the vast majority in French Polynesia – a remote collection of 130 tropical islands in the middle of the South Pacific, which together form France's sole remaining 'overseas country'. *Partula* snails wear their shells slightly off-centre – like a backwards-facing baseball cap nudged to one side. These shells are round in shape, gently sloping towards blunt, conical tips, and come in a wide range of sizes, patterns, and colours. Some *Partula* shells are strikingly vibrant, like the blood-orange blend of bright reds and yellows of a *Partula dolichostoma* shell. Most are more subdued, however: pastel shades of pink, brown, and yellow, or deep, earthy hues.

*Partula labrusca* was endemic not only to Raiatea, but to the Temehani Plateau. Just as *Plectostoma sciaphilum* had lived on a single limestone hill, *Partula labrusca* could only be found here, alongside the five-petalled flower Tiare Apetahi. *Partula labrusca*'s shell was distinctly earthy. Over its evolution, it had colour-matched to this specific setting, its body jet-black and its shell a perfect blend of the colours of all the plants, rocks, and leaf litter in this habitat.

If snails feel fear, Gerlach's *Partula labrusca* would probably have been absolutely terrified by the time they'd rolled out onto his bed. The altitude at which these snails live is 1,400–2,000ft, so they'd have undoubtedly sensed the colossal descent to sea level. What would it be like to have your world, in as much as your senses could understand it, turned upside down like this? How would it feel to suddenly exist outside the parameters your body had spent millions of years adapting to? Whatever fear these few snails may have felt as they charted the peaks and valleys of Gerlach's bedspread, they were in fact the furthest they'd been from harm in decades.

Carefully, Gerlach picked up each snail, checking its shell for damage and logging it in a notebook. Other scientists had previously visited the island looking for *Partula* snails, and Marie-Isabelle, who had hosted them for their stays too, had joked to Gerlach about those scientists treating the snails 'like they were platinum'. To Gerlach, though, they were much more valuable. These were, by now, one of the rarest species in the world.

Had he visited Raiatea thirty years earlier, Gerlach would have found *Partula* snails everywhere. On the undersides of leaves, on the trunks and branches of trees, or crunching underfoot – they would have been difficult to miss. But by 1992 *Partula* snails had all but disappeared from Raiatea and many other Pacific islands.

During his stay, Gerlach searched for *Partula* snails in many different places. He found an overgrown, disused road carpeted with millions of empty *Partula* shells. Then amongst hibiscus and coconuts he found 'the ground littered' with the small pine-coloured shells of *Partula garrettii*, and the larger brown shells of *Partula faba*, all empty. In his diary, later, he wrote, 'I hope, but do not really expect to find living survivors.' Clearly, something had gone drastically wrong for *Partula* snails. In just a fraction of a millisecond in geologic time, they had transformed from a thriving genus into a failing one. In fact, Gerlach had arrived on Raiatea just in time for *Partula labrusca*: his was the last sighting of the species in the wild, making the handful of snails he had rolled out onto his bed that evening the very last wild *Partula labrusca*.

WHEN I FIRST spoke to Justin Gerlach about *Partula labrusca*'s extinction, I couldn't shake the image of those last dusky-red snails rolling out of a lunchbox and onto a bedspread. There was something cartoonish about it all – this Cambridge University biologist counting snails on his bed while, downstairs, his landlady ushered in his dinner companions from the garden. But the events that led to biologists like Gerlach having to rescue *Partula* snails in the first place are even more incredible.

According to one version of the story (of which there are many), it all began in 1967 when a French border guard on Tahiti, a nearby French Polynesian island, developed a hankering for a certain creature comfort from his homeland: snail soup. At the time, governments around the world had yet to properly recognize the inherent danger invasive species posed to native ecosystems, so the border guard was able to secure a shipment of giant African land snails (*Achatina fulica*). He had hoped to establish a side business, offering French officials and citizens on the island some small respite from their homesickness. None of this panned out, however, and the border guard soon found himself with little to show for his venture but an expensive and ever-expanding snail collection. Perhaps short on time, or maybe out of some residual affection for the gentle giants he had taken into his care, he decided not to destroy these snails, but set them free instead. Gathering them up into a bucket the border guard simply tipped them over his garden fence, and off they went.

Giant African land snails are the largest land snail on Earth, some growing to around 20cm in length, with shells the size of an average man's fist. In a single year, one snail can produce up to 1,200 young, depositing as many as 400 eggs in each clutch. If they have access to a good food source they can live for up to a decade, and even without food they can survive for months, sometimes years, by resting inside their shells. A giant African land snail's radula (a mouthpiece used to grind food) contains 142 rows of teeth, with 129 teeth per row, in total around 4,000 more teeth than a common garden snail. It would be fair to describe the appetites of giant African land snails as 'indiscriminate'. They can eat everything from leaves, flowers, seeds, fruits, bark, wood, and fungi to bones, eggshells, other snails, and even man-made materials like stucco, concrete, and paint. But they're particularly keen on many of the plants we humans grow to feed ourselves. A single snail can consume 4.3g of grain per day and infestations of thousands are common, making them one of the most destructive agricultural pests on Earth. Recent history is replete with examples of these snails escaping from pet enclosures, or being purposefully released, and wreaking havoc on local agriculture and ecosystems that have no natural defence against such an alien species. Their introduction on Tahiti, purportedly by someone whose job, ironically, was to guard the island's borders, had devastating and far-reaching consequences.

Although French Polynesia's land area totals just 3,827km² – roughly twice the space occupied by London – this is a vast country. Its 130 islands are divided into five groups: the Tuamotu Archipelago, Gambier Islands, Marquesas Islands, Austral Islands, and Society Islands (the group of fourteen islands to which Raiatea and Tahiti belong). And the stretch of the South Pacific Ocean that these islands are scattered across is roughly equivalent in size to Europe. Colonizing the entirety of this disconnected landmass would have been difficult for a terrestrial animal like the giant African land snail. However, the country's human occupants had already solved this problem for them. Clinging to the undersides of trucks and cars, or tucked away inside shipping containers, giant African land snails boarded ferries leaving Tahiti. At just 45km away, Moorea was the closest stop; then, a little further west, there was Maiao, and then Huahine. Each of these islands was like a stepping stone, bringing the snails one step closer to their goal of total domination of the Pacific. Unfortunately for *Partula labrusca*, their island home was next on the itinerary. And so, giant African land snails arrived on Raiatea, alighted from whatever vehicle or piece of cargo they had bolted themselves to, and crept out into the rainforest, unnoticed.

Before long, the giant African land snail had established new populations throughout French Polynesia. At the height of the crisis, islanders described millions of these snails invading farmland and entire plantations being stripped bare; in one orange grove on Moorea, not a single blade of grass remained. There were so many snails that people's homes became infested; one man reported removing two wheelbarrows full of giant African land snails from his house, where they were most likely feasting on the paint on his walls. They weren't just a threat to crops and home decor, however. The shells of giant African land snails are so tough that the largest can puncture car tyres; otherwise, if they are run over and crushed, the slime from their bodies on the road can send cars skidding. In London Zoo's profile on them, they are even described as turning 'into projectiles when they encounter lawnmower blades.' And then there is the disease risk they pose, being vectors for the parasites that cause rat lungworm and eosinophilic meningitis in humans. The spread of the snails was more than just a nuisance: it was a danger to life. Something had to be done quickly, and it just so happened that the United States Department of Agriculture (USDA) was ready and waiting with a solution. *Euglandina rosea* is a predatory snail native to the state of Florida. Commonly known as the rosy wolfsnail, it also goes by another name, on account of its eating habits: the cannibal snail.

It isn't uncommon for snails and slugs to eat each other in the wild – lots of species do this, either out of preference or as a last resort when other food sources are scarce – but the rosy wolfsnail sets itself apart from other snails by virtue of its unique predatory effectiveness. Studies suggest they spend as much as 80 per cent of their lives hunting, and consume around 350 snails and slugs during their two-year lifespans. The key to this success is their speed (8mm per second, in the snail world, is far from sluggish) and their extraordinary palate. In the slime trails other snails leave behind them, a rosy wolfsnail can taste important details about its would-be prey. Its size and species, its health status, and even an idea of how far away it is, all of this and more are reduced into a kind of taste-signature within the trail it leaves behind. The rosy wolfsnail senses all of this using its lappets – hook-like features that protrude sideways from the head and that have chemical receptors dedicated to tasting and analysing trails. All the rosy wolfsnail needs to do is follow this trail, which it does at relative great speed, to find a meal. They are, in the words of Lead Keeper of Invertebrates and Fish at ZSL London Zoo Dave Clarke, 'the guided missiles of the snail world'.

It's easy to see why many governments and agricultural bodies are attracted to biological pest control solutions like the rosy wolfsnail. What simpler, more elegant solution could there be than introducing another species to the affected ecosystem to eat the pest? The approach has, in some instances, been enormously successful. In East Africa, a fungus that eats away at the bodies of locusts has been harnessed to help bring billions-strong plagues to heel, thereby freeing tens of millions of people from the leading natural cause of famine. The prickly pear cactus became a pest in Australia in the early twentieth century, infesting 30 million hectares around Brisbane alone. The problem was solved by the introduction of the moth *Cactoblastis cactorum*, whose larvae ate around 1.5 billion tonnes of prickly pear in under a decade. But, of course, even the most successful biological pest control programme is, by necessity, a game of Russian roulette.* We can never have enough knowledge about a natural ecosystem to perfectly predict the effects a fungus, a moth, or any other biological solution might have on local species. Even the most successful of these operations may have negative effects that we will never know about. Sometimes, such as in East Africa, the gamble is clearly worth it. Sometimes it is not. A careful calculation has to be made that weighs all potential outcomes. On 19 December 1974, the government of French Polynesia instead abandoned reason and stepped out into the dark. Following the advice of the USDA,

---

* The same moth introduced in Australia to control the prickly pear cactus, for instance, now threatens multiple species of prickly pear cactus in North America.

they purchased rosy wolfsnails directly from the department, released them on their islands, and hoped that nature would take to this augmented new course they had set out for it. It ended up being one of the most costly and counterproductive decisions a government has ever made under the pretext of pest control, and culminated in an ecological catastrophe that continues to haunt these islands today.

The rosy wolfsnails released on Raiatea and other Pacific islands largely ignored their intended targets. Rather than follow the thick slime trails left behind by giant African land snails across garden paths and fields, they were instead drawn to the thinner, more delicate lines that crisscrossed the countryside, each of which ended with a *Partula* snail. The banquet that followed would have eclipsed even the wildest fantasies of the snail-soup-loving border guard who had set these events in motion. Rosy wolfsnails swept across entire islands, consuming *Partula* species after species, leaving nothing but hollowed-out shells behind them, like discarded soup bowls. Or, if the species of *Partula* was small enough, they ate the whole of their prey, shells included, to obtain the calcium needed to build their own shells, spending a good thirty minutes on the meal. The same inter-island ferry networks that giant African land snails had used to spread enabled the rosy wolfsnails to hop to islands where they hadn't been introduced, and the cycle was repeated. In the world of *Partula* snails this was nothing short of a cataclysm, as if a giant, world-ending asteroid had suddenly fallen from space.* But, as Gerlach explains to me, it could have been

---

* Could anything else have been done to control the spread of giant African land snails in French Polynesia? The answer is yes. Children have been an effective control agent in Brazil, where there is a festival called 'C Day', during which local school children go out and catch as many giant African land snails as possible. At the end of the festival, the snails are culled.

much worse. '*Partula* were lucky, in a way,' he says, 'they were lucky that *someone* was watching them.'

IN HIS 1779 book on his travels through the Alps, the Swiss naturalist and explorer Horace Bénédict de Saussure coined the phrase 'laboratory of nature'. Elevated above the rest of the world, cut off, providing a 'pure' and 'immaculate' environment, mountains were places where, Saussure argued, 'all the phenomena of general Physics are displayed . . . with a greatness and majesty.' The metaphor persisted, changing over time to become the phrase 'natural laboratories', used today to describe not only mountains, but other environments too. Sebastian V. Grevsmühl describes natural laboratories as places where, 'instead of reproducing nature within the confines of laboratory walls, it could be equally or even more convenient and appropriate to introduce key aspects of the laboratory to the "natural" environment.'

Groups of islands; valleys separated by tall, hazardous mountains; hills surrounded by rainforest – on our planet, natural laboratories like these give us unique insights into the mechanisms behind evolution. 'Geographic isolation, often combined with low predatory and competitive pressures, provides a setting for indigenous species to undergo bursts of adaptive radiation and rapid evolution,' explains the science writer and broadcaster Elizabeth Murray. Birds scattered across one of these natural laboratories, for example a system of islands, might develop different colouring over time, or variations in beak shape, or begin to build slightly different-shaped nests and produce eggs of a different size or colour to their island neighbours. If the habitat is the laboratory, then these slow processes of change are the experiment, whose results have been drawn out over thousands or millions of years. There is huge value in observing these 'experiments'. Charles Darwin's theory of evolution by natural selection was partly inspired by his observations of Galápagos finches – a family of birds found only in the geographically remote Galápagos Islands, which were then mostly undisturbed by humans and almost devoid of predators. The islands of French Polynesia were another such 'natural laboratory', and the thing that made them special – in the eyes of Elizabeth Murray, and many other scientists – was *Partula* snails.

By the turn of the twentieth century, evolution was no longer a controversial idea amongst the scientific community. However, exactly how this process operated was still deeply mysterious. A key topic of debate was whether internal or external factors were the driving force behind evolution. Some believed that gene mutations determined the direction of a species' development, whereas others thought that a species' environment ultimately held the reins. Darwin's theory posited the latter, but crucially it hadn't yet been demonstrated in nature.

In 1906, Henry E. Crampton, an American evolutionary biologist, wanted to find proof for emerging ideas about genetic inheritance. With their varied shell shapes, colorations, and banding patterns, *Partula* snails were the perfect subjects for his experiments and the islands on which they resided the ideal laboratory. Over the next five decades, he visited French Polynesia twelve times to study *Partula* snails. Traipsing through valley after valley, on island after island, he collected specimens and meticulously recorded the conditions in which he found them. He collected hundreds of thousands of snails during this time, and described nineteen new *Partula* species (one of which was *Partula labrusca*), publishing his findings in a series of extensive monographs covering the islands of Tahiti, Moorea, Raiatea, Taha, Bora Bora, and Huahine.

Crampton dissected thousands of *Partula* to study the embryos inside adult snails and learn about the extent to which they had inherited their parents' shell characteristics. He also

*Henry E. Crampton's son (left) and a field assistant demonstrate the abundance of* Partula *snails on the island of Saipan in July 1920.*

recorded all instances where the habitats of *Partula* species were separated by natural divisions, such as the sea or mountains, and made meticulous notes on the ecological conditions he collected each snail from. Through this research, Crampton hoped to determine the extent to which diversity in shell pattern and shape was linked to plant life and other environmental features, which would demonstrate evolution through natural selection. However, to Crampton, the habitats in which he found *Partula* snails seemed much alike. He concluded that this uniformity meant that the differences between *Partula* species – the colour, banding, and forms their shells take – couldn't be the result of environmental influences or natural selection. It had to be genetic, each a kind of fluke, born by chance over long periods of time. Crampton claimed that his findings supported the idea that speciation could come about simply as a result of isolation, and that variation in *Partula* snails was random.

Crampton's conclusion about the evolution of *Partula* snails remained unchallenged for decades. That is, however, until a chance encounter in a library. In 1961, two young geneticists, Brian Clarke and Jim Murray, stumbled upon the work of Crampton as they scoured the bookshelves of a library at Oxford University. They realized that it could be expanded on, using a different approach. Rather than dissecting and studying the embryos of collected *Partula* snails, Clarke and Murray decided to breed them and observe how traits were inherited in real time, over generations of captive snails. They would go to French Polynesia themselves, in a way returning to Crampton's laboratory and picking up his work where he had left off.

It was the island's unique characteristic as a 'natural laboratory' that meant that *Partula* were, in Gerlach's words, 'being watched'. Concurrent to the border guard craving soup, his importing of giant African land snails to Tahiti, and the introduction of rosy wolfsnails from Florida, Clarke and Murray's field research was taking place. All over the islands, from Raiatea to Tahiti to Moorea, *Partula* snails were being collected, in the hope that they would help shed light on evolutionary biology's central concern: the origin of species. Elizabeth Murray, who was married to Jim Murray, was a research assistant on Clarke and Murray's project, along with Ann Clarke, Brian Clarke's wife. Together, these four traipsed the same valleys and mountains Crampton had half a century earlier, filling plastic lunchboxes with specimens to take back to a captive-breeding programme they had established at the University of Nottingham and the University of Virginia. There, the group – at this point oblivious to the new threat arriving into their subjects' habitats – studied the way certain traits were inherited, or not, within populations of *Partula* snails. Peering into the genetic make-up of their test subjects, the Clarkes and Murrays were able to observe *Partula* evolution in a way that simply wasn't possible in Crampton's time. As a result, they discovered that, contrary to Crampton's view, the characteristics of *Partula* species could be determined by environmental factors after all – plants that only grew in certain valleys influencing the patterning of a shell, or the amount of leaf cover affecting its colouring. Natural selection, not just isolation, drove their evolution.

It didn't take long for the threat facing the scientists' research subjects to become clear. Elizabeth Murray writes in her paper 'The Sinister Snail' that it would have been 'interesting' to continue sampling populations of *Partula* in the wild. 'We only wish we could,' she adds. On field trips on which previously the Clarkes and Murrays would have been able to find an abundance of specimens, they now found only empty shells. About Moorea, a neighbouring island of Raiatea, Murray writes: 'We have had the depressing distinction of being able to document very precisely the complete extinction of all Moorean species in the wild in just one decade.'

The Clarkes and Murrays devised a simple experiment to demonstrate what was now happening to *Partula* snails across the Pacific. They placed the three snails at the centre of this story – a *Partula* snail, a giant African land snail, and a rosy wolfsnail – in a lunchbox, and observed them over the next few hours. The giant African land snail lounged in a corner, untouched, while the rosy wolfsnail crept up on the *Partula* snail, flipped it over, and quickly ate it. It turned out that the French Polynesian government had already been warned, years before; the rosy wolfsnail had wreaked havoc on native Hawaiian land snails, and the American malacologist Yoshio Kondo had written letters to the French Polynesian government, pleading with them to not make the same mistake by introducing the rosy wolfsnail to French Polynesia. Malacologist John B. Burch had also warned the Service de l'Économie Rurale (responsible for agriculture in French Polynesia). Both had been ignored.

The focus of Clarke and Murray's project changed abruptly, and so did the professions of the scientists involved. Field trips collecting specimens to study became rescue missions, the geneticists became conservationists, and their research laboratories – where all Moorean species of *Partula*, apart from one, were held – became a kind of ark.*

Soon, zoos around the world – including ZSL London Zoo, Perth Zoo, Jersey Zoo, Edinburgh Zoo, and Berlin Zoo – joined the effort to save these snails under the coordination of the *Partula* Propagation Group, formed in 1986. Snails from Clarke and Murray's laboratories, along with newly collected wild snails, were sent to facilities around the world to establish 'insurance populations' – so that the continuation of that species' existence didn't depend on just one institution. The devastation wreaked by the rosy wolfsnail was so sudden and so rapid that the now-conservationists had to act quickly; specimens were kept in lunchboxes and transported carefully wrapped in tissue paper and tucked inside 25cm cardboard tubes that were packed neatly in crates. In the dark, *Partula* snails entered into torpor, a kind of sleep state, oblivious to how much their lives were changing, and of the fact they were travelling halfway around the planet. Once confined to remote Pacific islands, these snail species now spread out across Earth, ending up in universities and zoos – their neighbours no longer the 'itata'e (white tern) and flying foxes of French Polynesia, but

---

* The species not held already in captivity was *Partula exigua*. It became the first *Partula* to go extinct.

giraffes, bears, and tigers from disparate countries and climates. Back in French Polynesia, the rosy wolfsnails continued to predate on the remaining *Partula*, skipping from island to island and broadening their territory. Meanwhile, *Partula labrusca*, the subject of this chapter, remained undisturbed on the Temehani Plateau – one of the few spots on Raiatea not yet reached by the rosy wolfsnail due to its high elevation.

ONE OF THE first things I learned about *Partula* snails was the details of their captivity. I knew only that they'd previously lived in what looked to me like a tropical paradise, and that this paradise had become a kind of hell at the hands, or radula, of a cannibal snail.

At the time it surprised me how seemingly lo-fi and almost DIY everything about their captive lives was. The ark that had saved these species was made up of 12cm x 12cm x 2.5cm clear plastic sandwich boxes, stacked one on top of another like ready meals at a supermarket. Each had a strip of three-ply toilet tissue laid inside it like a rug, which was moistened and then allowed to dry out, to mimic the fluctuations in humidity that occur in the snails' natural habitats. For food, the snails were given porridge oats, grass, or nettle powder, and powdered chalk for calcium. Around five adult snails and their young could live in each box. The natural barriers between species in their island homes – mountains, the sea – were now replaced with plastic lids. This was not some hi-tech, state-of-the-art facility; these snails had narrowly survived thanks to the creativity of a handful of people who had never intended to hold the fate of an entire genus in their hands, and had suddenly had to improvise. None of this surprises me now. Conservation projects so often take this form: ad hoc, guerrilla but, ultimately, tenacious.

Later, I read a sixty-one-page document detailing the guidelines on how to keep *Partula* snails. It was issued by the International *Partula* Conservation Group in 2019 and shows how those initial, seemingly off-the-cuff approaches have been continually refined over time. I read in this document about how sandwich boxes had been swapped for larger glass aquariums topped with clingfilm that could house up to fifty individuals and their young, and that these aquariums were kept in a specially designated room – a *Partula* room – climate controlled to ensure a temperature of between 20 °C and 24 °C and humidity between 60 and 80 per cent. This ensured that the snails' 'micro-environmental conditions' were maintained at all times.

To prevent contamination by chemicals or other organisms, lab coats had to be worn, a strict cleaning regime adhered to, and researchers had to disinfect their hands with antibacterial soap before interacting with the snails. I read of the extraordinary lengths taken to save *Partula* snails from death or injury; how the floors of these *Partula* rooms had to be soft, so that snails were less likely to be damaged if accidentally dropped, and how the researchers should ideally handle the snails with bare hands, because a tiny infant *Partula* might be crushed in the folds of a plastic glove. In one part of the document, I came across a recipe:

- Grass pellets 300g (Drygrass Ltd – 25 per cent oil, 16 per cent protein, 25 per cent fibre, 9 per cent ash)
- Oats 300g
- Trout pellets 150g (Vextra Trout intermediate 3mm – 18 per cent oil, 45 per cent protein, 2 per cent fibre, 8.5 per cent ash)
- Cuttlebone 150g (only the clean inner is used)
- Stress multi vitamins 25g

The instructions called for these ingredients to be reduced in a coffee grinder, one by one, and the resulting powder mixed with filtered tap water, left for five minutes so that the liquid is absorbed, then mixed with more water to form a runny paste. The document described this as a 'complete food', and it is accompanied by a list of backup ingredients, should anything in the recipe be unavailable.

I found this document strangely moving. It had no doubt been devised strictly with utility in mind – an instruction manual of sorts, a simple set of directives to achieve a clear, scientific outcome. But there was something striking about seeing this level of care and precision poured into the welfare of snails. It felt to me like a parenting book, as though I might close the cover after reading and see the title: *It Starts with You: How to Give Your Snail-child the Best Start in Life*. One particular passage, warning researchers to be extra-careful when counting snails by hand, felt like a variation of something I'd heard every parent of a toddler say – 'They can be surprisingly speedy and zip off while your back is turned!'

This parent–child relationship wasn't entirely one-sided, either. Vivien Frame, a research assistant working at Nottingham University's *Partula* room, was someone the captive snails seemed especially attached to. 'The snails knew when it was Viv looking after them and when it wasn't,' Gerlach tells me. This much was obvious because, whenever she had a day off or, worse, went on holiday, *Partula* snails would die. The cause of these deaths was a mystery, but the team speculated that something as simple as the brand of hand cream Frame used might have been the cause. Tests were conducted, including having other assistants use the same hand cream when she was away, but nothing they tried made a difference. 'It was extraordinary. And no one ever found out what it was,' says Gerlach. Whatever the apparent bond between Frame and the snails, this incident demonstrates just how sensitive *Partula* can be to their surroundings. Other *Partula* keepers around the world have noted similar attachments to specific members of their team; 'I think this is mostly to do with the amount of care, the amount of time, and how precise they are,' explains Gerlach.

It was March 2021, during the Covid-19 pandemic, when I first spoke with Paul Pearce-Kelly, coordinator of the *Partula* Propagation Group at the Zoological Society of London. A national lockdown was in place across the UK, creating a surreal parallel between the lives of ordinary people and those of the *Partula* snails under his care. From my own enclosure in Kent, I hopped on Zoom to speak with Pearce-Kelly, who had been involved in the *Partula* snail programme since 1986 and – along with his long-time colleague Dave Clarke – had contributed much parental wisdom to the care guidelines I'd read.

'Sometimes, no matter who you are and how much experience you have, you just can't replicate what they may need,' Pearce-Kelly explains when I ask him how it feels to care for a species that then goes into decline. Subtle things might tip the balance for a species in captivity – parasites, disease, nutrition, the altitude compared with their natural range, and then of course things that are less understood: the carer on duty, their hand cream, the weather. '*Partula* don't respond well to change. And in the early days, the worry was: "Oh, if Viv has a day off or someone goes into the lab on a different day to usual or you change an element of their diet, they'll all die!"' says Pearce-Kelly. 'Sometimes, there were things we worked out in hindsight. You never know, at the time, all the information you truly need.'

Some of the species that entered the captive-breeding programme thrived, so much so that the guidelines even include a section on how to humanely euthanize a population that has grown too large. There have been remarkable success stories – Jim Murray collected only one specimen of *Partula gibba* in 1972 and yet, since this species finds self-fertilization easy (even in sandwich boxes), by 1996 there were 520 in captivity, all descended from the single snail Murray saved. Other species had a harder time adjusting, like *Partula aurantia*, which never adapted to captivity and quickly went extinct. Fluctuations were common, too. *Partula nodosa* was brought into the programme in 1984; numbers dropped to eight adults in the early 1990s, but the population recovered and by 2014 numbered 5,000 snails. *Partula faba*, collected in good numbers in 1991 and 1992 (Gerlach brought some back with him from Raiatea, along with *Partula labrusca*), steadily declined. By 1995 there were 180; and in 2013 just fifteen individuals remained. Pearce-Kelly and his team, and biologists like Gerlach, would frantically search for solutions when this happened, but often to no avail. In these instances, even with the best of care from a dedicated and experienced team, nothing could turn the tide. For instance, in 1994 *Partula clarkei* had a strong population numbering 296.* However, after just under two years, only one remained. This last individual died at 3.30 p.m. on Wednesday 1 January 1996 – 'the most precise moment of extinction ever recorded,' Gerlach explains in his book, *Snailing Around the South Seas*. A spokesperson for London Zoo said, soon after this extinction, that a tombstone for the species would read: '1.5 million years BC to January 1996'.

Gerlach delivered the last *Partula labrusca* snails to the *Partula* room at ZSL London Zoo on his return to England in September 1992. After counting them on his bedspread and feeding them the *Partula* Propagation Programme-approved food paste, he carefully packed them into cardboard tubes and

---

* Many of the people involved in *Partula* conservation have species named after them. *Partula clarkei* was named after Dave Clarke by Justin Gerlach; and a species discovered in 2016 by Gerlach was named *Partula pearcekellyi* after Paul Pearce-Kelly. The specimens Gerlach collected did not survive in captivity, and *Partula pearcekellyi* is now presumed extinct.

then into plastic boxes, and travelled with them, first by boat, then by plane – during which they were tucked away in his hand luggage in the overhead compartment – and finally by London Underground, via the Piccadilly line. 'On the flight, I got up to check that the bag wasn't being crushed in the overhead, that the lid was still on the container, and the snails hadn't woken up,' Gerlach tells me. 'There was the worry of what could happen at every stage of the journey: *What if you miss your flight? What if you lose your bag? Have you got your bag? Where's your bag?!*' At Regent's Park station, an exhausted and jetlagged Gerlach weaved through crowds of tourists who were, like him, on their way to London Zoo. But unlike them, he was carrying all that was left of an entire species on his back. He handed the last *Partula labrusca* and the other snails he had collected over to Paul Pearce-Kelly and, his load now significantly lightened, went straight home to bed.

Four years later, in 1996, the Temehani Plateau was finally invaded by the rosy wolfsnail. The landscape changed; in place of the perfectly colour-matched shells of *Partula labrusca*, the sharp, long, pale-brown shells of the rosy wolfsnail moved on the ground around the Tiare Apetahi flowers.

By 2000, three decades after the introduction of the rosy wolfsnail, only five of the Society Islands' sixty-one endemic *Partula* species survived. Once thought to be the solution to all of French Polynesia's snail-related problems, the rosy wolfsnail had wiped out three-quarters of an entire genus. On Raiatea, once home to thirty-three *Partula* species (the most of any island in the country), none survived in the wild. Those that clung on did so exclusively in tanks, alongside Gerlach's *Partula labrusca* – now the last remnants of their species.

WE DON'T KNOW much about *Partula labrusca*. Crampton's discovery of the species gives us the first written description, which compared its shell to the labrusca grape. However, the species wasn't paid much attention after that, until Gerlach's visit to the Temehani Plateau almost forty years later. What we do know is that *Partula labrusca* had an instinct to move towards the source of gravity and away from light. They were adapted to living at high altitudes. They had right-coiling shells, and their colour varied from grape-red to golden-brown. This was all conservationists knew about *Partula labrusca* at the time it entered into their care.

Soon, *Partula labrusca* settled into the rhythm of life in the *Partula* room. Each day, they crept slowly across a Perspex sheet, smeared with food paste, that was placed in their tank. They grew accustomed to the clear boundaries of their new vanishingly small habitat, and the impressions of other snails just about visible through them. Twice each week, they were lifted out of this tank with great care by their keepers, and placed on tissue or in a petri dish while their tank was cleaned. After a while, the individuals collected by Gerlach from the Temehani Plateau died, leaving behind their descendants who only knew of life in captivity.

When I spoke with Gerlach, I explained to him that I hoped to give an impression of what life was like for *Partula labrusca* in the wild. But, quite quickly, I learned of the impossibility of that task. 'We just don't know much about them at all,' he told me. We spoke about a section in his book, *Snailing Around The South Seas*, in which he describes the findings of a 1979 study on the courtship dances of *Partula* snails. It says of how, on rainy nights, they would come out en masse, pair up, circle one another, and then one would climb onto the back of the other. The snail on top – which was, for the moment, playing the role of the fertilizing 'male' – would move across the shell of its mate, slowly tracing a pattern with its body, over and over again. This pattern differed from *Partula* species to species, like a signature. Some were like figures of eight, others less complex, forming just slightly warped lines or circles. Then the dance would repeat, with the roles reversed, and the two snails would return to cover, back under rocks and leaves, eggs fertilized. 'I'd love to know more about this too,' Gerlach says. But these mating dances have never been observed in captive populations, meaning we don't know what *Partula labrusca*'s or many others' looked like. This behaviour, and so much more about their lives, Gerlach explained to me, 'might simply have been something the snails left behind when they left their homes.' This is one of the paradoxes of breeding extinct-in-the-wild species in captivity. When *Partula labrusca* was rescued, something intrinsic – specificities about the species' behaviour and role in its habitat in the wild – was lost.

The middle of the *Partula* story is, in this sense, missing. However, we do know the beginning. All *Partula* – every single one at London Zoo and other conservation zoos around the world – share the same origin.

AROUND 6 MILLION years ago, the ancestor of all *Partula* snails migrated west from either Southeast Asia or Australia – picked up and carried by strong winds or birds. These snails landed on the Pacific islands, where they split into three distinct genera, one of which was *Partula*. First, Huahine and Raiatea were colonized by *Partula* snails; then, carried by typhoon winds and sea birds, they landed on other Society Islands, becoming the many species of *Partula* snail we know today. This is the beginning of *Partula labrusca*'s story. At some point after *Partula* snails first

landed on Raiatea 2.71 million years ago, a snail arrived on the Temehani Plateau and evolution took care of the rest.

Much later, our own species emulated this epic tale of migration, and in the process became entangled in the story of *Partula* snails. Between 3000 BCE and 500 CE, *Homo sapiens* migrated eastwards from Southeast Asia, riding the winds – albeit, on far gentler winds than those that had transported the snails. As they sailed, these seafaring people settled on new islands. Raiatea, *Partula labrusca*'s island home, became the centre of their religious, cultural, and political lives, and was known to them as Ha'vaii, the ancestral homeland of all Polynesians.

Soon, stones began migrating across Raiatea, one travelling from each islander's home in order to build a temple, named Taputapuatea, which was dedicated to the gods that the islanders believed governed all aspects of life and death. The shells of snails started to move about the island too, collected by these new arrivals and given a new definition and role. The shells of *Partula* snails, in their varied and beautiful shades, were pierced and then threaded together to make necklaces and other jewellery. *Partula* had now become a part of what is known today as the ancient *hei* jewellery tradition, adorning ceremonial pieces central to Polynesian cultural and civic life. Some necklaces were gifts to signify friendship or romantic interest, while others demonstrated rank and status, or that the wearer belonged to the specific island to which the shells of a necklace were native. The small shells of *Partula* made perfect beads for these sculptural pieces of jewellery that were detailed and complex in design. Twisting this way and that, the whorls of *Partula* gave necklaces a sense of movement. The dusky-red shells of *Partula labrusca* would have almost certainly found their way onto *hei* jewellery. Not yet recognized as an individual species, not yet named, its existence not yet recorded in written word, *Partula labrusca* was recorded here, instead, on thread, as a shade

*A Tahitian woman wearing a traditional* hei *necklace and headdress made from* Partula *snail shells, circa 1920. (© California Academy of Sciences; this image has been clarified with AI image enhancement software.)*

on a palette of colours – a warm, rich burgundy amongst the pink, purple, and yellow of other *Partula* shells.*

It can be said that a celestial event brought the West into

---

* A 2007 study found that species with white shells were not restricted to only one island, as was the case with most *Partula* species. Live populations of *Partula hyalina*, a species with a pearlescent milky-white shell, were taken from their home on Tahiti to the Austral and Southern Cook Islands, so that a stock of the snail could be established for jewellery-making. Where before, storms and birds indiscriminately dispersed

contact with *Partula* snails. Occurring on average once every eighty years, the transit of Venus is a rare astronomical occurrence, during which the planet Venus is visible from Earth as it passes in front of the Sun. Thanks to the odd cadence of this phenomenon – in which a pair of transits, eight years apart, occurs either 121.5 years or 105.5 years after the last – its study has been an intergenerational effort spanning centuries. In 1691, Edmond Halley predicted that observations of the transit could be used to calculate the distance between the Earth and the Sun, and in turn the size of the solar system. Halley died in 1742, long before the next pair of transits came around, in 1761 and 1769. But thanks in part to his insistent lobbying for observations to be made, various European nations – some, like Britain and France, at war with one another – momentarily united in the spirit of scientific collaboration. For the 1761 transit, 122 observers were stationed at sixty-two viewing positions, located everywhere from Siberia to Calcutta, from Canada to South Africa, and even on St Helena. Then, eight years later, as part of the effort to observe the 1769 transit, the *Endeavour*, under the command of James Cook, was sent to observe it from the recently 'discovered' island of Tahiti.

This voyage – the first of Cook's famous three – was ostensibly a scientific mission. Arriving in Tahiti on 13 April 1769, the *Endeavour* carried both a full complement of scientific instruments as well as the scientists necessary to operate them. These included the astronomer Charles Green, the botanist Joseph Banks, and the naturalists Daniel Solander and Herman Spöring. Their observations of the transit of Venus, on 3 June 1769, turned out to be inaccurate, rendering this lavishly expensive scientific mission a failure – or so it had seemed. Cook had left England with a set of sealed secret orders from the British

---

snails across different islands, now humans were distributing them based on the popularity and usefulness of a certain colour.

Admiralty, to be opened only after the observation of the transit. Now, with that goal completed, at least in spirit, he unsealed the secret orders. The *Endeavour* was to sail south from Tahiti in search of Terra Australis Incognita, the hypothesized southern continent, long sought by European powers.

One of the first places Cook landed after receiving his new orders was Raiatea. 'I then hoisted an English jack, and took possession of the Island and those adjacent in the name of His Britannick [sic] Majesty,' reads his journal entry on 21 July 1769. Meanwhile, while Cook focused on the land, Solander set about procuring its wildlife, including a handful of small shells from a genus of snail never before recorded. These were from the species *Partula faba* (in Latin, 'faba' means 'bean' – although, curiously, the shells don't look at all like beans). This handful of *Partula faba*, also known informally as Captain Cook's bean snail, would introduce the European scientific community to the incredible world of *Partula* snails.

Cook's first voyage was an expedition of profound scientific significance, but it also left a greedy trail across the Pacific for European powers, hungry for new colonial acquisitions, to follow. Soon, Polynesian traditions, religions, and cultures were suppressed. Native islanders were subjected to violence, imprisonment, and exile, as officials and missionaries from European countries wrested control of Raiatea and other islands from local monarchs and religious leaders. Christianity was imposed, and the temple on Raiatea, which had been built with stones donated by everyone in that community, was abandoned.

There are many points in this story where the fate of the *Partula* genus was tilted towards extinction: a French border guard's homesickness; the USDA's indifference to the risks of exporting rosy wolfsnails; the French Polynesian government's dismissal of warnings from scientists. But in the long run, the deconstruction of a society that not only lived harmoniously alongside *Partula* snails but had incorporated them into its culture might just have been the first domino to fall. It was at this time, too, as Polynesian customs were being dismantled piece by piece, that – on the other side of the world in 1821 – the *Partula* genus was named after the Roman goddess of childbirth by the French biologist André Étienne d'Audebert de Férussac. There is irony in the name *Partula*. Not only was the Roman goddess Partula responsible for overseeing childbirth, but she was also responsible for weaving death into a newborn baby's future, and turning it from something immortal into something mortal. Of course, Férussac wasn't thinking about this when he named the genus. Nonetheless, unbeknownst to him, the diverse species he grouped together under the name *Partula* were almost all doomed to die, suddenly, within a few years, less than two centuries later.

The *Partula* room at London Zoo is tucked away near the Reptile House, in a space that used to be an office. The door to enter it stands next to a gigantic fish tank full of endangered corals. I visit in late summer 2022. Paul Pearce-Kelly, one of the founders of the *Partula* Propagation Group, gives me a tour. Visitors to the zoo watch us through a large window. I am told that, often, people briefly stop here to observe the team at work. Inside the room are rows of stacked-up tanks that, on first glance, look empty, until you notice the feeding plates smeared with brown-green pulp and the small, dark flecks dotted about on the glass. These flecks are *Partula* snails. On the glass walls of the tanks that the snails have left winding trails on are labels, each with the name of a *Partula* species and the phrase, 'Extinct in the wild'.

Today, not much is happening. In the corner, while Pearce-Kelly and I talk, a team member diligently scrubs some Perspex plates on which *Partula* feed is smeared each day. The snails in the containers are all small enough to fit on a fingertip; some shells are topped with the tiniest of dots – barely visible, these colourless specks are baby *Partula*. Like beads of glass, they are so small and their shells so thin that they are translucent. Little of this is perceptible from the other side of the window. While I'm here, a boy pulls his dad away, urging him on to something fiercer perhaps, or cuter. And I get it: had I visited as a child, I probably would have been just as underwhelmed by these little dots. But in this room, each day, at all hours, incredible things are happening; species that are on the brink of extinction are nursed back to life, while others slowly dwindle in number, despite the researchers' best efforts.

Two decades before my visit, on 6 July 2002, there would have been people visiting the zoo to see the resident penguins; others would have come to the Reptile House, which had been made an impromptu tourist attraction thanks to a scene in the film *Harry Potter and the Philosopher's Stone*. Children might have peered through glass at pythons, trying their hardest to conjure a greeting in Parseltongue. And as visitors rushed this way and that, they would have been oblivious to the fact that, metres from where they stood, the long story of a species was quietly coming to an end.

*Partula labrusca*'s extinction was not a surprise to those taking care of it. According to Gerlach, the snail never adapted to captivity, and a big part of that is probably down to its natural habitat. 'What we've found is that the high-altitude species have proved very difficult to keep,' he says. 'None of them have done well.' The reason for this isn't fully understood. Compared with the more generalist *Partula* found at lower elevations, high-altitude species possess certain specialized attributes, such as thinner shells due to reduced access to calcium at higher elevations, but it's unclear how this might have disadvantaged them in captivity. Despite this, says Gerlach, '*Partula labrusca* probably lasted the longest out of any of the high-altitude species. They adapted to the diet well, they were easily transported, they even bred.' For ten years, the species he had rescued from the Temehani Plateau in 1992 held on at London Zoo's *Partula* room, at one point their population even slightly increasing. 'The species was definitely on the up,' says Gerlach, 'but then it just failed, and we don't know why.' Without much ceremony, on 6 July 2002, the last *Partula labrusca* faded away.

For anyone working with critically endangered species, moments like these are difficult. Pearce-Kelly describes the experience of losing a species in your care, no matter how expected the loss, as 'devastating'. 'It's impossible not to feel something for them,' he says. Gerlach describes something similar. 'When you get down to one, you know it's almost certainly going to happen, so there's a sense of inevitability . . . But, it is hope extinguished,' he explains, 'even if you had known it was a vain hope.'

Fourteen years after *Partula labrusca*'s extinction, *Partula faba* – the species that travelled back from Cook's voyage and became the first *Partula* known to science – also fell into decline at ZSL's *Partula* room. Gerlach, who had collected the last wild specimens on the same trip on which he'd rescued *Partula labrusca*, told me how he'd worked feverishly to find some way to help the species, frequently calling the *Partula* room staff to suggest adjustments to their care that he thought might help. He also conducted a study on *Partula faba* specimens that had been captured in the wild and preserved in alcohol. Cutting open the stomachs of these snails, he discovered dozens of decapitated ant heads inside. That ants might be a part of the natural diet of *Partula* snails, which were now being fed a plant-based mix, was something of a revelation. But despite adjusting the ailing snails' diets to account for this discovery *Partula faba* continued to decline, until, in February 2016, it went the way of *Partula labrusca* and thirty-two other *Partula* species before it. This extinction – and many others caused by the rosy wolfsnail – has, at the time of writing, yet to be declared by the IUCN.*

In an office adjacent to the *Partula* room, a looming cabinet stands against one wall, containing dozens of small drawers. Each is labelled in the same way as the tanks of live specimens: with a species' name and a status. But this chest of drawers, unlike the tanks, is a catalogue of deaths and extinctions. Each drawer contains the preserved bodies and shells of specimens that have died in captivity. A drawer just like all the others, near the top, is labelled *Partula labrusca*. It slides out like a mortuary cabinet: I almost expect it to be cold inside. Small glass vials are neatly lined up within it, each a tiny coffin for an individual *Partula labrusca* specimen, labelled with the date of its death. I hold one – the date on its side, 20 November 1995. I realize that this is likely one of the individuals collected by Gerlach from the Temehani Plateau, one of the last wild *Partula labrusca*. It's a slightly lighter colour than in descriptions. This one is yellower, almost golden, perhaps faded from age; but I still see it – how much it resembles a beautiful bead, how much it looks like a precious stone. As I turn the vial, the shell clinks against the glass and I worry that it might break, that it might shatter and turn to dust.

WHAT IS LOST with the extinction of a *Partula* species? Dissections of *Partula* collected and preserved in the 1960s suggest that they are detrivores, meaning that they feed on rotting leaves, algae, lichen, mould and mildew, and it is speculated

---

* Another species, *Achatinella apexfulva*, a Hawaiian land snail, also recently went extinct. The last of the species was named Lonely George, after the last Pinta Island tortoise, which is the subject of Chapter Nine. Lonely George's species had also fallen victim to a disastrous introduction of rosy wolfsnails. Lonely George died at the age of fourteen on 1 January 2019.

that each snail may consume as much as a third of its own body weight in detritus per day. 'Given the numbers of *Partula* snails that there used to be, they must have been removing a phenomenal quantity of detritus and dead plant tissue from the environment,' says Gerlach. If this is true, these small snails are like the cleaners of their habitats, taking care of the health of the land, looking after plant life and maintaining balance in the ecosystem. Without them, native plants are likely under greater stress from disease, which may explain why they are often being outcompeted by more hardy introduced species. 'The disappearance of *Partula* snails has probably contributed quite significantly to the deterioration of the forests,' says Gerlach.

Then there is the human cost of *Partula* extinctions. The Polynesian artisans who make *hei* jewellery are almost always women. *Hei* jewellery 'mamas', as the women who run each workshop are called, would often provide for their whole family. Mamas play many important roles in Polynesian society. Anna Laura Jones, an archaeologist at Stanford University, explains that 'mama' is a name that 'is used to honour women wise in the ways of tradition, oral history and craft'. Before Polynesian islands were colonized by European powers, women occupied positions of authority; the idea that women should not partake in politics was imported from Europe. Leaders of artisan associations, many of which have no male membership, however, have historically been an exception. These women wield political influence, since politicians often seek to publicly support traditional crafts to emphasize their Polynesian heritage. In this way, the mamas have long represented 'a positive image of Polynesian womanhood that contrasts with the image of the tourist brochure beauty who is forever young, overtly sexual, and easily exploited,' writes Jones. Since the introduction of the rosy wolfsnail, however, the mamas' main resource, the shells of *Partula* (or *areho*, the Polynesian name for *Partula* snails) has largely disappeared. On Huahine, where *Partula* snails have been completely wiped out, serious economic hardship befell the *hei* artisans of the island. Many women were thrown into poverty and the island's artisan association was shut down. The extinctions of *Partula* snails leave gaps not only in the ecosystems each species has left behind, but also in the cultural heritage and economies of Polynesians.

ON RAIATEA, THE temple made of stones brought from islanders' homes, Taputapuatea, faces out towards the sea. Traditionally, this place was seen as a jumping-off point for the dead, where they leave the island for the afterlife. Taputapuatea was dedicated initially to Ta'aro, the creator of the world in the Polynesian creation story; then later to Ta'aro's son, 'Oro, the god of fertility and war, life and death. Here, it was believed that people and gods could interact and that te'pö (the world of the dead, ancestors and gods) met with te'ao (the world of the living). The temple was seen as being the jaws of the gods; human and animal sacrifices were offered here.

After the colonization of Raiatea, Taputapuatea was abandoned. By the time Crampton climbed the Temehani Plateau in 1934 and made the first collection of *Partula labrusca*, Taputapuatea was a ruin, neglected and in disrepair.

In his 1964 book, *Vikings of the Sunrise*, the Maori–Anglo-Irish anthropologist and politician Te Rangi Hiroa (also known as Peter Henry Buck) wrote of his first visit to Taputapuatea. No longer the temple it once was, it was now 'speechless . . . and inanimate' stone. 'The dead,' he writes, 'could not speak to me.' He continues: 'The bleak wind of oblivion had swept over [the village of] Opoa. Foreign weeds grew . . . stones had fallen from the sacred altar . . . The gods had long ago departed.'

By the time the last *Partula labrusca* died, parts of Taputapuatea had been restored, first in 1968, and then in 1994.* There once again existed this place where the worlds of the living and the dead were said to intersect. Now, on platforms where sacrifices to the gods were once laid out, mounds of bananas and yams were offered. And, at a stone shrine where 'Oro's god-house once stood, *hei* jewellery necklaces made from local shells – some pieces including the distinctive walnut-brown of rosy wolfsnail shells – lay in offering.

In Polynesian culture, the boundary between life and death is permeable. Death is not always the end. After speaking with Gerlach and Pearce-Kelly about *Partula labrusca*'s extinction, I read about the ghosts that are believed to populate the valleys and forests of the snail's island home. Upon death, a person's soul was said to travel out of the body, through the tear ducts and journey around the island, following the trail of spirits that had gone before, to Taputapuatea. Some people, I read, were once believed to possess the power to capture the soul of a person who had died before their time, and revive that person by inserting their soul back into the body via the deceased's foot, under their big toe.

During the autopsy of the last *Partula labrusca*, no pathogens or any other likely cause of death were found. The species had simply slipped quietly into extinction. This was the end for *Partula labrusca*. But for *Partula faba*, Captain Cook's bean snail, extinction was not the end of its story. A thin sliver of the last snail's foot was cut off post-mortem, frozen, and then added to The Frozen Ark – a 'biobank' at the University of Nottingham, founded in 2004 by Bryan and Ann Clarke and Dame Anne

---

* A local cultural association of 'Opoa people called Na Papa E Va'u Raiatea was formed in the 2000s to protect Taputapuatea, restore connections between the people of the Pacific triangle and beyond, and work towards World Heritage classification for Taputapuatea. The site was classified as a UNESCO World Heritage Site in 2017.

MacLaren, to preserve the tissues, cells, and DNA of the world's endangered animals.* Much like seeds in a seed bank, these samples are kept as a safeguard against extinction. It's even possible that some will one day be used to resurrect extinct species. Cloning techniques don't yet exist that would enable scientists to bring snails back, however; so for the time being at least, *Partula faba* remains adrift, waiting to cross back over the divide between the world of the dead and the world of the living.

THE ROOMS THAT *Partula* snails are kept in were always intended to be a temporary solution, holding rooms where the snails would wait until it was safe for them to return home. However, over the years they have spent in captivity, new threats have emerged in French Polynesia that have made reintroductions less and less feasible. Forests and natural vegetation have been cleared to make way for farmland and building projects; a carnivorous flatworm (*Platydemus manokwari*), introduced as another attempt to control numbers of the giant African land snail, now predates on endemic snails along with the rosy wolfsnail; and climate change means the rosy wolfsnail will be able to invade higher altitudes that were previously too cold for it.

However, in 1994, the opportunity arose to create special predator-proof exclosures for *Partula* on Moorea. Rosy wolfsnail numbers had dropped in the absence of their favourite

---

* Had the Clarkes and Murrays not been studying *Partula* snails, The Frozen Ark might not exist. The threat posed to *Partula* encouraged them to ensure that tissue samples were preserved. At the time, institutions around the world stored animal material, including tissues and DNA; however, there was little global coordination. The Frozen Ark was founded to provide it, with particular emphasis on the preservation of material from endangered species.

food source, but not significantly enough that *Partula* snails could be reintroduced into the same environment. In 1993, Elizabeth Murray wrote, '*Euglandina* [the rosy wolfsnail] is still present, living as a cannibal and perhaps on carrion as well.' While not a sustainable solution to the plight of *Partula* snails, these small reserves were a step in the right direction. But the researchers faced a problem; many of the captive *Partula* whose numbers were large enough to be considered for release had spent generation after generation in plastic lunchboxes and glass tanks. It was feared that these snails and their forebears might have selected their mates on the basis of how good they were at surviving in boxes above any other talents, which might have rendered them incapable of surviving in their natural habitats. To assess this problem, an experiment was devised. In the Palm House at Royal Botanic Gardens, Kew, sixty *Partula taeniata* were released onto a *Pandanus tectorius*, a tree native to most French Polynesian islands.

Researchers watched these snails closely, around the clock. Over six- to ten-hour shifts, teams of two observers recorded each snail's position every thirty minutes. These sixty *Partula taeniata* became, as Gerlach puts it, 'probably the most intensively monitored snails ever', with keepers making 15,000 observations by the end of the experiment. After fifteen months, researchers found that the snails had dispersed and, most crucially, among them were sixteen juvenile snails, meaning they were breeding. Concerns that the snails might have adapted to captive life were dispelled, and the plan to create an exclosure in French Polynesia went ahead. A 20m × 20m area of land on Moorea was designated for this, and surrounded by a 75cm-high barrier of galvanized iron sheeting. To keep rosy wolfsnails out, a plastic trough filled with salt was laid around the base of the barrier and a 12-volt car battery was used to create an electric fence. In September 1994, three hundred *Partula* snails were released into this veritable snail fortress, one of the smallest wildlife reserves in the world, and their progress monitored weekly. Four years later, however, the experiment was brought to an end. The exclosure, battered by wind and rain and overgrown with plants, had failed to keep the rosy wolfsnail out. Only two *Partula* snails had survived.

In 2012, however, there was finally some good news. Trevor Coote was an English field biologist stationed on Tahiti, whose position overseeing the monitoring of *Partula* snails was funded by the various international institutions involved with the *Partula* programme, including the French Polynesian government. While performing fieldwork in a Tahitian valley, he was met with a surprising sight. Mape trees (*Inocarpus fagifer*), also known as Tahitian chestnuts, are a striking tree native to French Polynesian islands, with thin, sail-like roots that can stand at a height of up to 2m weaving and folding around their trunks. On one of these trees, Coote spotted some small light-coloured dots. These were the pearlescent shells of *Partula clara*. Here, the snails seemed to have escaped predation. The dry and dusty trunks of mape trees were, it turned out, a natural barrier against rosy wolfsnails, which are mostly terrestrial and prefer moist habitats, and for which the bark is too dry. This was a native tree species that would, Coote realized, provide a natural sanctuary for *Partula* snails reintroduced into the wild.

Coote's discovery also coincided with the realization that rosy wolfsnail populations were starting to decline naturally. Though the cannibal snails had not completely disappeared from French Polynesia, their numbers had dropped dramatically. While the introduced flatworm *Platydemus manokwari* poses a new threat to *Partula* snails, the trees seem to be a natural barrier against them, too. The flatworms have only been observed killing *Partula* snails when the snails have descended the trees to around 30cm off the ground.

Reintroductions started in 2015. In *Partula* rooms around the world, snails were again packed into cardboard tubes and plastic boxes and loaded onto planes. This time, however, these *Partula* snails were returning home. Coote oversaw the reintroduction of over 15,000 *Partula* snails, before the Covid-19 pandemic abruptly halted the programme as lockdowns prevented travel and local fieldwork. In early 2021, Coote took one of the first available flights back to the UK from Tahiti. After a career of nearly twenty years in *Partula* conservation, he tragically caught Covid-19 and died in February the same year. In an obituary published in the IUCN/SSC mollusc newsletter *Tentacle*, Paul Pearce-Kelly writes: 'In addition to a wealth of fond memories, Trevor's conservation impact is a truly outstanding legacy for which we are all grateful and inspired to carry forward.'

Gerlach took over field monitoring duties after Coote's death, overseeing the largest ever reintroduction of any extinct-in-the-wild species, when over 5,000 *Partula* snails were released in April 2023. At the time of writing, 21,000 snails from the captive-breeding programme, including eleven species extinct in the wild, have made the trip back to French Polynesia. Before these snails were released, small coloured dots (colour-marked by year) were painted onto their shells. They were then left to their own devices – wild again. When I speak with Gerlach again in January 2024, he tells me that the conservationists are now waiting to find an adult *Partula* snail in the wild without a dot on its shell – evidence that, once reintroduced, the snails are breeding successfully and that their offspring are surviving in their natural habitats.

With these reintroductions, there is hope not only for these species but also for the restoration of Polynesian cultural heritage. In 2019, a seminar was held in French Polynesia bringing together politicians, researchers, conservationists, and Polynesian artisans. A statement was made by what are collectively called the Pacific Countries, which has been recorded in the deeds of the seminar. Its final three lines read: 'Our snails move

# A Tale of Three Snails

*One of the* Partula *snails that has been released on the island of Moorea as part of the reintroduction programme. To help conservationists track the snail's progress, its shell has been marked with luminous red paint. (© London Zoo.)*

slowly but are dying out rapidly. We must act fast to save them. Together, we can!'

Back in 2017, Coote, Pearce-Kelly, and a local scientist named Rodrigo Navarro were recorded in the BBC radio documentary 'SOS Snail' as they searched for *Partula* snails that they had reintroduced in previous years. There is wonder and excitement in their voices as they look up into the canopy of mape trees. Although such a solution could not be found for *Partula labrusca*, many of its cousins now have a second chance to stay in our world at least a little while longer.

'Oh, look, LOOK!' shouts Coote in the documentary, '. . . that's one of the releases from last year, it has to be.'

'That's a sight that no one's seen for nearly thirty years,' says Pearce-Kelly.

'To imagine that these had disappeared from the wild,' adds Navarro, 'to see them back here again, it's truly a success.' And then, speaking to the snails themselves up in the mape tree canopy, he says, 'Welcome home, guys.'

# 4 The Little Masked Bird

Poʻouli

*Melamprosops phaeosoma*

THERE IS A place on the Hawaiian island of Maui where it was once possible to travel back in time. Clouds roll up the steep slopes of the 10,000-foot Haleakalā volcano towards its crater. In amongst the murk of these clouds, on the north flank of the volcano, is a dense knot of rainforest where the rain rarely stops, cycling between drizzle and torrential downpours. 'Ōhi'a lehua trees (*Metrosideros polymorpha*) – a native species with distinctive flowers comprising hundreds of stamens, found all over Hawaii – grow unusually large here, and are covered from root to branch tip with mosses and lichens. In this mysterious place, magnetic anomalies caused by the terrain mean that a compass would lead its bearer in circles. For millennia, this rainforest – now known as Hanawi Natural Area Reserve – was unexplored by humans. With all of its quirks of perpetual rain, magnetic anomalies, swampy land, dense vegetation, and steep terrain laid out like booby traps, it was as though the forest itself was trying to protect something. And perhaps it was.

The year is 1973 when the story of this chapter begins. The rainforest is still unexplored. And beneath the blanket of clouds that covers it are species of native Hawaiian plant, bird, and insect that are absent or rare elsewhere across the archipelago. Spiky, sea-urchin-shaped plants the size of basketballs, called greenswords (*Argyorxiphium grayanum*), found only in very small numbers elsewhere, pack themselves into clearings in their hundreds here. Above, swooping through the mist like ghosts, are native birds that are dwindling in number everywhere else. The air is alive with their calls and colours, just as, at one time, all of the island's forests would have been. There is the orange of the Maui 'akepa (*Loxops ochraceus*), and the yellow-green of the Maui nukupu'u (*Hemignathus affinis*) that creeps up and down tree trunks hunting for spiders and insects using its strikingly curved beak to peck at the bark. And there is the green of the kiwikiu (*Pseudonestor xanthophrys*), also known as the Maui parrotbill after its parrot-like beak. It is a landscape and ecology that looks less like modern-day Hawaii, whose native wildlife has been decimated since the arrival of humans, and more like a snapshot of its distant past. This small slice of land has remained pristine, as if transported from a time before people had reached the archipelago.

In amongst the hubbub of the rainforest, cradled in the V-shaped fork of an 'ōhi'a lehua tree branch, there is a nest. Inside, a bird that humans have yet to lay eyes on sits on an off-white egg, speckled with brown and grey. Small, round, and with a stumpy, almost non-existent tail and conspicuously large feet, this bird looks like nothing else in Hawaii. Its feathers are brown and grey, apart from the streak of black it wears across its face like a bandit's mask. While every other bird adds its looping call to the chorus of bewitching melodies that fill the forest air, the little masked bird is mostly silent. 'Chik-chik-chik' is the quiet, understated greeting she gives her mate whenever he returns to the nest carrying in his beak a beetle fished out from under a cushion of moss.

The two masked birds have been preparing for this day for weeks. They had started by perusing countless branches until they'd found the perfect foundation, and they'd then assembled their nest, piece by piece, from twigs, moss, and bits of fern scavenged from the forest floor. Once their egg was laid, the female had carefully lowered herself over it, encasing it in a skirt of ruffled feathers, and gently warmed it against her skin. Now and then, to ensure no part of the egg got cold, she'd stand and turn the egg before retaking her place on top of it. The male, meanwhile, had started working double shifts to keep them both fed, combing the forest's thick understorey for insect larvae, beetles, and small snails. Now, the female feels a stirring beneath her, and the speckled egg the birds have worked so hard to protect finally cracks.

Featherless and with eyes still sealed tightly shut, the chick can barely hold its own head up. But, day by day, under its mother's feathers and with the steady stream of food delivered by its father, it grows stronger and stronger. And as the chick grows, its mother slowly expands the nest, flattening the doughnut-shaped rim with her large feet. Occasionally, the chick's parents fend off other birds. A far more serious threat, however, arrives when the sky darkens and fierce winds rock the V-shaped branch back and forth. The clouds moving up through the forest blacken and burst, and a deluge of rain pummels the nest, the parents, and their chick from all sides. This storm is the greatest test yet of the parents' judgement. If the branch they've selected for the nest is not sturdy enough, or the nest not wedged firmly enough in place, it could be tossed up into the air and reduced to a pile of twigs and moss. The chick might become hypothermic and die if the shelter of the nest and its parents' plumage fail to insulate it from the wind and rain. The rain continues for days, drowning out the sound of all life in the forest, reducing visibility to near-zero, until, eventually, one morning the clouds part and a rare spell of sunshine warms the air.

All is quiet in the forest, and the nest, storm-battered and frayed but intact, is still in the branches. A small, downy head pops up over the nest's rim, and an expectant 'chik-chik-chik' issues from the chick's beak. To live here means to pick up the pieces after a storm and continue. Greenswords immediately get to work, growing new leaves and repairing the damage left behind by rainfall. Kiwikiu chisel for insects in moss-covered branches that now hang from trees at odd angles. And the pair of little masked birds, in the wreckage of their home, continue with their parenting duties. Just as before, the father leaves the nest to look for food, while the mother keeps the chick warm.

A few weeks later, having overcome the most treacherous ordeal of its short life, the chick leaves the nest and its parents behind. Fully grown and donning its species' signature black bandit mask, the bird sets off to claim a territory of its own. It joins a flock of a variety of species of birds – the odd one out, being brown amongst vibrant greens and yellows. And then, one day, as it is flying between the trees, from branch to branch, moss cushion to moss cushion, it comes across an unusual sight: animals it has never seen before. Three humans – all young, drenched and muddy, one in jeans he has tried to make waterproof using melted candle wax – pointing up at the bird, excitedly taking notes and making sketches. They are university students making the first ecological survey of this area after having accidentally stumbled into it last year while lost on a trek up to the Haleakalā crater. And, in this moment, they have discovered one of Hawaii's oldest birds.

So it was that on 26 July 1973, the bird that became known as the poʻouli (*Melamprosops phaeosoma*) was discovered. High up in their ridge-side forest, they had avoided the attention of humans since our species had first arrived in Hawaii, roughly 800 years earlier. In that time, an entire civilization had risen and fallen, and the ecology across the rest of the archipelago had been transformed almost beyond recognition. And yet, somehow, this species had never been recorded or even named, not by Europeans or native Hawaiians before them, until now.

'THEY DIDN'T REALLY know what humans were, I think,' ornithologist Tonnie Casey tells me. Casey was twenty-two years old at the time of the poʻouli's discovery. 'Until we got there, they'd never seen one before.' It was Betsy Gagne (née Harrison), a student botanist, who spotted the po'ouli first. Three strange little brown birds 'just materialized' in front of her, she later said, flying in a mixed flock of bright green kiwikiu, yellow-green Hawai'i 'amakihi (*Chlorodrepanis virens*) and Maui 'alauahio (*Paroreomyza montana newtoni*) – the latter a 'little yellow marshmallow' of a bird, according to Casey. Gagne led Jim Jacobi, the project's student mammalogist, to see the birds, followed by Casey – all of whom were stumped.

When I talk with Casey, she speaks warmly about those initial moments of getting to know this 'chunky little bird' with its 'very, very short tail . . . and huge feet'. She describes spotting them hanging from the undersides of branches, upside down and hopping – always hopping, rather than flying – from branch to branch or, occasionally, across the forest floor. She tells me how they'd 'come slowly hopping into your area' if you sat in just the right spot (preferably amongst lots of *Cheirodendron trigynum* trees of the ginseng family and the endemic *Vaccinium reticulatum*, a flowering shrub of the heather family, both of which, for some unknown reason, the birds seemed to like) and how, if all else failed, she could summon them by imitating their call. 'Chik-chik-chik,' she demonstrates to me over Zoom, as if hoping to call the birds back to her now.

'Friendly' is the word Casey uses to describe the po'ouli. 'They were wild,' she adds, with a chuckle, 'but they weren't very scared of you, they were always inquisitive . . . you could put your hand out and they'd be just beyond your fingertips.' Over several weeks, Casey and Jacobi followed these friendly little birds around the forest, recording their movements and behaviour in detail. Amongst the notes, photographs, and other materials Casey shares with me from this time, there is a painting. This image of the bird, which Casey painted on the evening of its discovery, is the first ever recording of the species' likeness. In it, a po'ouli perches above us on a thin branch, legs

*'Ōhi'a lehua trees shrouded in mist in Maui's high-altitude rainforest. (Courtesy of MFBRP.)*

sharply bent as if ready to spring away from us, its plump belly and masked face on full display.

Casey and Jacobi returned to Honolulu buzzing with excitement about 'the mystery bird'. It certainly looked like nothing else in Hawaii, but could it really be that a group of undergraduates had stumbled upon an entirely new species? One that countless naturalists and ornithologists had overlooked? A few academics who the students consulted were dismissive, suggesting that they had merely mistaken a known bird as a new species. However, they were encouraged by Dean Amadon, ornithologist at the American Museum of Natural History, who told them to gather evidence to prove that their discovery was in fact new. The only catch was that this proof would come at the expense of the species itself – Casey and Jacobi would have to travel back to Hanawi, not to study the po'ouli but to track and kill them in order to obtain scientific specimens for a formal species identification.

'That was a really challenging trip, emotionally,' says Jacobi. 'It was something I wasn't very much behind doing in the first place, but that I convinced myself was necessary in the circumstances.' Jacobi, who is now a biologist with the US Geological Survey, explains to me that after he and Casey returned from Hanawi word began to circulate about the new bird species they'd seen there. Amadon, state authorities, and others feared that without swift legal protection the po'ouli would be disturbed by birdwatchers and other people eager to catch a glimpse of it. 'A price was on its head,' says Jacobi.

The two students travelled back into the forest with Dave Woodside, a biologist working for the state who had the permit needed to kill wild native birds. They shot three po'ouli, but could only locate two bodies, so were forced to leave one behind in the forest. Jacobi prepared one of these specimens himself, skinning the bird and preserving its internal body parts in alcohol, before stuffing it with cotton and stitching it back up. Head positioned facing forward, wings tucked back and legs straight, the finished specimen looked 'sort of like a popsicle,' says Jacobi.

Casey then hand-delivered the birds to Amadon at the American Museum of Natural History and he was soon able to confirm that not only were they a new species, but the little masked birds that Gagne, Casey, and Jacobi had found were so different from any other bird that they warranted the creation of an entirely new genus within the group of birds known as Hawaiian honeycreepers. The species was given the scientific name *Melamprosops phaeosoma* (which translates from the Greek as 'black-faced bird with brownish body'). Some academics proposed an English common name, since the bird hadn't been known by native Hawaiians. Casey and Jacobi, however, believed it should be given a Hawaiian name. And so, Casey asked Mary Kawena Pukui, the Hawaiian author and scholar responsible for preserving and cataloguing Hawaiian folk tales, songs, and the Hawaiian language, to name the bird. 'We felt that this was important,' says Casey. The names given to native birds in the Hawaiian language are often onomatopoeic – in this way, the calls of the kiwikiu and 'i'iwi (*Drepanis coccinea*) can in some sense be heard when speaking their names aloud. For the quiet new bird, however, Pukui chose a name which reflected its appearance instead: po'ouli, meaning 'black-head'. It soon became clear that the po'ouli was actually an ancient resident of the archipelago. One of the oldest Hawaiian honeycreepers, it had diverged from its closest ancestor species millions of years ago, when Hawaii's current islands did not yet exist.

A FLOCK OF birds, lost at sea, 5.8–7.2 million years ago – this is how the story of the Hawaiian honeycreepers begins. As a forest-dwelling species, originally from Asia, life for the ancestor of honeycreepers was all about short flights between branches, picking from an endless buffet of seeds and insects. Or, at least, that's how it was meant to be. Now, the birds found themselves lost over the Pacific Ocean with nothing to eat, nowhere to stop, and nothing to do but fly on and on, over the salty blue expanse beneath them.

Before long the flock began to thin. The weakest amongst them fell quietly into the sea, exhausted. This would have been the fate of them all, were it not for the smattering of tiny dots that appeared on the horizon ahead. 3,800km west of the nearest landmass – today, the continental United States – lay Hawaii.

In all volcanic archipelagos, islands are created by the convergence of two geological processes: a volcanic hotspot where molten rock forces its way through the Earth's mantle and up to the surface of the sea, and the movements of tectonic plates (in the case of Hawaii, the north-westerly movement of the Pacific Plate). Together, these two processes create something like a conveyor belt. At the start of the production line, molten rock from the hotspot bubbles up to form an island. This island is colonized by plants, animals, and other life forms. Then, slowly, over tens of millions of years, the island and its inhabitants are dragged across the ocean in the direction of plate movement, making space behind it for another new island, and another and another, and so on. Finally, on reaching the end of the belt, the island – now the oldest in the chain – is subsumed again by the ocean. When the lost flock of forest birds reached Hawaii, none of today's eight main Hawaiian islands, including Maui, yet existed. In their place were nameless islands that have long since been pulled back down into the sea.

At long last, the lost birds could rest, drink, and feed. Once recovered enough to investigate their new surroundings, they found an ecosystem stranger than any they'd encountered before. Instead of cows or goats, a tribe of flightless geese called moa-nalos (literally 'lost fowl' in Hawaiian) were top of the herbivorous food chain. On the youngest islands, 'ōhi'a lehua trees sprouted straight up out of cooled lava, their flowers little explosions of long bright-red filaments, as if the trees had somehow imbibed some of the magma's essence. Perhaps more notable than all of these strange inhabitants, however, was what was missing from this ecosystem; no amphibians, lizards, or even mammals lived anywhere in the archipelago (the only native land mammal, the Hawaiian hoary bat, arrived only 10,000 years ago). As with St Helena, Hawaii's extreme isolation had worked as a kind of evolutionary trial that few organisms could overcome. The resulting ecological hodgepodge might have seemed alien to the lost flock of forest birds; however, it was a land packed with opportunity.

As the birds spread, they found predator-less habitat everywhere. Over time, each subpopulation adapted to its particular corner of this new world. Some developed special tongues to sip nectar from specific plants, and long curved beaks so perfectly shaped they could almost have been cast directly from their corresponding flowers; the corollas of certain Hawaiian lobelioid flowers, for example, fit like a glove around the beak of the Hawaii mamo. Other birds, such as the kiwikiu, grew broad beaks ideal for digging insect larvae out of wood. And one species, the ʻaki, developed a curious double-purpose beak with a straight, firm lower mandible suitable for drilling holes into the trunks of trees, and a curved, flexible top mandible for probing and scraping these holes for insects and sap. Before long, the lost flock of forest birds had spawned fifty-five species that collectively spanned the entire archipelago.

Today, we call these birds the Hawaiian honeycreepers. Like the *Partula* snails of Chapter Three and Darwin's Galápagos finches, their evolutionary story is an example of adaptive radiation – where one ancestor species rapidly diversifies into many distinct species. Adapting to fill a broad range of ecological niches, these species developed unique plumages, bill morphologies, foraging behaviours and diets. And yet, even within this uniquely broad church, the poʻouli ended up being a misfit.

The Hawaiian honeycreepers are typified by canary-like songs and bright colours. With vibrant hues of cadmium yellow, electric blue, crimson, lilac, turquoise, and violet, many have to be seen to be believed. Birds are associated with the divine in Hawaii. It was from them that the demi-god Māui – and, in turn, all humans – learned how to make fire, according to Hawaiian myth; in the islands' creation story, the archipelago is fished up out of the ocean by Māui using a magical hook baited with birds. In another myth, Māui is said to have painted the bodies of honeycreepers himself, making of their feathers and beaks a vast fragmented canvas filled with colour. Then, to impress a deity visiting from another island, he had the birds appear and sing, overwhelming the visitor with their beauty. This myth, and the fact that the feathers of Hawaiian honeycreepers have been used in traditional Hawaiian dress for centuries, testifies to their striking colours and songs.

And then there was the po'ouli. With its dull brown, grey, and black feathers, it was as if Māui forgot to paint this bird. Its call was understated and rarely used; its tongue turned out to be nothing like the rest of the honeycreepers', too, and its stumpy tail was so short that when Dean Amadon examined the first specimen, he presumed that its actual tail had been ripped off. Strangest of all, however, was its scent; or, to be more precise, what its scent lacked. All Hawaiian honeycreepers smell, with varying degrees of pungency, like old canvas. But not, for some reason, the po'ouli.

The various ways in which the po'ouli differed from other Hawaiian honeycreepers led some to question whether it belonged to the group at all. But, in the early 2000s, DNA analysis confirmed its place as the oldest living Hawaiian honeycreeper, having diverged from its closest ancestor 5.7–5.8 million years ago.

We often think of the moment when European nations 'discover' a place like Hawaii as being the point at which a natural order is destroyed. However, it's an unavoidable fact that our species is, seemingly by its very nature, one of the most potent agents of ecological destruction, regardless of time, place, or culture.

Polynesians arrived in Hawaii around 800 years ago. And, tucked in the hulls of their boats, nestled amongst stores of food, were rats – the first rodents to set foot on the archipelago. Geckos and skinks also arrived, and were the first reptiles (other than the native sea turtles). Small pigs came ashore as a source of food for the Polynesians, as did yams, taro, sweet potato, breadfruit, and bananas – all plants the islands hadn't seen before. Alongside them, the seeds of weeds, ferns, grasses, and trees also made the journey, attached to clothes and hair, and tangled in the fur of livestock. As Polynesian society grew, homes and temples were built, and forests were felled for fields and timber; meanwhile, these introduced species made Hawaii their home too.

Many native species were squeezed out of their habitats, and others were hunted during this time. The moa-nalos were among the first to feel the presence of the Polynesians. Like dodos, these plump, clumsy birds lacked the means or even the instinct to flee hungry humans, and were quickly eaten to extinction. Just as the moa-nalos had never considered humans, the native plants of the archipelago had never encountered pigs. As a result, they lacked the kinds of adaptations – thorns, spines, or chemical defences, for instance – which plants elsewhere rely on to fend off grazers. This made them easy pickings for the pigs and, to a lesser extent, the chickens that arrived with the Polynesians.

By the time James Cook showed up to 'discover' Hawaii in 1778, Polynesians and the species they'd introduced had driven over half of the archipelago's native birds to extinction. The arrival of Europeans, then, rather than kick-starting the destruction of Hawaii's natural environment, was more like a passing of the baton in an ecocidal relay race. But, of course, the next runner to take this baton was a far more dangerous prospect. Whatever ills native Hawaiians had inflicted on the land were made to look like child's play by the incoming Europeans. Labourers were shipped in from East and Southeast Asia, as agriculture, which had previously operated on a localized scale in Hawaii and without the concept of private land ownership, became industrialized. Sugar was soon the dominant monoculture; fields of cane raced far inland, consuming native forests as they advanced. Wood was needed to fuel the boilers in sugar mills, driving yet more deforestation. Eventually, the situation became so dire that the authorities had to bring in protections to prevent further degradation, although these decisions were driven not by any concern for wildlife but out of the need to protect Hawaii's watersheds – the rainforests that produced water to feed fields of thirsty sugarcane.

New invasive species were introduced as well. While Polynesian rats had certainly made an impact on the environment, the black rats (*Rattus rattus*) that swam ashore from European ships outperformed them on every front. Larger and more aggressive, they had a particularly detrimental effect on native birds, feasting on eggs, chicks, and females. Whereas Polynesians had confined their pigs to pens, the much larger pigs that Europeans brought to Hawaii were allowed to roam free, uprooting trees and shredding native vegetation.

The American science writer William Allen uses the metaphor of sound to characterize the cumulative effects of threats to native Hawaiian species: it is 'a modern-day crescendo of extinction,' he writes, 'that began with the original human settlement and grew with Western contact, large-scale ranching, and a rising tide of development.' The Hawaiian historian Samuel Kamakau turned to the metaphor of an invasive plant. In his 1866 book, *Na Mo'olelo a ka Po'e Kahiko* ('Tales and Traditions of the People of Old'), he describes Cook's landing on Hawaii as being like a dreadful seed that 'sprouted and grew, and became trees that spread to devastate the people of these islands'. From this seed also came 'the spread of epidemic diseases', 'changes in plant life', and, significantly for this story, the arrival of 'fleas and mosquitoes'.

In 1826, Hawaiians started to notice small flies that visited them mostly at night. They were described by one Hawaiian as presenting themselves by 'singing in the ear'. To Europeans and the Asian labourers living in the archipelago, this 'singing in the ear' was familiar, but to native Hawaiians mosquitoes were an entirely new phenomenon. Thanks to its geographical isolation, the archipelago had remained completely free of mosquitoes over the millions of years it had existed – until now. Soon, a new word entered the Hawaiian lexicon: 'makika', a loanword from the English 'mosquito'.

A year earlier, the British Royal Navy ship HMS *Wellington* had stopped at Maui to resupply. The ship's water barrels, which had last been filled in Mexico, weeks earlier, needed to be cleaned before they were refilled, which the assigned sailors decided to do in a freshwater stream. They found 'wrigglers' – slang for mosquito larvae – in the stagnant water at the bottom of these barrels. Making no attempt to contain the larvae as they tipped out the dregs, the sailors refilled their barrels and were on their way. The wrigglers, meanwhile, seized the opportunity they'd been given, and soon people all over the archipelago were encountering flies that 'sung' in their ears. In and of itself, the arrival of mosquitoes in Hawaii would not have been a significant event for the native birds, beyond the occasional itchy bite or two. However, there were other, far more dangerous things that had long been present in the ecosystem, waiting for something like a mosquito to come along.

*Plasmodium* are protozoan parasites that spread avian malaria, a disease which, like its human variant, infects red blood cells, causing them to rupture, leading to anaemia, coma, and eventually death. *Avipoxvirus*, also known as bird pox, is a genus of debilitating and often fatal viruses that, as well as causing the lesions that all poxes are named after, can affect a bird's ability to consume food and water, or even to breathe. Both of these organisms had long been present in the Hawaiian archipelago, likely for thousands of years, in the bloodstreams of migratory birds like Pacific golden plovers and sanderlings (over a million of which visit Hawaii from North America each winter). However, without transportation, they were essentially stranded in their hosts. The mosquito changed all that. Like an inner-city metro network, mosquitoes connected them to new destinations all over Hawaii. Suddenly and disastrously, it was now possible for bird pox and deadly protozoa to move freely from a migrating plover to a native honeycreeper.

Imagine a bird sleeping. What you are picturing is most likely the sleeping position of birds that are native to places with mosquitoes: the head and bill tucked into fluffed-up back feathers, legs folded beneath the belly. Compact and neatly contained, a bird sleeping like this has no featherless part of the body exposed. Native Hawaiian birds, however, sleep in an entirely different position – standing, with legs exposed, heads untucked. At night, these birds became easy prey for mosquitoes. Quietly singing, drifting along in the air, they would perch on a sleeping honeycreeper and bite around the bill, on the forehead or on the legs. Birds started to die en masse. In 1902, the American ornithologist and ethnologist Henry Wetherbee Henshaw wrote about the eerie and desolate atmosphere he encountered in some of Hawaii's forests, from which native birds had vanished:

> The 'ōhi'a blossoms as freely as it used to and secretes abundant nectar for the iiwi, akakani and amakihi. The ieie still fruits, and offers its crimson spike of seeds, as of old, to the ou. So far as human eye can see, their old home offers to the birds practically all that it used to, but the birds themselves are no longer there.

Hawaiian honeycreepers, in particular, started dropping like flies. Extinction followed extinction: the Oahu nukupu'u in

1892, the ula-'ai-'hawane, lesser koa-finch and Oahu akialoa in 1893. Numbers continued to plummet, with two species going extinct in 1894, two more in 1895, and then another pair by the end of the century. 1901, 1907, 1918, 1923, 1940 and 1963 – in each of these years, a species disappeared.

IN 1939, THE American painter Georgia O'Keeffe visited Hawaii on a trip funded by the Dole Pineapple Company. O'Keeffe was to capture something of the wildness and beauty of Hawaii in a series of paintings, which Dole – whose canned pineapple chunks were produced in the archipelago – would use in advertisements. 'It is really a beautiful world,' O'Keeffe wrote to her husband, Alfred Stieglitz, during the trip: 'The country is very paintable.'

O'Keeffe produced many paintings during her visit – some landscapes, others close-ups of plants she'd seen: *Hibiscus*, *Bougainvillia*, *Plumeria*, white lotus (*Nymphaea lotus*), pineapple (*Ananas comosus*). In her painting *Crab's Claw Ginger*, the eponymous species (*Heliconia wagneriana*) is front and centre, as in a portrait, its bright-red bracts backdropped by clouds. The white bird of paradise (*Strelitzia nicolai*) is the focus of another painting, its flowers zig-zagging across the canvas, pointing this way and that like the beaks of birds. O'Keeffe wrote to Stieglitz: 'So many of the flowers just simply seem unbelievable', then of the crab's claw ginger in particular: 'it is really a wonderful flower – You will think I made it up – You will not believe it is true . . .' O'Keeffe told Stieglitz that her very idea of what nature was had been transformed by the trip, writing that, in Hawaii, 'many things are so beautiful that they don't seem real'.

Dole was satisfied. In her paintings, O'Keeffe had captured the Hawaii they hoped would help to sell their pineapple. But O'Keeffe's written descriptions of Hawaii – 'fantastic', 'unbelievable', 'you will not believe it is true', 'don't seem real' – unintentionally intimate something closer to the truth. None of what O'Keeffe painted – in fact, not one of the plants she wrote home about, including the pineapple at the centre of it all – was native. The 'fantastical' crab's claw ginger is from Central and South America, and the white bird of paradise is native to South Africa. O'Keeffe's paintings, together, are like a map of historical plant migration to Hawaii: the *Hibiscus* from China; the white lotus from India; the *Bougainvillia*, pineapple, and *Plumeria* from South America. Below 800m elevation, Hawaii's native ecosystems had by now – thanks to the work of mosquitoes, invasive species, and humans – simply vanished.

True Hawaii existed only in the rich green of the mountains, which O'Keeffe also painted: the green she described as 'different than [the colour of] any mountains I have seen'. There, hidden in that 'different' hue, rare and endangered plants and birds clung to life, while the introduced plants carpeted the landscape below. Hanawi was one such green refuge, and deep within it the po'ouli survived, just above the altitude ceiling of mosquitoes, long after much of Hawaii's native wildlife had vanished. And, so it was that in 1973, when the po'ouli was discovered, it was the bird's pristine forest home that seemed alien, amongst acre after acre of land covered with thriving foreign introductions that had themselves become emblematic of Hawaii. After decades of extinction playing out across the archipelago, the discovery of a new bird was like a miracle.

ON 2 AUGUST 1974, Casey and Jacobi's paper announcing the discovery of the po'ouli was published. In it they describe the species' morphology, plumage, behaviour, diet, and genealogy, alongside photographs of their specimens, and diagrams from

a then-unpublished study of its unique tongue by the ornithologist Walter Boch. Concluding with a frank discussion of the po'ouli's conservation prospects, the authors write '*Melamprosops*, with a very localized range, might well have become extinct before being discovered.' Casey and Jacobi estimate that only 200 individuals remain, and recommend that the po'ouli 'be included in the Department of the Interior's List of Endangered Species, and the State of Hawaii's Endangered Bird List, and should be given complete protection . . . and strict preservation of the remaining vital habitat should be accomplished.'

The po'ouli was added to the US Fish and Wildlife Service (USFWS) list of endangered mammals and birds on 21 April 1975. What followed, however, is perhaps best described as a long period of silence. Despite further studies noting the precariousness of the species' situation and frequent calls to do more, most vocally from Casey, who returned to Hanawi every summer to study the bird, no concrete action was taken to protect the po'ouli. Then, in 1981, the Hawaiian Forest Bird Survey – a seven-year project to assess all of Hawaii's forest birdlife – returned from Hanawi with bad news. Pigs, it emerged, had discovered the Hanawi Natural Area Reserve at pretty much exactly the same time as humans during the 1970s, and the result was not far off what we had been doing to lowland Hawaii for centuries. The pigs cleared vast swathes of the understorey in Hanawi, ripping up plants to feast on their roots, and flattening other vegetation to create mud wallows. And with this sudden erosion of the understorey, many of the insects, snails, and larvae that the po'ouli fed on disappeared too.

By 1981, just five years after Casey and Jacobi had described the po'ouli to science, pig activity had increased in Hanawi by approximately 475 per cent and the po'ouli population had plummeted by 80–99 per cent. This discovery wasn't the wake-up call that many hoped it would be, however. A plan to create pig-proof exclosures was proposed in 1984, but the US government took over a year to approve the project. Disagreements with state authorities on whose land the exclosures were to be built, along with the realization that the cost of the project had been significantly underestimated, caused years of further delay. It wasn't until 1990 that the first exclosure was fully constructed, by which point pigs and, to a lesser extent, goats had been deconstructing the Hanawi understorey for well over a decade.

Despite the exclosure now providing a protected space for the po'ouli, 1990 and 1991 passed by without a single po'ouli being spotted. Over the next few years, there were only sporadic sightings until a 1994 search turned up five, possibly six po'ouli. This, it seemed, was the extent of the species' population; that three of these birds were a family unit (a male, a female and their chick) was the faintest of silver linings.

Hope was raised further when a new effort to save the po'ouli, headed up by British biologist Paul Baker, caught six birds, two of which were pairs, and fitted them with identification bands. However, by mid-1997, this population had halved, with three of the birds vanishing without a trace. Crucially, these disappearances had broken up the two couples who were thought to be the species' best chance at recovery. As the project concluded, a sombre Baker wrote: 'We presume the po'ouli to be on the brink of extinction.'

At this crucial juncture, with only three birds left, the conservation efforts became mired in indecision and infighting. Some saw the po'ouli as a lost cause, suggesting, before the species had even become extinct, that a funeral be organized and that members of the conservation team attend grief counselling to deal with the loss. Others wanted to continue to try and save the species, but frequent disagreements about how to do this stymied progress. As the twentieth century drew to a close, time was running out for the po'ouli.

'ACTUALLY, THAT PHOTO is a bit of a cheat,' says Jim Groombridge, Professor of Biodiversity Conservation at Kent University. On the wall in front of us are framed photographs of birds he has worked with over the years. And, in amongst this disjointed flock – alongside Mauritius kestrels and a host of exuberantly colourful tropical birds – is the plump, brown-grey body and unmistakable black mask of a po'ouli. The photograph in question appears to show the bird perched on a branch – 'as if it's in the wild,' says Groombridge, 'but what you can't see, just behind this branch, is the hand of the person holding it.'

In June 2000, Groombridge was hired as Coordinator of the Maui Forest Bird Recovery Project. Two other Hawaiian honeycreepers – the akohekohe and the kiwikiu, 1,000 and 400 of which remained, respectively – also fell under his remit, but the overwhelming priority for Groombridge and his team was the three remaining po'ouli. The first step was to hold a meeting with everyone who had been involved in the conservation of the species, from organizations like the USFWS and the Hawaii Division of Forestry and Wildlife to individuals like Casey and Gagne who had been there since day one. 'I felt it was important to get everyone in the same room, even if they didn't agree on what to do,' Groombridge says.

The disagreements or 'fault lines', as Groombridge describes them, lay in things such as whether or not intensive control of invasive species, like rats, would resolve the situation, or if the birds should immediately be brought into a captive-breeding facility. On either side of these debates, people clashed so fundamentally that progress was nigh on impossible. 'There was a lot of frustration and sadness,' Groombridge tells me. 'Sadness at the situation, but also I think everyone took ownership as well, in the sense of: "Well, we are part of the problem because we can't agree on what to do."' It took a long time, but gradually the various options were whittled down to just one: translocation.

By now, the greatest hurdle facing the poʻouli was a matter of proximity. The three remaining birds were spread out across Hanawi, living independently within distant and well-defined territories. As a result, like three singletons registered on different dating apps, these birds were almost certainly never going to cross paths. Instead, if the species was ever going to stand a chance at recovery, humanity would need to play Cupid.

The plan was simple: catch one of the two male poʻouli, fit it with a radio transmitter, and then release it in the territory of the last remaining female. With any luck, this simple matchmaking gesture would create a breeding pair that might rekindle the species.

In early 2002, Groombridge and his team set up aerial mist nests throughout poʻouli territory. These special nets have an ultrafine weave that makes them almost imperceptible when hung, and they are used to catch small birds without harming them. To lure the poʻouli in, speakers played a looped recording of the calls of kiwikiu, with which they were known to flock.

'Have you ever had as much trouble catching a bird?' I ask Groombridge. 'No,' he replies, 'this was the record!' The team spent 157 field hours, over the course of twenty-one days, trying to catch a male poʻouli. 'I think I saw the male nineteen times . . . he would fly from one side of the net to the other, over the top,' says Groombridge, 'and it was just the most frustrating thing in the world.'

On Valentine's Day, the team gave up on the male and hiked across Hanawi to try their luck with the female, who they eventually caught on 4 April 2002. Finally, Groombridge and his team had their poʻouli, and their translocation attempt – involving a whole team of specialists, including a vet prepared to give the bird emergency life-saving support using the bird-intensive-care unit they'd created in a carry-box – could begin. 'I remember just trying to condition myself to think, "It's not rare in any way, it's not special, you've done it a thousand times in the same environment, same materials, same process . . . It's *not* one of the last three po'ouli!"' says Groombridge. 'It was nerve-wracking, and you can see my hands shaking in the video.'

The video he's referring to is a documentary that the USFWS produced to chronicle the translocation attempt. In it, members of the team stand nearby, silently watching as Groombridge handles the bird, which looks tiny between his fingers and surprisingly calm while he trims its feathers, gently glues a radio transmitter to its back, then blows the glue dry. A moment later and the bird is in a specially prepared cage lined with cloth to prevent it from injuring itself if panicked, happily gobbling down mealworms and snails. Live footage of this is piped from inside the cage to a nearby monitor set up in the unlikely

*A po'ouli gazes down at the ID ring conservationists have fitted around its leg. (Courtesy of MFBRP.)*

surroundings of Hanawi, and around which people gather. At one point, the camera turns to a watching team member, who silently mouths 'Unbelievable!' back into the lens.

Groombridge and his team released the female po'ouli at sunset, in the spot where one of the males had been seen a few hours earlier. Since birds like the po'ouli roost overnight, it was hoped that she would settle down for long enough in the male's range that the two birds would cross paths, or at least become aware of each other's existence. As dawn broke, however, the signal from the bird's radio transmitter grew fainter and fainter. She was returning to her own home range alone, seemingly oblivious to the male whose territory she had slept in. Later, she was spotted flocking with her old friends, the kiwikius.

Although translocation had ultimately failed, the effort was game-changing in a number of ways. Groombridge's team were able to track the female po'ouli for ten days, learning an enormous amount about its behaviour. More importantly, however, the experience helped allay the concerns many people had about the distress that handling and captivity might cause the bird. 'I think a lot of people were surprised we got as far as we did,' says Groombridge. 'We got the bird in the hand and it didn't die. We hiked it two kilometres in a box and it didn't die. We glued a transmitter to its back and it didn't die. All those things that had given us [such a feeling of] foreboding didn't happen.' Particularly encouraging was the fact that, after everything she had experienced, during the two hours the female po'ouli had spent in a cage, she had happily fed on mealworms, which were the food of choice for captive-breeding programmes.

Groombridge left Hawaii soon after the translocation and, although the project hadn't brought them any closer to saving the po'ouli from its plight, the learnings from it energized the remaining conservationists. 'Rather than people becoming despondent about it, it was more, like "What next?"' says Groombridge.

The only course remaining, which had long been rejected, at least in part for fears that the stress caused might kill the bird, was captive breeding.

As far as strange experiences go, there can't be many stranger for a bird than riding in a helicopter. On 9 September 2004, the Maui Forest Bird Recovery Project – led by ornithologist Kirsty Swinnerton, who had replaced Groombridge as the project's coordinator – finally caught another poʻouli. It was the female, and conservationists were shocked to discover that she only had one eye. Packed inside a box, then sealed inside a soundproof carrier, this little one-eyed bird, who was more accustomed to hopping than flying, soon found herself soaring high into the air and leaving her forest home behind.

This evacuation was the culmination of eighteen months of work tracking the birds' positions through Hanawi and attempting to snare them in mist nets. The plan had been to establish a captive-breeding colony at the Maui Bird Conservation Centre, in the hope that, amongst the last three poʻouli, a breeding pair would form. 'We'd been tracking the last three birds,' Swinnerton tells me. 'But in early 2004, two big storms had hit and two of the birds just disappeared. By the time we caught the [one-eyed female], we pretty much knew that this was the last poʻouli.'

Nonetheless, at the captive-breeding centre, great efforts were made in anticipation of the bird's arrival, led by the facility manager Mary Schwartz. A special cage had been built, furnished with 'ōhiʻa lehua tree branches and generous helpings of mealworms, insects, berries, and snails. In the event that the bird was in any way perturbed by her first experience of propeller-driven flight, padded walls would minimize the risk of self-injury. And, to help the poʻouli feel more at home, Swinnerton and her team had scoured the understorey of Hanawi for one of its favourite foods, gently unfurling the fronds of ferns to collect the insect larvae inside. However, when the bird arrived at the captive-breeding centre, she seemed to have taken the last twenty-four hours – no doubt the most bizarre of her life – in her stride. 'It was a very calm, very resilient little bird,' says Swinnerton. 'A lot of bird species would just go rigid and not do anything, a sort of fear response. But this little bird, it was like, *oh, yeah, this is nice . . . this is my new home.*'

Like in Casey and Jacobi's descriptions of the species from 1973, this po'ouli seemed more curious than afraid. Hopping around her cage and feasting on insects and berries, she was as calm as could be. Even with the staff at the breeding centre, who carefully examined and weighed her on check-in, the bird was fairly relaxed, cocking her head to one side and peering at them through her one good eye. This bird had been captured and taken to a strange new place but it was the people, rather than the po'ouli, who were nervous.

Rich Switzer was just thirty years old in September 2004, when he was hired as the new manager of the Maui Bird Conservation Center. The team working with him were younger still: a flock of three fledgling aviculturists in their twenties, all of whom suddenly found themselves with the fate of a species resting – sometimes literally – in their hands. 'We had to be *very* regimented,' Switzer tells me when I speak with him in 2024. 'Our main goal was to minimize the stress felt by this bird.' This meant eliminating human contact wherever possible. Instead of observing the bird directly, the team gathered around a TV in a nearby room and watched it through a video feed. When, at feeding time, the little po'ouli hopped up excitedly onto its feeding plate, little did it know that its weight was being recorded by a set of scales stealthily positioned underneath this plate. And preventative medication was, wherever possible, injected not into the bird itself but into its favourite foodstuff, live waxworms – the avicultural equivalent of a cat owner wrapping a tablet in chicken.

It was also the team's job to keep mosquitoes out of the captive-breeding centre. When the one-eyed po'ouli had been helicoptered away from Hanawi in its dark soundproof box, the bird had crossed an invisible boundary. The 'mosquito line', so called because it is the uppermost boundary of the habitable zone of Hawaii's mosquitoes, could for the purposes of this story also be called the 'end of the line'. Beneath it, a native Hawaiian bird was at the mercy of mosquitoes and the malaria they could transmit. And the po'ouli's new home at the Maui Bird Conservation Center placed the bird well within what was essentially a death zone. 'In Hawaii, our main consideration from a biosecurity perspective isn't just rats, or mice. It's mosquitoes,' says Switzer.

While the po'ouli settled in, trips to capture a male bird continued. However, each time, Swinnerton and her team returned empty-handed. The last sighting of either of the other birds had been nearly a year ago by now, and the odds of one suddenly materializing seemed vanishingly small. Meanwhile, back at the conservation centre, the one-eyed po'ouli continued to gorge itself on an expansive menu of live and partially crushed insects, snails, berries, mealworms, and waxworms. Roughly once a week, Swinnerton and her team delivered a fresh shipment of succineid snails that they had foraged from the rainforest. 'We'd give these snails to the po'ouli and they'd just be devoured in the space of three minutes!' says Switzer. By all accounts, the bird seemed to be doing well at this stage. 'When I visited, it looked perfectly happy, pottering around,' Swinnerton tells me, '... but then it just went downhill.'

At the end of October, the po'ouli was moved into a larger, 12ft by 4ft aviary. Just one day later, however, it was clear that something was wrong. Despite flapping her wings, the one-eyed po'ouli struggled to lift herself off the ground. She was weighed, and it was discovered that she had lost 10 per cent of her body weight, which in peak condition was already just 25–27g. Decline for a small bird can be swift. 'It can all be over in just a couple of days,' says Switzer. So the po'ouli was quickly moved back into her smaller cage with a heat lamp. For over a week, the bird sat with her feathers fluffed up next to this heat lamp, looking visibly less alert. Broad-spectrum antibiotic and antifungal medications were administered in the hope that she might recover.

On 15 November, a blood test revealed that the one-eyed po'ouli had malaria. The following day, there was another revelation: lab results showed that this bird, who everyone had thought was female, was actually male. With this news, a possibility dawned on the conservationists. Perhaps the bird that Groombridge and his team had translocated had also been male (sexing po'ouli was especially challenging with so few specimens for comparison). Had the last three po'ouli steered clear of each other's territory because they were all male, and each was waiting for a female that no longer existed? Perhaps the species had in fact become functionally extinct as far back as 1997, when the last confirmed females had vanished without a trace.

With the malaria diagnosis, Beth Bicknese, an avian vet from San Diego Zoo, was immediately dispatched to give the po'ouli specialist care. Examining the bird in an avian intensive care unit – a plastic container in which light, heat, humidity, and oxygen levels are carefully controlled – she used a transilluminator, a torch-like instrument typically used by opticians, to peer through the skin at his internal organs. Inside, everything seemed normal: a tiny fluttering heart; a liver – where malaria parasites typically gather – of normal size. It turned out that it was not a new infection, but one the bird had been fighting for months, if not years. Bicknese administered subcutaneous fluids and several shots of the antimalarial drug chloroquine, and let the bird rest. He seemed to perk up, fluttering around to avoid capture, and even pecking at Bicknese's finger when eventually caught. This uptick in mood was felt to be a sign that the bird was on the mend, so Bicknese returned to San Diego. Switzer now took over, administering another antimalarial drug called primaquine by gavage (via a small plastic tube inserted down the bird's throat). Nonetheless, the bird's health continued to deteriorate.

From 22 November, Switzer barely slept. He and his team were now checking on the po'ouli every ninety minutes. But once the work day was over, this responsibility fell predominantly to Switzer, being the more senior of two staff members living on-site. 'Typically, if you're caring for a sick bird, you want to leave them alone in peace and quiet,' says Switzer. With the last of a species, however, he explains to me, there is more at stake. It was hoped that, after the death of the last po'ouli, the body would provide scientists with tissue samples from which to develop cell cultures for study and cryogenic preservation. That way, the species might in some sense live on beyond extinction. The viability of this hinged on how quickly tissue samples were taken, refrigerated, and transported. 'As well as making sure the po'ouli was comfortable,' says Switzer, 'I was checking on him so frequently so that if he did die I could pick him up, place him in a Ziploc bag, and then put him in the fridge.'

During this time, Switzer and Bicknese kept in close contact via telephone. By 25 November, the bird was extremely underweight, and they decided that a liquid diet should be administered by gavage. That evening, Swinnerton hosted Thanksgiving dinner at the Maui Forest Bird Recovery Project for staff. 'I can't remember many details about the meal,' Switzer tells me. 'Everyone was so invested in the po'ouli, but it was almost as if people didn't want to talk about it, for the sake of Thanksgiving,' he says. Immediately afterwards, Switzer returned to the facility to resume his watch over the bird.

The following day things got worse. The one-eyed poʻouli was unable to even perch. Switzer rolled up a towel into a doughnut-shaped ring around the bird to support his body and prop up his head. Slumped in this makeshift nest, with oxygen piped in near his nostrils and fluids injected under the skin at the top of his leg, the poʻouli clung feebly to life. 'It got to the point where it just wasn't being a bird any more,' Switzer tells me. He recalls how he'd told his then-supervisor, Alan Lieberman: 'If this was my pet budgie, I would have taken it to the vet to have it euthanized.' It becomes clear to me during our conversation how conflicted Switzer felt while caring for the last poʻouli. 'By this stage, we were essentially keeping this bird alive to prolong the existence of its species and for the sake of preserving its cells, which were both very admirable goals in the bigger picture,' he says. 'But for that one individual little bird, I did feel sorry for him.'

That night, Switzer found the poʻouli lifeless, propped up in the towel he'd arranged for it. The last poʻouli was dead and its species was extinct. I ask Switzer about this moment, and he begins to tell me: 'It was around eleven o'clock at night. I remember going in to check on the bird and . . .' But his voice cracks. For a moment we are silent. 'Sorry,' he says eventually. 'This is why I don't think about this much. I can't believe that twenty years later I'm still . . .' He wipes his eyes.

In the early morning after the poʻouli's death, Switzer phoned Swinnerton to tell her the news and ask if she would perform the final phase of its care – the extraction of tissue samples. 'It was very dark, I remember that. And nobody else was in the building, which made it all the stranger,' she recalls, before adding: 'You don't forget these things, these moments in time.' With the poʻouli gently transplanted from the refrigerator and onto a surgical table, Swinnerton got to work extracting tissue from the bird's liver, eye, musculature, and any other part of its body that was likely to yield high-quality cells. Packaged up and frozen, these samples were then sent to the Institute for Conservation Research, a branch of San Diego Zoo, where an examination of the bird's body revealed that the last po'ouli had been very old and that a systemic fungal infection, which typically impacts birds under physiological stress, had exacerbated its decline. From the tissue samples Swinnerton had prepared, geneticists at the institute worked to develop cell cultures, eventually succeeding with tissue extracted from the bird's eye. Now, the po'ouli's DNA could be stored in the Frozen Zoo. This is an American parallel to Britain's The Frozen Ark. It is 'the world's largest collection of species on ice,' explains Elizabeth Kolbert in her book *The Sixth Extinction: An Unnatural History*, where in a 'windowless room', the cells of numerous species 'are kept alive – sort of.' Back in Hanawi, neither of the two remaining po'ouli were seen again, and in 2019 the species was declared extinct.

'Where does the blame lie?' It's a question that often feels too simplistic to apply to extinctions. Except in rare instances, extinction is a slow and creeping process, often with many causes. Habitat destruction, hunting, disease, predation and competition from introduced species – the roots of the po'ouli's demise are about as varied as they come. However, many people have expressed to me that there were missed opportunities to save it. Some felt more should have been done to remove pigs and rats from Hanawi. Others highlighted the slow pace of the state and national bureaucracies, and the fact that captive breeding was only attempted once the population had dwindled to just three ageing birds.

In amongst all this were instances of simple human error, too. Casey recalled one such moment to me, which occurred right at the tail end of the 1973 expedition. While in Hanawi, she and Jacobi 'had spent months putting out rat traps, cutting up rats, and removing their stomachs'. But this grisly work ended up being for nothing. When Casey and Jacobi returned from Hanawi, they discovered that whoever had unloaded the helicopter carrying their equipment had mistaken these specimens for rubbish and thrown them out. Years later, rats would become Casey's prime suspect in the decline of the po'ouli, and this lost box of guts would have gone a long way to establishing the extent of their culpability, she tells me. 'It would have been really important data,' says Casey. 'I'm sure there would have been bird feathers and all sorts of things in the rats' stomachs.' Crucially, this data would have arrived right at the point of discovery of the po'ouli, when its population was larger, and any conservation actions would have stood a far greater chance of success.

'One doesn't have to be a scientist to know that something is wrong with how this country [the US] finances endangered species programs,' writes Alvin Powell in his book *The Race to Save the World's Rarest Bird*, an exhaustive account of the story of the po'ouli. Powell looked at how funding was split in 2004 between 1,340 endangered and threatened species, discovering that half of it went to just twenty species. The top beneficiary of this funding, the Chinook salmon, received $161.3 million and was not even endangered; whereas, in the same year, as it slowly petered out into extinction, the po'ouli was allocated just $67,203. According to Powell's sources, the reason funding disparities like this exist simply comes down to geopolitics; a small, isolated state like Hawaii has far less political relevance than continental states, and is therefore overlooked.

Another factor in the po'ouli's extinction was what can only be described as a reluctance to act. 'Perhaps the most important lesson from the case of the po'ouli is that action must be taken early, before a time when a species' future rests on a single risky endeavour,' wrote Groombridge and others in a 2006 paper on the handling of the po'ouli's conservation. I've been told that fear of failure, of inadvertently causing an extinction, permeated many of the institutions working to save the poʻouli. When there are so few of a species left in the wild, there is almost no margin for error, and every decision can become paralysing. In the case of the poʻouli, the thirty-year effort to save the species was pockmarked with delays, sometimes years long. And when it was finally accepted that captive breeding was the only thing left to try, it was simply far too late.

Regardless of the decisions that were made, and the reasons behind them, it is possible that the roots of this extinction actually lay somewhere else. Fossilized remains of poʻouli, found on the western side of the Haleakalā volcano in the 1980s, show that the species was once more widespread. Crucially, the lowland area where these fossils were found was once dry forest – a far cry from the wet, murky conditions in Hanawi. If the species had originated in dry forest, but later moved to the rainforest, it might explain the poʻouli's odd characteristics. Its dull brown and grey plumage would have been the perfect camouflage in dry forest, whereas in the ever-present green of Hanawi, it made far less sense. Also, its lack of a canvas scent correlates with it having evolved to live in dry conditions, where the less pungent Hawaiian honeycreepers live. It might also explain why, for the poʻouli, surviving in Hanawi was such a struggle. Why, if they did not belong in Hanawi, was this last population living there? Perhaps they were ousted from their original habitat by pressures wrought on them by humans, such as land-clearing, invasive species, and mosquitoes. If this is true, from the moment we'd first laid eyes on the poʻouli, it had never really been at home in Hanawi, and the life it had been living there was as artificial as O'Keeffe's paintings of 'Hawaiian' plants.

'Hawaii is a fascinating place to work because of its unique ecology. But, on the other hand, it's a frustrating and sad place to work because of what is happening to that ecology,' Jim Jacobi tells me when we speak. Jacobi is keen to stress that what happened to the po'ouli is just one part of a broader degradation of Hawaii's native ecosystems. Sometimes dubbed 'the extinction capital of the world', Hawaii is home to nearly a third of all species and subspecies listed under the US Endangered Species Act (as of October 2023). 'It's not just about the loss of individual species, it's also about how they connected with the other species in their ecosystems,' says Jacobi. 'We need to think beyond just that focal point.'

Nonetheless, I sense that the grief of losing the po'ouli is still fresh for many people today. Having been one of the students who discovered the bird, Casey in particular took on the role of caretaker for the species. In the years following the discovery, she returned to Hanawi frequently by herself to learn about the bird. She even took her mother to meet it, she tells me. Later, when time was running out for the species, Casey learned how to pilot a helicopter so that surveying the habitats of the species she was working with would be easier and cheaper. I don't ask her how she feels about the extinction of the po'ouli; I don't have to. It's there in every syllable of her story, audible in her tender descriptions of the then-nameless little bird hopping up to her in 1973, and her recollections of the many frustrations she felt during the years that followed. 'Do you miss Hawaii?' I ask Casey, who moved back to the US mainland not long after the po'ouli went extinct. 'Sometimes, yeah,' she replies, before adding: '... I miss the birds.'

Today, in a lecture room at Kent University's Canterbury campus, the story of the po'ouli takes on a different hue. Each year, a fresh batch of around thirty undergraduate students cram into this small room to learn about the po'ouli. They hear about that first flock of lost birds that miraculously found Hawaii, and the astounding array of species that descended from them. They learn about the pristine forests of Hanawi and the gaggle of lost students – undergraduates like them – who were the first to experience this rare place. They hear of how, during the 1973 survey, Casey, Jacobi, Gagne, and their cohorts gathered at 'Frisbee meadow' (the nickname, still used on maps of Hanawi today, where the students played frisbee during downtime) to catch the potatoes, cabbages, oranges, and other foodstuffs that were periodically dropped from their support helicopter. And they learn about the bird itself, the most ancient and elusive of the Hawaiian honeycreepers, the desperate attempts to save it and, finally, the death of the last individual.

As the lecture finishes, amidst the bustle and chatter and squeaking of chairs, there is electricity in the air. 'I *loved* that!' one of the students says to another. This sad story, which, despite the efforts of so many people, peters out into nothing by its end, has become something new in this setting. The students hearing it here for the first time will one day be heading up conservation programmes of their own. And their lecturer, Jim Groombridge, believes that teaching the story of the po'ouli to the next generation of conservationists might help save other species (in fact, one of his former students is now working for the Maui Forest Bird Recovery Project).

'I think that if there's one uplifting note from my own involvement with the po'ouli, it's that the story really resonates with students,' Groombridge tells me. The story of the po'ouli has become a kind of cautionary tale for many conservationists, an example of what can go wrong if fear and uncertainty are allowed to guide decisions. But for Groombridge's students, I wonder if the reason it resonates is more because its human protagonists were themselves students when they discovered the bird and pushed for its conservation – the same age as the students in the classroom, only just adults, still full of hope. With their own careers ahead of them, Groombridge's students can perhaps imagine similar stories for themselves, only with different, happier endings.

'I've learned from the po'ouli that we should get used to rolling our sleeves up, getting invasive and hands-on with endangered species, not being reluctant because of the weight of responsibility. And that's what I tell my students,' says Groombridge. 'Because there's nobody else out there doing this stuff. The probability of extinction is kept at bay by us, and that's it.'

It's evening on 9 August 2023 and Maui is in the headlines. In a banner flashing red at the top of my news feed is a slideshow of photographs – landscapes of grey ash and blackened trees and, close by, the conspicuous orange glow of fire against what looks like a night sky. Above, the headline tells of 'unprecedented' wildfires on the Hawaiian island of Maui. I click through – it's not a night sky in the photographs, I realize: it's smoke.

Satellite images of Maui, created by NASA using shortwave infrared light, paint the island and the sea around it in calm blue tones, against which the fires burn bright yellow and orange. It looks like a disaster movie, as if Maui's twin volcanos have suddenly erupted not from their summits but from random spots about the island. The historic town of Lahaina, on the western side of the island, is entirely engulfed by this sinister glow, and there are more fires closer to the centre of the island. One, partway up the north-west side of the Haleakalā volcano, I recognize as the exact location of the Maui Forest Bird Recovery Project headquarters and Maui Bird Conservation Center – where the last po'ouli had briefly lived before its extinction, and where many other critically endangered species such as the 'alalā, or Hawaiian crow, now cling on to life.

As news spreads of the human toll taken by the fires – the

entire town of Lahaina lost, thousands of Maui residents displaced, over a hundred people dead – CCTV footage emerges from the Maui Bird Conservation Center. The video, which quickly goes viral, shows the forest across the road from the centre in flames. Then a staff member rushes towards the encroaching fire, brandishing a garden hose.

'The fire could have destroyed half the 'alalā population . . .' says Hanna Mounce, Project Program Manager at the Maui Forest Bird Recovery Project – adding, after a pause: 'I don't know what we'd even be talking about now if that had happened.' Luck was on the centre's side: the fire stopped just short of the facility.

Mounce arrived a year after the last po'ouli died. We're talking on Zoom, and behind her, I see a small cabinet with taxidermied birds positioned as if alive on an arrangement of branches. She tells me that they are Hawaiian honeycreepers, listing their species names. I recognize a few that the po'ouli used to flock with in Hanawi. 'Some are extinct now,' she says. 'Others . . . well . . . we're trying. I take these [taxidermies] into communities on the island, so people can see the birds, recognize what we're losing and what we're trying to save.' We move on to talking about the po'ouli: 'I never actually saw the species alive. Only the aftermath of its extinction.' Many of the staff quit at this point, according to Mounce. 'They actually talked about disbanding the project altogether,' she says. Eventually a decision was made not to give up, but to refocus the project's efforts on the kiwikiu, the next most imperilled of Maui's native forest birds and the bird the po'ouli was most frequently seen with. Despite their work, however, the total kiwikiu population declined from around 544 birds in 2001 to just 157 in 2017.

In October 2019, fourteen kiwikiu were transported to an area on the southern slope of Haleakalā known to be free of malaria-carrying mosquitoes. However, what had initially seemed like a successful translocation ended in failure. 'The disease landscape basically shifted under our feet while we were planning the translocation,' says Mounce. 'By the time we opened the aviary doors, the area was no longer safe.' By November 2019, all but one of the fourteen birds had either disappeared or died, and those that were recovered tested positive for avian malaria.

Morale within the project plummeted again. 'We did not have anything that we felt would make a meaningful difference,' Mounce says. During this time, she tells me, she finally understood why people had left the project after it had failed to save the poʻouli. 'Conserving the kiwikiu was all I had been doing for twenty years . . . it was almost like my whole identity, in a way,' she says. As if history was repeating itself, the project was once again left with no clear idea of what to try next, and how to move forward. 'If someone had said to me then, "If I gave you five million dollars right now, could you save the kiwikiu?" – I'd have had to say "Nope!"' says Mounce. 'There was just nothing we could do – we had no tools to deal with malaria.'

In 2023, showers of a different kind fell on Maui's near-perpetually-drenched rainforests. *Wolbachia*, a naturally occurring bacteria carried by many of Hawaii's insects, has been found to wield an enormous influence over the reproductive systems of mosquitoes. 'Say you are a female mosquito with strain A of *Wolbachia*, and you go and mate with a male carrying strain B, the two strains are incompatible, so your eggs will not develop,' says Mounce. The 'Birds, Not Mosquitoes' project, a collaborative scientific initiative between fifteen institutions, aims to exploit this natural contraceptive, with Mounce and the Maui Forest Bird Recovery Project leading the charge on Maui. 'Right now, we're throwing half a million male mosquitoes out of helicopters, twice a week,' she tells me – meaning that, for the mosquitoes of Hawaii, it is now literally raining men. By flooding an area with millions of males infected with an incompatible strain of *Wolbachia*, a male-to-female ratio of ten to one is created, the net result of which is an almost immediate suppression of the mosquito population. And with this, at long last, Mounce and her colleagues have something with which to fight back against avian malaria.

As encouraging as this new technique is, it is not a cure-all. 'On the one hand, it's really good to be able to do this without using a GMO [genetically modified organism], because the public are much more comfortable with it,' says Mounce. 'However, at the same time, because it isn't a genetic technique, it's limited.' *Wolbachia* isn't passed down between generations of mosquitoes, so this project relies on the continuous breeding, infecting, and releasing of vast numbers of host insects. 'I don't know how sustainable it is to think we are going to be releasing mosquitoes over the entire archipelago forever. Plus, the cost is huge!' says Mounce, before adding: 'Though it's not even a day of our country's military budget.'

The day before we speak, Mounce and her team had a rare win. The one kiwikiu that had survived the 2019 translocation attempt had been caught a month ago and a blood sample was taken. 'We got the results back yesterday and he is chronically infected with malaria,' says Mounce, 'but he's doing fine!' Migrating back to the wet, northern side of the volcano, this bird has now found himself a mate. 'This is the first kiwikiu known to have had malaria for more than a year and seemingly recovered,' says Mounce.

Speaking with Mounce, I get a sense of the lingering impact left by the fires of August 2023. The road leading up to the Maui Forest Bird Recovery Project, now charred and blackened, is a constant reminder of the human loss in places like Lahaina, Mounce tells me. 'Maui has a very, very long recovery ahead of it,' she says. I also get a sense of something else, however – of the determination of the people fighting for Maui's critically endangered species. After our call, Mounce tells me she and her colleagues will be venturing out into the rainforest once again to try and capture a kiwikiu. This bird, they hope, will be the first of a captive-breeding population at the Maui Bird Conservation Center. Perhaps, just like its old friend, the poʻouli, the best chance at survival for this species lies in captivity. And with an estimated 100–200 birds left, captive breeding stands a far better chance of success, this time around.

The issues facing the kiwikiu and birds like it – habitat loss, invasive species, pox, malaria – can themselves seem almost like a fire. The climate is warming, and as it does so, the habitable zone of mosquitoes is rising. Metre by metre, these insects are creeping up Haleakalā and other Hawaiian volcanoes, like rising flames, bringing avian malaria with them. In 2022, the US Department of the Interior predicted that the kiwikiu, along with three other Hawaiian honeycreepers, may go extinct within the next ten years. In the face of such a catastrophe, it's hard to imagine feeling anything approaching hope. 'A lot of people ask me why I do this – because it's so depressing,' says Mounce. 'It's a huge emotional rollercoaster for everybody involved, but we have a lot of very dedicated people, and we *all* believe in what we are doing.'

Though the plight facing Hawaii's native birds shows no signs of abating, it gives me hope knowing there are people fighting tooth and nail to protect them. As I turn the page and leave the poʻouli behind, I think of that video of fearless conservationists rushing to beat back a wildfire with nothing but a garden hose. And then, on 1 February 2024, I get an email from Mounce. Against all odds, there is good news from Hanawi: 'We have brought eight kiwikiu into captive care!'

# 5 Driftwood Stowaway

Bramble Cay Melomys
*Melomys rubicola*

I FOLLOW ROBERTO Potela Miguez, Senior Curator of Mammals at London's Natural History Museum, through corridors and rooms, up and down staircases, twisting and turning back on ourselves as we go. Everywhere I look there are cabinets, cupboards, shelves, and crates, stashed with skeletons, fossils, and other remains. A Bronze Age dog rushes past on our left; next, a cabinet of bears – polar, grizzly, and panda – followed by the headless skeleton of a famous racehorse from the 1800s, his skull resting in a wooden box at his feet. Above, below, and all around us, the museum's vast collection of 80 million objects – including 29 million animal specimens – sprawls over eight floors in one of the building's towers. Eventually, somewhere in the middle of all of this, Potela Miguez leads me through a door and into his office.

I put on a pair of latex gloves and sit down at a desk as Potela Miguez darts off to fetch the specimen I'm here to see – the first collected example, and the holotype, of an extinct species that is, even amongst the many millions of specimens here, especially unusual. The Bramble Cay melomys made headlines around the world when in June 2016 conservationists announced that it had gone extinct. A large, fat rat with a long, scaly tail, it was an unlikely animal to attract this kind of attention, usually the preserve of the so-called 'charismatic species', like lions and elephants. Rats – an informal, catch-all term that describes not only 'true' rats of the *Rattus* genus, but also hundreds of species of long-tailed rodents – are perhaps the group of animals most reviled by humans; still, something about this one in particular was undeniably newsworthy. The Bramble Cay melomys was, in the words of its conservationists, 'the first recorded mammalian extinction due to anthropogenic climate change'.

Potela Miguez emerges holding a wooden box with a glass viewing window in its lid. The object he then presents me with is so carnivalesque, it takes me by surprise. Straight and stiff, and mounted on a thin rod that runs the length of its tail before disappearing into its anus, the Bramble Cay melomys holotype looks like roadkill ready for a spit roast. Its cotton-wool stuffing, some of which protrudes from its open mouth, gives it a slightly bloated and unnatural shape compared with photographs I've seen of the species when alive. Its eyeless face is stretched over a cotton and wire simulacrum of its head, like an animal wearing another animal's face as a mask. Altogether, the impression is less 'scientific specimen' and more 'exhibit from a Victorian cabinet of curiosities'.

As instructed by Potela Miguez, I pick up the holotype by its reinforced tail. It is light, and its fur is bristly but slightly greasy – residue, Potela Miguez tells me, from the liquid it was preserved in in 1846. 'You can see where the trap came down . . . here,' he says, pointing to a break in the animal's skull, which is kept as a separate specimen. In taxonomy, Potela Miguez tells me, the skulls of mammals are often key to distinguishing between different species and subspecies. In the case of the Bramble Cay melomys, however, its species designation, its name, and the very matter of its extinction, all hinge less on *what* it was than *where* it was found.

Bramble Cay is a small coral cay in the Torres Strait, a narrow sea passage that connects the Indian and Pacific Oceans. It is the northernmost of 274 small islands (the vast majority of which are uninhabited) that make up the Torres Strait Islands. Despite being 227km away from mainland Australia, it is a territory of Queensland and is Australia's most north-easterly point. It is, however, much closer to Papua New Guinea, which lies just

53km north-west. Satellite images of the island give the illusion of a place of moderate size: an oval of turquoise surrounded by the deep blue of the ocean, with a hint of yellow sand at its north-west. However, the turquoise is coral – the northern extremity of the Great Barrier Reef. The cay itself is only the tiny area of yellow sand, which spans roughly 3.6 hectares (the size of three Trafalgar Squares), and its only geological features are lumps of phosphatic rock that run along its south-eastern edge and, at their tallest, stand 1.9m above high tide. The sediment that forms the cay is foraminiferal sand, made from the shells of microscopic sea creatures called foraminifera. Look closely, and each grain is a magnificent shape – tiny stars and spirals and discs with swirling patterns. This sand, along with the compacted guano of the millions of seabirds and turtles that have made this little cay their rookery, sits on top of a coral reef shelf that expands under the sea around it. The cay is in a state of constant flux, which is why any description of its size can only be rough. Wind, waves, and tides continuously mould and remould the island's periphery, like spinning pizza dough. Its shape and size change so much that from one year to the next it can be 100m smaller or larger.

Apart from a spotty covering of short vegetation, very few species live on Bramble Cay. As a rookery, it is a stopover for birds and turtles. It is also in the path of migration for butterflies; miles-long clouds of wandering white butterflies have been observed passing overhead, travelling on the wind, between the Fly River delta and mainland Australia. Bramble Cay melomys, however, lived out their whole lives on the cay, on this shifting patch of sand, and were distinguished as the Great Barrier Reef's only endemic mammal species.

In April 1845, the British Royal Navy Lieutenant Charles Bamfield Yule was sailing through the Torres Strait on a survey mission when he came across Bramble Cay. Yule named it after his ship, HMS *Bramble*, and then hopped ashore to chart the cay and claim it for the British Empire. There, Yule encountered hundreds of Bramble Cay melomys, which he referred to in his journals simply as 'large rats'. Perhaps an indication of both humanity's disdain for rats and Yule's particular scientific ignorance, he and his crew took this first encounter with the species as an opportunity not to collect or study it, but to practise their archery skills. A month later, a Scottish naturalist called John MacGillivray collected the first specimen, which I hold in my hands today.

In Potela Miguez's office, I turn the Bramble Cay melomys holotype slowly, and watch as the dull scales on its tail dimly catch the light. Across the room, a stern-looking man observes me from an oil painting, cradling an antelope's skull in one hand. This is Michael Rogers Oldfield Thomas, a prolific zoologist and former Curator of Mammals at the Natural History Museum. Oldfield Thomas named and described more mammal genera, species, and subspecies than anyone in history – over 2,000. In 1924, he added the Bramble Cay melomys to this list, seventy-nine years after MacGillivray had collected the specimen, naming it *Melomys rubicola* (the genus name *Melomys*, coming from 'Melansia', the name given to the region extending from New Guinea to Fiji in the early nineteenth century, and 'mys', the Greek for mouse; and the species name 'rubicola', meaning 'bramble dweller'). As with other *Melomys*, this Bramble Cay species had scales on its tail that didn't overlap (in other genera of rat, they do overlap) but neatly tessellated, like a mosaic. Oldfield Thomas noted, however, that the Bramble Cay melomys was 'comparatively large . . . with a very long tail' that was 'more prehensile than usual'. He also described its colouring: on top, a 'dark buffy brown'; its nape and middle, 'strongly ochraceous' (a light brown colour); and its belly and feet white.

And so, without much fanfare, the Bramble Cay melomys was introduced to science; another rat, amongst many, in the sprawling collections of the Natural History Museum.

LOOK AT THE island of New Guinea on a map and you'll see a straight line running down its middle from north to south. This line separates the western half, which is part of Indonesia, from Papua New Guinea, to the east. Around halfway down, however, it deviates from its course and lurches to the west, as if the cartographer drawing up this border had momentarily slipped before recovering their footing. This lurch is the Fly River, which for 150km is itself the border between New Guinea's two halves. At 1,120km, it is the world's longest river to remain uninterrupted by dams, and the twentieth largest in the world by discharge volume. Every second, around 6,000m$^3$ of water flows out to sea from the Fly River delta – the equivalent of 207,360 Olympic-sized swimming pools being released over the course of a single day. At the point in this torrent where it weakens just enough for coral to grow is the northernmost tip of the Great Barrier Reef, the tiny island of Bramble Cay.

Clumps of earth, branches, entire tree trunks: these are some of the things that litter the shores of Bramble Cay. Torn up and washed away from somewhere along the vast, uninterrupted expanse of the Fly River system, debris like this arrives in a near-constant flow. Sometimes, far larger flotsam washes up, too. Whole chunks of riverbank, which peel away during severe floods, arrive complete with trees still standing upright. There have even been instances of entire human-made huts – built on the banks of the Fly River – making their way to Bramble Cay on these earthen rafts.

It isn't known exactly how the Bramble Cay melomys found its way onto Bramble Cay. It may have arrived via the vast land bridge that once stretched from the north coast of Queensland to Papua New Guinea, only to be stranded there between 9,000 and 5,800 years ago, when sea levels rose and engulfed it. If so, the Bramble Cay melomys could have been a relict – an ancient species that was once widespread, but now survives only in small isolated populations. In this scenario, it is estimated to have diverged from its closest relative, the Cape York melomys (*Melomys capensis*), 900,000 years ago. The other hypothesis is that the melomys originated from somewhere along the banks of the Fly River. At some point in time a piece of driftwood or other debris, perhaps even an entire chunk of riverbank, was washed out to Bramble Cay with a group of melomys stowed away on board.

On the face of it, there doesn't appear to be much on Bramble Cay from which a group of castaway rodents can build a life. There are no trees, other than those washed up by the tide, and only eleven plant species have ever been recorded growing there. Several depressions, including a large one in the south-east, are routinely filled with rainwater, but empty again after a few days. Only the small winding tunnels inside the phosphatic rock, piles of driftwood, and the shells of dead turtles provided shelter for the species. Yet, marooned, the Bramble Cay melomys had no choice but to make this place its home.

Burrowing into the sand under the herbs and grasses on the cay, and into the miniature caves in the rock, the melomys nested and mated. Studies of related *Melomys* species – the fawn-footed melomys (*Melomys cervinipes*) and the Cape York melomys – suggest that litters of one to three pups were born after a gestation period of thirty-eight to forty days. Born blind, the small, helpless pups clung to their mothers' undersides until they grew too large. Spending the day asleep in their burrows, the melomys emerged at night to feast on the plant life.

In many ways, life was peaceful for the melomys on Bramble

Cay. They had no predators or direct competitors and seem to have sustained themselves on a mostly plant-based diet. They had some fearsome neighbours, however, in the thousands of nesting birds that squeeze onto the cay to breed. Many dwarfed the melomys significantly; for example, the slate-grey or white-feathered Pacific reef heron, which stands up to 60cm tall, and the spotted harrier, an Australasian bird of prey, with its wingspan of 121–147cm. Sacred ibis, terns, and silver gulls, as well as kingfishers and the red-chested frigate bird, also frequent the cay. The melomys avoided nesting birds, who would violently peck at any intruders. However, contact was sometimes inevitable. Whenever vegetation cover on the island decreased, the melomys were forced to broaden their palate, venturing into turtle and bird nests to feast on the eggs inside – with the latter incursions sometimes proving fatal.

Every night, for much of the melomys' existence, their cay home was plunged into a deep darkness. After Yule's 1845 discovery of the cay, it began to appear on navigational charts as

*Birds fill the air above Bramble Cay. (Photo by Ian Gynther.)*

the entrance to the Great North East channel, through which European vessels traversed the dangerous waters of the Torres Strait. Bramble Cay and its surrounding reef were a significant hazard to these ships, especially at night. The only sign of land for passing boats was the din created by nesting birds. By the early twentieth century, Bramble Cay had become 'a veritable graveyard for ships', according to an article in *The Bulletin*. And so, in 1924, a lighthouse was finally built. Groups of two flashes with six seconds in between became Bramble Cay's 'characteristic' – the specific sequence of flashes that identifies a lighthouse to mariners. Now, for those navigating the treacherous shallow waters of the Torres Strait, Bramble Cay was defined by its light rather than its darkness. What the nocturnal melomys thought of this is anyone's guess.

Bramble Cay was, to Europeans, a hazard and then a lamp in the middle of the ocean; it was also an emergency larder. Accounts of shipwrecks and of boats anchoring on the cay also read like recipe books – replete with descriptions of omelettes, boiled and scrambled eggs, with sides of 'spinach' (the vegetation growing on the cay, one of which is the edible herb purslane). At the same time, however – and for centuries before European 'discovery' – the cay was visited by other people. The Erubam Le are indigenous Torres Strait Islanders from Erub Island, 50km south. With expert knowledge of the surrounding waters, the cay had never been a hazard for them. Even in the dead of night, they would skilfully navigate the coral reef and land on the cay to catch birds and gather eggs to take back to Erub.

The Bramble Cay melomys features in only very few accounts of the cay. In 1906, a boat transporting 'boys' (i.e. native labourers) to Daru Island, as well as 'an odd murderer or so going to the same place "to do time"', anchored on the cay, and the 'boys' and murderers 'caught some rats ... singed them and ate the lot practically raw.' There is anecdotal evidence of a Papua

New Guinean visiting the cay, and catching melomys by 'hitting them with sticks'. More recently, Aaron Ketchell, an Erubam Le man, recalls trips to Bramble Cay as a child with his father in either 1983 or 1984, and seeing several melomys during the day, running from clumps of vegetation to flee the humans. A mackerel fisherman, Al Moller-Nielson, also recalls seeing them in the 1980s and 1990s; the fishing fleet would anchor on the cay and let their Jack Russell terriers loose to chase melomys for exercise. When turtle researchers later visited the cay, the inquisitive rodents clambered up onto equipment and were photographed by scientist David Carter. In one image from 1979–80, a plump Bramble Cay melomys sits, happily perched, on top of an anemometer (a device used to measure wind speed and direction). Tucked beneath it, clutching onto its belly, is a pup. Tiny – the size of a button – it clings on, both mother and pup seemingly oblivious to the human taking photos. These images of melomys clambering over equipment suggest that

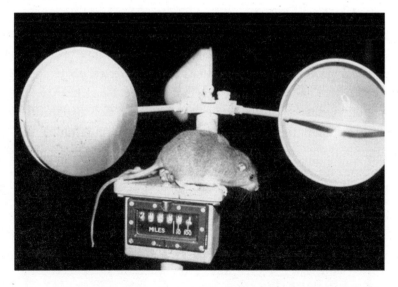

*An adult Bramble Cay melomys with a pup clinging to her underside sits on top of an anemometer. (Photo by David Carter.)*

although the species lived on an almost entirely flat island, it may have been scansorial (adapted to climb), just like the fawn-footed melomys and the Cape York melomys that nest in trees. Carter later wrote of the Bramble Cay melomys: 'One of my strongest recollections is dawn light revealing a sea dotted with flood debris right to the horizon in every direction: trees, great rafts of nipa palms and tangles of grass and reeds . . . [and] there were always rats climbing on these beach washed items even as they tossed in the surf.'

Very little is known about the lives of Bramble Cay melomys beyond these broad strokes, despite the fact that researchers have tried various methods to better understand them. The most novel was in 1998 by two conservationists working for the Australian Government's Department of Environment, Andrew Dennis and Daryn Storch. They attached bobbins of thread to the backs of captured Bramble Cay melomys, which were then released: as the rodents went about their business, the thread unspooling behind them would catch on plants, rocks, and other surfaces – leaving a trail across Bramble Cay that traced their every movement. However, when the frightened test subjects were released they immediately burrowed down into the sand, knocking their bobbins off in the process. A similar rat-tracking attempt during the same trip saw melomys dusted with fluorescent powder, which dropped off them as they moved about, to form a trail visible under UV light in their wake. This failed too, the trails petering out, along with the researchers' hopes of following them home, after ten or so metres.

Dennis and Storch did manage to catch forty-two Bramble Cay melomys on their trip, however. Individuals were trapped overnight, and then transferred into calico bags at dawn. At their camp, the two men then took out each specimen one by one to study it before releasing it. It would have been a surreal scene. The island was densely packed with not only nesting birds but also unpaired and non-nesting birds and, at night, returning mates that were out fishing during the day. Working at sunrise, to the 'racket of calls' from these 'screaming, squawking, flapping, running' birds, of which there were thousands, Dennis and Storch weighed each melomys with a Pesola spring balance. They then took measurements using callipers. It was exhaustive work – head length, head width, ear length, tail length, testes length for the males, all were measured. Hunched over their specimens, they even counted the number of nipples on the females, and noted their arrangement. From thirty-eight individuals, thin slivers of ear tissue were surgically removed, and stored in 70 per cent alcohol, to be analysed alongside all of the measurements. Passive integrated transponders – tiny tracking tags that responded each time a melomys passed an antenna on the island – were implanted under the skin near the shoulder blades. The two men also attempted to find melomys in the vegetation and follow them – a final attempt at tracking the species and learning about its behaviour. However, the sight of humans creeping around in the grass only disturbed the relative peace of the birds, causing a cacophony of sound that likely scared any nearby melomys away.

It was established from Dennis and Storch's study that the Bramble Cay melomys was the 'most morphologically distinct melomys in Australia', significantly heavier than other melomys and with distinctively large feet and small ears. The Bramble Cay melomys was also confirmed to be declining in number. Two decades earlier, another scientist, Col Limpus, had estimated that the population was at 'a few hundred individuals'. He had also noted the erosion of the island, and the Bramble Cay melomys had been listed as endangered in 1992. Now, in 1998, Dennis and Storch estimated that only ninety-three remained.

At the end of their paper, Dennis and Storch include a list of 'conservation options' in response to the decline of the melomys. They run through several possibilities, including: 'do nothing', 'captive breeding', 'introduction to another sand cay', 'stabilise Bramble Cay' (by building a cement wall, or using sandbags, to protect the cay from flooding), 'monitor cay erosion', and 'establish a recovery team'. Under 'do nothing', they write that 'extinction, if it occurs is likely to be natural.' They come to this conclusion based on the rate of sea-level rise in the Torres Strait between 1993 and 1998 (2.1cm per decade), and suggest that coral growth will keep up with this rate of sea-level rise, thereby protecting the cay. Due to this, they say there is an argument for not interfering; storms and high tides would, eventually, erode the cay away or inundate it, meaning that the extinction of the melomys would be both natural and inevitable. 'However,' they continue, 'there is an equally strong argument that although the process may be natural, humans have caused so many extinctions and lowered the world's biodiversity so significantly that we should do what we can to preserve biodiversity where-ever possible.'

BEFORE VISITING BRAMBLE Cay for the first time, Natalie Waller made an offhand joke to a friend. 'I remember saying, "Imagine I get to the cay and there's no vegetation there. How ironic would that be?" and then I just sort of laughed,' she recalls. Waller, an Australian mammalogist, had heard about the Bramble Cay melomys from her supervisor while working on a rodent eradication programme in the Torres Strait. She'd read about how population estimates for the species had plummeted between 1983 and 1998, and how surveys in 2002 and 2004 had caught just ten and twelve individuals respectively. A Queensland government 'recovery plan' for the melomys had also been created in 2008; but she knew that, while this plan existed, there was nothing like captive breeding in place.

And, lastly, she knew that the melomys' dependency on plants as both a habitat and a food source left it extremely vulnerable to any variation in the vegetation cover on Bramble Cay. So, in December 2011, she hitched a ride with some turtle researchers heading to the island, to check on the species herself.

'Desolate' is how Waller remembers the state of Bramble Cay on that first trip. 'I saw it from the boat,' she says. 'There was very little vegetation and it looked brown and just . . . trashed.' This desolation was not at all what Waller had expected. In all the photographs she'd seen, taken on prior surveys and expeditions, the cay had looked 'lush', she tells me, with large parts of the island covered in plants – particularly the succulent species *Portulaca oleracea*. What she found herself staring at now was anything but lush; while not entirely devoid of vegetation – as Waller had innocently quipped to her friend, days earlier – the cay was in a far more barren state than it had ever been before, and she knew what that might mean for the melomys. 'I got there and I thought to myself, "Holy shit, you shouldn't have said that,"' Waller says. '"You jinxed it, you idiot."'

Despite the state of the cay's vegetation, Waller still held out hope for the melomys. 'Rodents are pretty resilient,' she thought to herself. Once ashore, she set to work laying traps, baited with an irresistible putty formed of oatmeal, peanut butter, and golden syrup. However, she quickly noticed two problems. Seabirds in the middle of their nesting season started hopping up onto her traps and settling down to roost, which could deter even the hungriest melomys from entering them; and the green turtles, heaving their heavy bodies across the sand, paid absolutely no heed to Waller's traps, trampling and crushing them. 'I was just so worried about a Bramble Cay melomys getting squashed in a trap,' says Waller, 'and then you've just killed a critically endangered species!'

In order to keep the traps clear of interlopers, and ensure they didn't – with the help of the turtles – accidentally cause the extinction of the melomys, Waller resolved to stay on Bramble Cay and check her traps every half hour. So it was that, while the rest of the expedition returned to the relative luxury of a support vessel just offshore, she and two rangers from the Torres Strait Regional Authority (TSRA) spent the next two nights on the cay, perched on deckchairs and wrapped in bin liners, their only protection against the storm that was sweeping across the island. Still, after all her efforts, checking and rechecking her traps, she found absolutely no trace of the melomys.

'After that first trip, I really wanted to go back,' Waller says. In order to understand what had happened to the species, and answer the urgent question of whether or not it was still even extant, a far more thorough search would be needed. But putting together an expedition to Bramble Cay – distant and expensive to access – was fraught with difficulties. 'At the time, the government had this whole "green army" thing going on, where they funded volunteers to plant trees around suburbia,' Waller says. 'Those kinds of things were more a priority at the time, but things like rodents? People didn't care so much.' Here, Waller touches on an issue facing endangered species that lack 'charismatic' appeal. How people feel about a species can directly impact the level of conservation support it will receive, and all melomys species have the unfortunate distinction of being considered types of rat, i.e. a pest in most people's minds. With her supervisor, Waller applied for funding to survey the Bramble Cay melomys; but, despite the extreme peril facing the species, her application was rejected.

Eventually, the TSRA agreed to her joining a 2012 expedition to Bramble Cay, on one condition: that she conduct a programme of rodent extermination on a nearby island beforehand. Ironically, to save one rat, she would first have to eradicate another. After a grisly two weeks laying snap traps for rats, which she then decapitated by hand so that the skulls could be studied, she was finally ready to go to Bramble Cay. However, at the last minute, the TSRA called off the trip due to poor weather, and Waller returned to the mainland with nothing but a bag of severed rat heads to show for it.

IAN GYNTHER, SENIOR Conservation Officer at the Queensland Department of Environment and Science, was the next person to search for the melomys. Waller had warned him that the cay and its vegetation were in bad shape. However, this did nothing to prepare him for what he arrived to. 'When I saw the cay, I felt sick to my stomach,' he tells me in 2024. The native vegetation on which the melomys depended had diminished dramatically, from an estimated area coverage of 1.1 hectares in 2011 to just 0.065 hectares (a reduction of 97 per cent), and the cay itself had significantly shrunk. 'It hit me like a sledgehammer,' Gynther says. 'This animal, if it was there at all, was on its very last legs.'

Gynther had just two and half days to find the Bramble Cay melomys. He laid traps along the 'little shot' of vegetation that remained and near the phosphatic rock outcrop, and also conducted nightly 'headlamp searches'. Peering into cracks and crevices in the rock, and under piles of timber that had accumulated during the daytime, he scoured the cay for the melomys but found nothing. 'In my heart, I felt that this animal was gone and it was already too late,' he says. 'But I couldn't rule out the possibility that they might be hanging on somewhere.' Back on the mainland, Gynther worked with the TSRA and Waller to organize another trip to Bramble Cay as soon as possible.

In this instance, 'as soon as possible' meant five months later. This delay was partly to avoid the turtle-breeding season, during which trapping would be virtually impossible. Gynther described the cay as looking 'like a warzone' when turtles dig up their eggs, creating huge craters in the sand. The delay also reflected the scale of the intervention the Bramble Cay melomys now needed. Given the scarcity of the species, it seemed unlikely that Bramble Cay held much of a future for the melomys, so Gynther and Waller decided that any they found would be caught to establish a captive-breeding population. This plan came with its own logistical hurdles. 'We had to get special authority from the Federal Environment Minister to fly them back to the mainland, and a captive-breeding agreement signed by the University of Queensland,' he tells me. Suitable enclosures and food also had to be sourced, and transportation via multiple boats and planes meticulously planned. With all of this arranged, Gynther and Waller made a special plea to the Erubam Le elders for permission to take the melomys into captivity. 'They didn't make a decision on the spot. They said they wanted to think about it and talk to other members who couldn't be present at the meeting, and that they would decide when we actually caught some melomys,' Gynther explains. With all the arrangements made, in late August 2014 Gynther and Waller finally set off for Bramble Cay again.

With the last recorded sighting of the melomys now an entire decade ago, Waller and Gynther knew that finding the species alive, and in sufficient numbers to breed in captivity, was the longest of long shots. 'There wasn't much hope by the time we went up there,' says Waller. 'It was more like, "*If* we find it . . . hopefully we do . . . but we probably won't."'

They tried to think optimistically. 'Perhaps it's just trap shy,' they pondered. To address this possibility, they decided to bring night-vision cameras equipped with motion sensors, which they mounted on poles. Unlike the traps, the cameras didn't require a curious melomys to venture into a strange metal box in order to be triggered. Instead, bait was stuffed into spherical metal tea-strainers, secured in place by sturdy pegs and positioned in front of the cameras. Any movement would result in a burst of photographs. This way, even if they failed to capture melomys, they could at least confirm their continued existence.

There was an early moment of drama when, before they'd even set foot on Bramble Cay, Gynther fell out of the boat, completely soaking his equipment and possessions. But once dried off and ashore, he and Waller got to work. Twenty traps and twenty cameras were set up around the now minuscule vegetated area, where the melomys were most likely to be. Over the six nights of the survey, this equated to 120 'trap nights' (twenty cameras multiplied by six nights) of each trap type. The researchers also performed a two-hour search of the cay each day, to check for signs of the melomys – combing over every inch of vegetation, overturning logs, and crawling into cavities in the phosphatic rock. This was a far more extensive survey of the species than had ever been conducted.

The results of all this work? 'Hundreds and hundreds of photos of birds,' says Waller, 'and no rodents.' The only thing the traps caught was a thick coating of guano, which had to be wiped off each evening, and the in-person searches of the island were equally fruitless. The survey left no doubt about it – the Bramble Cay melomys was gone. When I speak with Gynther, I ask how it felt to witness this extinction; he and Waller are unique in this book for having never seen a live individual of the species they were trying to save. He says, 'I remember flying home and just feeling very empty. We both felt like we had witnessed the demise of the species, even though we didn't actually see it happen. It felt like we had just waved the melomys goodbye, and that was the last we were ever going to see of it.'

IT WAS A night like many others when, in early 2006, the sea began flooding cemeteries and graves on Masig, a Torres Strait island 80km south-west of Bramble Cay. 'The tide had been rolling in all afternoon, bigger than usual, like surfing waves. But we kept thinking it's going to stop,' one Masig resident recalled. A confluence of weather events – low pressure over Australia's Northern Territory, north-westerly monsoon winds, a high spring tide – had produced a devastating storm tide. In some places, sea levels rose to 0.5m above the predicted astronomical tide, and 0.3m above the maximum. Like Bramble Cay, Masig is a low-lying sand cay, its highest point less than 8m above sea level. As the water swelled, it reached inland, breaking through sea defences and inundating homes, roads, community buildings, and burial grounds.

Events like this have become increasingly common in the Torres Strait. They can result from a cocktail of causes similar to the 2006 inundation events on Masig, or from a single, severe weather event such as a cyclone. When they happen, they touch nearly every aspect of islanders' lives. Reports have emerged of seawater withering away coconut trees and turning drinking wells brackish; breaking through the bedroom walls of sleeping children; washing away ancestral graves and spreading bones across islands. A 2019 *New York Times* article about the plight of Torres Strait Islanders includes an interview with a Masig resident named Yessie Mosby. He describes walking on the beach with his family. Here and there on the sand are shells and fossils – but also his sixth great-grandmother's bones, disinterred and scattered by a flood. Mosby, his children, and his wife collect the bones and move them beneath a coconut tree. From afar, crouched down on the beach, they appear like any family around the world, beachcombing for shells and pebbles. In moments like this, any dignity ancestors had in death is washed away.

The increase in inundation events in the Torres Strait has been demonstrably linked to climate change. In Dennis and Storch's 1998 paper on the melomys, the average sea-level rise per year was 2.1mm. By 2010, however, it was 6mm; in 2016, it had increased to 6–8mm. During the same period, the intensity and frequency of cyclones and other extreme weather events increased substantially. The causal link is thought so strong that in a 2022 ruling the UN Court of Human Rights found in favour of a group of Torres Strait Islanders, including eight adults and six children indigenous to Boigu, Poruma, Warraber, and Masig, who argued that the Australian government's failure to address climate change had 'violated their rights to enjoy their culture and be free from arbitrary interference with their private life, family and home'. Under 'substantive issues' in the court's decision, 'right to enjoy own culture; privacy; right to life' are listed.

When Gynther and Waller surveyed Bramble Cay in August 2014, they found clear signs of seawater inundation. Patches of vegetation that Gynther had recorded just five months earlier were lying flat on the ground, dead and coated in salt. The flattened vegetation, all of which was pointing in the same direction, revealed the motion of the water as it had swept over the cay. Large logs and other heavy debris had been picked up and moved, along with masses of pumice, timber, and hundreds of dead birds' eggs, which were left strewn over the island in long drifts. Multiple inundations, over the previous four years, had almost certainly caused the enormous die-off of 97 per cent of Bramble Cay's vegetation, they asserted; and it was probably this – coupled, perhaps, with one or more mass-drowning incidents – that had killed off the Bramble Cay melomys.

In a 2016 paper, Gynther, Waller, and their colleague Luke Leung detail the findings of the 2014 survey, concluding that the Bramble Cay melomys is probably extinct. They write that 'available information about sea-level rise and the increased frequency and intensity of weather events producing extreme high water levels and damaging storm surges in the Torres Strait region over this period point to human-induced climate change being the root cause of the loss of the Bramble Cay melomys'.

In 1998, Dennis and Storch had asserted that if the melomys went extinct, it would be a 'natural' extinction, based on the rate of sea-level rise at that time. The extinction ended up not being natural, however; it was brought forward by us humans. Extinction in this case was a natural process, unnaturally sped up by climate change.

The news that climate change had claimed its first mammalian victim quickly resonated around the world. When the IUCN declared the extinction in 2016, articles announcing this news appeared in *The Guardian*, *The New York Times*, *National Geographic*, *The Washington Post*, and many more. 'Here's the sad part,' said Stephen Colbert, paying tribute to the Bramble Cay melomys in a segment on *The Late Show with Stephen Colbert*, 'we're all responsible for climate change, therefore we all killed this animal.'

This narrative, however, doesn't quite tell the whole story. John Woinarski, Professor of Terrestrial Ecology at Charles Darwin University, Australia, and Deputy Director of Australia's Threatened Species Recovery Hub, co-authored the Red List assessment of the Bramble Cay melomys. When we speak, he tells me, 'The dice was so loaded against this species that it was never going to have a chance.' The 'dice' he is referring to here are factors like the melomys' extremely limited distribution and small population size, along with the particular qualities of its habitat, all of which affected how well the species coped with threats. Inhabiting a small, flat island just a couple of metres above sea level at its highest point, the melomys had nowhere to seek refuge from a storm or freak high tide. Any that managed not to drown in their burrows or be swept out to sea during such an event relied primarily on a food source that was also vulnerable to seawater. 'I think it was probably doomed where it was, even if climate change wasn't happening, given the stochastic process of tidal inundation of that teeny island,' says Woinarski.

In July 2016, Woinarski and his colleagues at the Threatened Species Recovery Hub put forth an alternative explanation for the melomys' extinction, which was published in the journal *Nature*. 'Neglect and inaction caused this extinction. Simple management interventions could and should have saved this rodent.' While acknowledging the role of climate change, they criticize the lack of any substantial effort to conserve the species, honing in on the fact that there hadn't been an earlier attempt at captive breeding – even in 2008, when the Queensland authorities developed a recovery plan. Reading this recovery plan now, with the melomys long gone, its silence on the topic of captive breeding is surprising. The entire document, which runs to twenty pages and discusses, in detail, the threats facing the species, along with many possible solutions, contains not one mention.

'It was just one of those flukes of circumstance,' says Woinarski. 'The senior executive in charge of conservation management at the time just happened to be averse to captive breeding.' He goes on to explain how – at least as he understands the situation – the person responsible for writing the recovery plan was overruled by a senior executive, who excised any mention of captive breeding from the report. 'It's just tragic, really,' says Natalie Waller, 'that things like that could be missed, because somebody in the threatened species department was against captive breeding.' Exactly why it was so strongly opposed is a question no one has been able to answer, but the consequences of the resulting inaction are now crystal clear.

Although it omitted the one action that probably stood the greatest chance of saving the Bramble Cay melomys, the 2008 recovery plan did recommend some conservation objectives. These included the establishment of a monitoring programme, and increased efforts to facilitate community participation and education on the recovery of the melomys. However, the recovery plan was never implemented, and the Bramble Cay melomys was left to fend for itself.

In 2009, not long after the plan was written and then quietly shelved, a mackerel fisherman named Egon Stewart stopped at Bramble Cay. Walking along the shore, he came across a 'heap of sticks and a smashed-up dug-out canoe', which he overturned. A pair of terrified melomys zipped out from underneath it. There wasn't anything unusual about this sight, the fisherman thought. He had visited Bramble Cay many times and so had his grandfather, who had fished the waters around the cay for decades before him. Both had seen the melomys before, many times, in fact; sometimes, he'd even let his dogs come ashore and chase them about for exercise. There was something more significant

*Sculptures of Bramble Cay melomys constructed from ghost nets. This work was made as a collaboration between Erub Arts, the Ghost Net Collective, and Lynnette Griffiths. (Photo by Lynnette Griffiths.)*

about this particular sighting, however. Unbeknownst to the fisherman, this was the last time he or anyone else would ever see the Bramble Cay melomys alive again.

THERE IS A myth about the formation of Bramble Cay. It has been passed down from generation to generation of Erubam Le for centuries. And it begins with what is in essence a plan to conserve wildlife.

Long ago, Rebes and Paiwer – the leaders of two Erubam Le clans – were concerned that their people were overexploiting the natural resources around them, over-hunting the seabirds and turtles and taking too many of their eggs. They decided that to protect these animals a new island should be created, far enough out to sea that people could not easily reach it. Loading up a bamboo raft with earth dug from the ground on Erub, Rebes and Paiwer set off with a small party to build the new island. A group watching from a hilltop back on Erub communicated telepathically with the leaders, guiding them to the perfect spot. This was how the location of Bramble Cay, or Maizab Kaur as it is known by the Erubam Le, was decided. A shallow reef acted like a table, out onto which the islanders tipped the earth they had brought. The result was a perfect rookery for seabirds and turtles which, with no obstructions, could easily come ashore to nest. From Erub, elders would be able to oversee the comings and goings to Bramble Cay, so that the birds and turtles there would be safe from overexploitation. With the island built, Rebes and Paiwer prepared to return home. But just as they were leaving, fierce waves kicked up and a magical wind blew in, turning the two leaders, their expedition, and the watchers back on Erub into stone.

Today, six stone statues on a hilltop on Erub are believed to be the petrified remains of the watchers, while the phosphatic rock on and around Bramble Cay is said to be Paiwer, Rebes, and their companions. In 2004, the Erubam Le were finally granted native title rights (to possess, occupy, enjoy, and use the land, 'at the exclusion of others') over Bramble Cay by the Australian High Court. Now the legal guardians of the cay, anyone hoping to visit it must first seek permission from Erubam Le elders.

The legend of the cay makes no mention of the Bramble Cay melomys. However, in recent years the species has been adopted into Erubam Le culture. In 2021, seven local artists worked tirelessly to unpick an enormous tangle of netting in a workshop on Erub Island. This dense bundle was comprised of 'ghost nets' – abandoned fishing nets that litter the world's oceans and pose a significant threat to sealife, especially turtles. Hauled out of the water across the Torres Strait, these nets were separated by colour and cleaned, after which the artists patiently unravelled them to form thread. Coral pink, turquoise, orange, black, bright yellow and ocean blue – from abandoned ghost nets suddenly came a vibrant palette. Next, the artists gradually recombined these threads, to build three-dimensional sculptures. Out of tangles of colour came bellies, heads and tails. Flecks of orange, brown and yellow gave the suggestion of wiry fur, and balls of black thread became eyes. These were sculptures of Bramble Cay melomys, or 'Maizab Kaur mukeis' as they are known by the Erubam Le – twenty or so young pups surrounding an adult, perhaps their mother, who inquisitively sniffed the air.

And so it was that, over a decade after the extinction of the Bramble Cay melomys, its image was recreated in ghost nets. In this new form, the species is now a part of the effort to clear the Torres Strait of these hazardous nets, and an affirmation of the spirit in which its island home is said to have been built. 'We create things from the sea to make people aware of what the nets do to the environment – what they destroy,' said one of the artists, Racy Oui-Pitt. 'We're just on a small island, living in the big ocean, and we're trying to protect what is there.'

# 6 Bird of the Evening

Christmas Island Pipistrelle
*Pipistrellus murrayi*

THE STRETCH OF the Indian Ocean we've arrived at is a vast, empty canvas. In the middle is Christmas Island – a small but riotous blot of colour. Rising steeply out of the sea, a series of terraces lead up to the central plateau, a structure that's often described as like a wedding cake. There is no sign of other land from here; Indonesia is closest at 360km away. Bramble Cay, where we have just been, is 4,000km east – yet both it and this island are territories of Australia, 1,400km south.

From above, the ground appears to move. Miles-long streams of bright red weave this way and that through the tropical rainforest that covers the island, around the coast, and down the roads. All these streams move in one direction, towards the sea. What we are witnessing is one of the world's natural wonders. Every October to November, tens of millions of Christmas Island red crabs (*Gecarcoidea natalis*) emerge from their rainforest burrows and march to the sea to breed, perfectly timing their movements with the phases of the moon to ensure that they always spawn on a receding high tide, before dawn, and during the last quarter of the moon.

It is 1966 and, today, that march is drawing to its close. The many other cycles of life that take place here are in full swing. Suspended from the canopy above the crabs are the seeds of lantern-fruit trees, each surrounded by a spherical array of white petals, like miniature paper lanterns. Between branches, Christmas Island flying foxes hang upside down, slumbering after a busy night pollinating the forest. Then, far above, Abbott's boobies – large, majestic seabirds and the most ancient of all gannets – warm their eggs at the tops of emergent trees that break through the canopy and tower over everything.

Meanwhile, the crabs pass over the sail-like roots of Tahitian chestnut trees, forming long ribbons of crimson, all their minds set on the sea.

By morning the next day, the march is over. The flying foxes return again to their upside-down slumbers, the Abbott's boobies warm their eggs just as they did yesterday, but the crabs have gone. It is as if the forest has reabsorbed the red that carpeted the ground mere hours ago. On the coast, the crabs' eggs (100,000 per female) have been laid, colouring entire shorelines coral pink; amongst the trees, the crabs themselves have gone back into their burrows. Christmas Island has returned to its normal daily routine; it seems, now, to be a day like any other.

*A road swarming with Christmas Island red crabs. During the annual crab migration, islanders depend on a 'crab forecast' to get from place to place. (Photo by Brendan Tiernan.)*

Soon, however, over the cries of seabirds and the scuttling of the last crab claws on the ground, a new and foreign sound punctures the Christmas Island air. From the edge of the only human settlement – simply named 'The Settlement' – in the north-east corner of the island comes a growl, and then splutters and coughs. A fleet of bulldozers is chugging to life. Then, sleds down, engines roaring, and plumes of black smoke billowing from their exhausts, they push forward into the rainforest. In front of them, whole trees fold flat, as if bowing down before their new mechanical overlords. Crabs and lizards scamper off in all directions, and birds abandon their roosts, as the forest is torn up by its roots. Over the next three years, the bulldozers carve out a 506km island-spanning network of 3–7m-wide 'exploration lines' – 354 parallel cuts dissecting the island's rainforest into long, linear strips, as if the entire island had been passed sideways through a paper shredder. Thousands of workers are here on the island, many the descendants of majority Chinese and Malay indentured labourers brought to the island half a century earlier.* They will be doing the 'exploration' these lines through the rainforest are intended to facilitate. Soon, huge drills will be employed to dig into the ground in search of phosphate.

Christmas Island is unusual among remote tropical islands for just how long it managed to remain free of humans. Portuguese and Dutch navigators might have charted it in the late sixteenth and early seventeenth centuries, although these maps are so imprecise it's impossible to say with certainty. It was first definitively sighted by an English merchant in 1615 and then

---

* A Chinese businessman, Ong Sam Leong, supplied indentured labour to what was then the British company 'Christmas Island Phosphate' and made a significant fortune. The labourers he supplied were mostly from very poor villages in southern China, and they were known as 'coolies' – a name that comes from the Chinese characters for suffering and strength.

named, on Christmas Day 1643, by Captain William Mynors of the English East India Company; but the steep and almost continuous cliffs frustrated attempts to land until 1688. Various parties visited sporadically over the next two centuries, but no one seemed too keen on the idea of actually living there. It was seen by the ocean-faring powers of the world as an impenetrable rock, hardly worth the bother of colonizing at all.

The 1887 discovery of phosphate on Christmas Island swiftly changed all that. Phosphorus was an extremely valuable chemical – gold dust to the agricultural sector, which had used it in fertilizers since the mid-nineteenth century. Within months, Christmas Island with its vast deposits of phosphatic limestone was snapped up by the British Empire. Over the next seventy-nine years, possession of the island passed between British, Japanese, and then Australian hands. Eventually, in the mid-1960s, with the island now a part of Australia, the authorities rushed to open up the island's entire phosphate reserves, ordering exploration lines to be bulldozed through its rainforest. And, so, in 1966, the bulldozers arrived.

As catastrophic as this might sound for the island's wildlife, it seemed to have upsides for some. At night, the long barren scars left by the bulldozers teemed with dark shapes that swooped and barrel-rolled like dogfighters. Little brown bats, with red-blonde flecks in their fur, blunt-snouted faces, and fine leathery wings, these were Christmas Island pipistrelles (*Pipistrellus murrayi*), the smaller of two endemic bat species found on the island – the other being the Christmas Island flying fox. The word 'small' is an understatement here. The pipistrelle measured just 34mm long and weighed just 3–4g (the weight of four raisins). Nevertheless, despite its diminutive size, the species' widespread distribution and abundance made it one of Christmas Island's most prominent species. Like all small bats, its turbocharged metabolism meant that it needed an enormous amount of food – over half its own body weight in invertebrates, every night – to sustain itself. As an 'edge specialist', the pipistrelle foraged in clearings or at the forest's edges; and so, drawn by the masses of flying insects now hovering in the newly opened air above exploration lines, the pipistrelle feasted like medieval royalty.

The earliest population estimate we have for the Christmas Island pipistrelle is around 5,000–10,000 individuals, in 1984. Early scientific records, produced soon after the settlement of Christmas Island, suggest that it had been flourishing for a long time before this. But anecdotal accounts of Christmas Island residents paint the picture of this species' past abundance most vividly. In an interview for *We Were the Christmas Islanders*, a 1988 book by Marg Neale, lifelong island resident Gladys Randell recalls her childhood during the 1910s. 'Tiny bats used to fly in through the open doors, and now and then you could see one fall off the light, or off somewhere onto the table, or into someone's soup.'

Today, however, the air is still and empty. Dark shapes no longer dart around in the exploration lines, or anywhere else in the forest. The Christmas Island pipistrelle, a once common and widespread species, is nowhere to be seen. Today, your soup is safe on Christmas Island.

'BIRD OF THE evening' – this is how the Latin word for bat, 'vespertilio', translates into English. It is from this source that the Italian for bat, 'pipistrello', was derived, and thus the genus *Pipistrellus*. True to this ancient epithet, almost all bats are nocturnal, which is likely why they were historically associated with evil or danger. Equally, bats were widely considered to be a type of bird – perhaps in no small part thanks to Moses' description of bats as 'unclean birds' in the Old Testament – until 1758, when Linnaeus finally reclassified them as mammals.

*Pipistrellus* is a widespread genus, with one or more of its twenty-four species living on every continent, except Antarctica. Some of these species span multiple continents all by themselves, like the common pipistrelle (*Pipistrellus pipistrellus*), Britain's most common bat, which ranges from southern Finland to parts of North Africa, and from Continental Europe to Central and South Asia. However, the Christmas Island pipistrelle, as the name suggests, occurred only on Christmas Island. Fluttering, almost butterfly-like, on tiny, 3cm-wide wings, it – or a recent ancestor – probably made its way there from Indonesia, although when exactly this happened no one knows. This wasn't its intent, most likely, but the result of strong winds blowing it off course and stranding it on the island.

The pipistrelle soon became part of an extraordinary tapestry of island life. 'The most striking factor is the peculiarity of the fauna,' remarked J. J. Lister, the first naturalist to visit Christmas Island, in 1887. To date, over 250 endemic species have been recorded on the island. This includes 186 endemic invertebrate species – among them fifty-six beetles, thirty-one bugs, twenty-nine moths, ten crustaceans, eight snails, seven wasps, three bees, two each of butterflies, cockroaches, and cicadas, and even a silverfish – all found nowhere else on Earth. There are also sixteen endemic plants, five birds, five mammals, two geckos, two skinks, and a blind snake. Twelve endemic species live in the subterranean caves, including some other-worldly creatures like *Humphreysella baltanasi* – a translucent, blob-like 'seed shrimp', measuring just 0.5mm long – and a cave shrimp called *Procaris noelensis*, thought to be a relict species originating in the Tethys Sea, an ancient ocean of the Mesozoic Era (65–220 million years ago). Alongside these endemic species are many more native non-endemics, including at least fifty species of crab.

It is for all of this that the island is sometimes nicknamed 'the Galápagos of the Indian Ocean'. Amongst this diversity, the Christmas Island pipistrelles thrived. They could be found, most easily, in the nooks and crannies in tree trunks or the dense foliage of palms, in which they roosted. Peel back a sheet of loose bark on a tree during daytime hours and you could find yourself face-to-tiny-face with as many as fifty of these bats, bunched together like grapes, resting ahead of a busy night hunting. They had round faces engulfed in short brown fur, right up to the tip of their blunt snouts, which grew longer and stood upright from the brow up, as if blow-dried. In photos, the bats all wear the same tranquil expression, their pistachio-shell-shaped ears perched upright behind small, beady eyes.

Hidden away in their roosts, the pipistrelles tucked their soft wings – described by John Woinarski as having a 'strange texture somewhere between fine leather and balloon fabric' – against their sides, and huddled. They gripped the tree trunk with the claws on their feet and midway up their delicate wing bones. At times, they entered the semi-hibernation-like state known as torpor. Reducing their body temperatures to lower than normal, torpor enabled the pipistrelle to use far less energy during periods of inactivity.

Although some preferred solitary living, the bats usually roosted in sex-specific groups. Once a year, however, they'd come together and form mixed-sex roosts, in order to mate. Female pipistrelles could store the males' sperm within their bodies for months after mating. This enabled them to perfectly time the moment of conception so that they would give birth at the beginning of the wet season, when there was an explosion in numbers of beetles, moths, and other flying insects. This was crucial because a female pipistrelle required an almost continuous stream of food each night during lactation, in order to sustain herself and her pup.

In the evenings, the pipistrelles left their roosts to feed, flying as much as 2km to reach their chosen foraging sites.

Like almost all bats, they used echolocation to effectively 'see' with sound. By contracting the larynx as it exhaled, a pipistrelle could fire loud, high-frequency sound pulses into the air, which bounced back off anything in their path as an echo, thus providing precise information on the location of obstacles. To avoid being deafened by its own calls, the pipistrelle reflexively contracted a middle-ear muscle called the stapedius, which dampened the sound by temporarily pulling apart the hammer, anvil, and stirrup bones that relay sound vibrations from the eardrum. All of this happened with frightening speed, the echo from an insect a metre away taking just six milliseconds to return to the bat hunting it. Wherever flying insects gathered, pipistrelles whipped through the air acrobatically; and, after humans arrived on Christmas Island, the bats would even hunt inside buildings.

With its tropical island home devoid of native predators and abundantly stocked with food, it's easy to understand why the pipistrelle was flourishing. So what happened? The answer is complex. Whereas, for some extinctions – like that of the tiny Malaysian microsnail *Plectostoma sciaphilum* – it's easy to see the smoking gun, in the case of the Christmas Island pipistrelle, it's more a question of which member of the firing squad got the killing shot.

IT WAS SOME time around 1669 when Hennig Brandt, a German alchemist living in Hamburg, discovered one of the keys to life on Earth. Brandt had long been in search of a very different prize. The 'philosopher's stone' was a legendary alchemical artefact that could supposedly turn base metals into gold, and Brandt, like many alchemists of his time, had set his heart on its creation. The recipe he had come up with, however, might best be described as piss-poor. First, Brandt boiled a bucket of human urine, reducing it to a thick syrup. Eventually, a red oil formed, which he skimmed off the surface before allowing the rest to cool into a 'black spongy upper part and a salty lower part'. Brandt discarded the 'salty' part, then mixed the red oil back in and heated it all at high heat for sixteen hours, until white fumes billowed off this concoction, like a witch's cauldron, and an oil collected on its surface. Eventually, a strange new material emerged, which he removed and solidified under cold water. The object Brandt was left with looked like a rock; but, in the dark, it emitted an eerie green glow and it ignited spontaneously in the air the moment it was exposed to temperatures over 37 °C. Brandt had discovered the pure form of phosphorus – its name derived from the Greek 'phōsphoros', meaning 'light-bearing'. Not the philosopher's stone, it turned out, but something equally magical.

The chemical element phosphorus is essential to all life on Earth. The very backbone of DNA – the outer frame of the twisted, ladder-like structure we call the double helix – is formed of phosphorus and sugar. In human bodies, phosphorus is used to make bones and teeth, and plays a direct role in many processes, from energy transfer and enzyme activation to regulating the normal functioning of nerves and muscles. Phosphorus has found other ways to sustain us, too; before electricity, the safety match – still to this day tipped with a red phosphorus head – brought light and heat within cheap and easy reach. However, by far the most crucial benefit phosphorus brings us is the role it plays in the lives of plants. Without it, plants would be unable to respire, photosynthesize, and ultimately survive. It is for this reason that phosphorus fertilizers are critical to agriculture today, just as they were in 1887, when the first settlement was built on Christmas Island all in the name of phosphorus.

By 1887, Brandt's buckets were thankfully far in the rear-view mirror. From the mid-nineteenth century onwards, phosphorus was extracted from phosphate (rock and ore that contain the element). It was John Murray, the pioneering Scottish oceanographer, who opened up Christmas Island to phosphate mining. After serving as a naturalist on the *Challenger* expedition of 1872–6 – a landmark scientific study of the world's oceans, the findings of which laid the foundations of modern oceanography – Murray had developed an irrepressible hunch that untold reserves of phosphate lay hidden under the island's rainforest. In 1887, after much bullish petitioning from Murray, the British government investigated and confirmed his suspicions. Within the same year, Christmas Island became an annexed possession of the British crown; a small (by British imperial standards) flag-raising ceremony made it official.

It wasn't only Christmas Island's phosphate that intrigued Murray, however. As a biologist, he appreciated the importance of documenting the novel life forms that this little-understood island harboured. Murray contracted Charles Andrews, a thirty-one-year-old assistant curator at the British Museum, to travel to the island immediately and create a comprehensive record of its biodiversity. From 1897 to 1898, Andrews scoured Christmas Island's rainforests and coastal fringes, hiking up through its 'wedding-cake' layers and down to its beaches, recording in detail everything he could find. His 1900 monograph on Christmas Island was widely celebrated as a rich portrait of a remarkably abundant ecosystem, making it, at the time, one of the most extensively studied oceanic islands in the world.

There was also something else that Andrews noticed during his time on the island. Coconut leaves, imported to create thatch for homes and other buildings, sometimes arrived on Christmas Island with other things wrapped in their long fronds. A flash of red, there one moment and then gone the next: the long, segmented body of a giant centipede. Highly venomous, these arthropods are formidable hunters of invertebrates, but can also target small mammals and lizards. They are the last thing you want to see high-tailing it off into your pristine rainforest, and Andrews recorded seeing two or three during his 1897–8 visit.

But as far as introduced species went, these centipedes were just the tip of the iceberg. Tamarinds, oranges, limes, guavas, mangos, cocoa, bananas, coconuts, pepper, sugarcane, tea, pawpaws, date palms, chillies, pineapples, gourds, nutmeg, tobacco, maize, custard apples, pumpkins, pomegranates, and bamboo were just some of the crops being grown on Christmas Island to feed an over-200-strong workforce that Murray – now Chairman of the Christmas Island Phosphate Company – had brought to the island.* Murray noted that much of this 'flourish[ed] with great luxuriance in [the] virgin soil'. Unfortunately, many of these plants did not confine their flourishing to agricultural lands and spread into the rainforest. 'In a few years, a number [of plants] will have established themselves at the expense of the native flora,' Andrews warned.

In 1900, with the ink on his monograph barely dry, Andrews' worst fears about introduced species began to bear fruit. That year, a ship called the SS *Hindoustan* stopped at Christmas Island to, amongst other things, deliver hay. Inadvertently, it also dropped off some stowaways – black rats (*Rattus rattus*), carrying fleas, which were vectors for disease-causing parasites, the likes of which Christmas Island had never seen. Soon afterwards, residents began to spot the island's endemic Maclear's rat (*Rattus macleari*) and the bulldog rat (*Rattus nativitatis*) during the daytime, behaviour that was at odds with their strong nocturnal instincts. Stranger still, they weren't scampering about, as they usually would, but were – in the words of the island's medical officer – 'in a dying condition'.

By the time Andrews returned to Christmas Island, in 1908,

---

* This workforce included over 200 Chinese and Malay labourers, eight European managers, and five Sikh policemen, all brought to the island in 1898.

the Maclear's rat – which, during his last visit, he had described as having 'swarmed over the whole island' – and the bulldog rat were gone. A third endemic species, the Christmas Island flea (*Xenopsylla nesiotes*), was collateral damage, vanishing with the island's endemic rats, on which it depended. These three were the island's first recorded extinctions, but there were signs of more to come. Introduced plants had continued to spread, and the number of giant centipedes had risen. In tandem, the endemic Christmas Island frigatebird (*Fregata andrewsi*) – whose species designation, *andrewsi*, was a tribute to Andrews himself – had noticeably declined. This may have had something to do with another introduced species, known the world over for its ravenous appetite for birds. Since his last visit, the house cat (*Felis catus*) had come to Christmas Island and multiplied, with some becoming feral. All of this Andrews had explicitly cautioned against in his 1900 monograph.

With Andrews pointing out the alarming trends developing on the island, and Murray, the man responsible, being a respected scientist, it's tempting to assume that this would have been a turning point for Christmas Island's beleaguered ecosystem. Murray's scientific credentials were formidable, as evidenced by his ability to steer the British government towards the exploration and annexation of Christmas Island, but his interactions with policy-makers also reveal a darker side to his relationship with the island. After annexation, Murray successfully lobbied the government to grant him half of Christmas Island's mineral rights. Whatever instincts he might have possessed as a biologist were now tempered by those of a phosphate magnate. This is perhaps why, in the prologue to Andrews' 1900 monograph, Murray makes this alarmingly indifferent statement about the island's fate:

> It has not hitherto been possible to watch carefully the immediate effects produced by the immigration of civilized

man – and the animals and plants which follow in his wake – upon the physical conditions and upon the indigenous fauna and flora of an isolated oceanic island. I hope to arrange that this shall be done in the case of Christmas Island.

By positioning spectatorship as a scientific response, Murray rationalized his complicity in the island's ecological unravelling. As a bonus, he also became extraordinarily wealthy off the back of Christmas Island's phosphate – and the literal backs of dead workers. Between 1900 and 1904, 600 indentured labourers died because of disease, malnutrition, and inadequate healthcare, stemming from the horrendous conditions on Christmas Island.

In 1914, Murray died in a motorcycle accident, after a decade of sitting back and watching invasive species freely take over the island's habitats. He left the bulk of his fortune to science. He also left Christmas Island peppered with references to himself, including Murray Hill, its highest point, and many of its coves and beaches, which he'd named after the female members of his family. Ironically, his most indelible mark was left on Christmas Island's wildlife, with a click beetle, capsid bug, multiple fossil coral and foraminifera species, an entire family of marine sponges, a stinging tree, an earwig, and a scorpion all named after him. This included the pipistrelle, which in 1900 Andrews had named *Pipistrellus murrayi*.

Murray set the tone for what followed after his death. The exploration lines created in 1966–9 kicked off an era of extensive rainforest destruction, transforming an island that was more than 90 per cent forested into one 'with large, bare, mined-out fields, access roads, dumps and [an] air-field', to quote one group of visiting scientists. Poaching, particularly of endemic wildlife, had also become a concern. As soon as mining began, labourers were encouraged to, and indeed had to since conditions were so poor, find ways to sustain themselves using the island's resources.

Surprisingly, all of this – the establishment and then expansion of phosphate mining, land-clearing, hunting, and introduced invasive species – seemed to have had little effect on the Christmas Island pipistrelle. In 1984, Australian ecologist Chris Tidemann completed the first ever ecological study of the species, and found the pipistrelle everywhere he looked, even within The Settlement itself, 'feeding on small insects inside the Christmas Island [Social] Club'. Tidemann returned four years later, and came to the same conclusion; the nearly ninety years of ecological upheaval inflicted upon the island since its settlement seemed not to have affected the pipistrelle at all. In his 1988 paper on his findings, Tidemann wrote, 'there seems little reason to be concerned about their continued survival'.

'ONE OF THE things I often chuckle about is that somebody working on birds just has to pack their binoculars and disappear off, but I've got to pack bat detectors, traps, night vision gear – just a huge amount of equipment to take with me,' Lindy Lumsden tells me. 'It'd be so much easier working on birds! But I think it wouldn't be nearly as much fun.' One of Australia's foremost bat researchers, Lumsden has been studying bats – 'my job, my hobby, my passion, my life, all rolled up together', as she describes it – for forty-five years, ever since the cold wintry night, on the outskirts of Melbourne, when she put up her very first mist net and a bat chewed a hole right through it. 'I thought: I have caught this great big bat and it got away from me – they're never going to get away from me again!' she tells me. 'I got hooked on them then, and I've been hooked on them ever since.'

Lumsden stumbled upon the Christmas Island pipistrelle in 1994. Her friend Richard Hill was studying the endemic Christmas Island hawk-owl, just as Lumsden was mulling over how to spend some hard-earned time off work. 'I remember thinking: Where's somewhere interesting to go on holidays?' she says. 'Oh, I'll go and visit Richard!' Naturally, Lumsden packed a few essentials that no self-respecting bat expert would leave home without: a bat detector and a large bat trap called a harp trap. She arrived in June and quickly became curious about the little-studied pipistrelle that was apparently so widespread here. Lumsden set up her harp trap and bat detector outside The Pink House – a salmon-coloured bungalow in the middle of the island that operated as a research centre. Then, as night fell, she sat on the ground nearby and waited for the silhouettes of hungry pipistrelles to emerge from their roosts.

Contrary to what she had heard from Chris Tidemann, who had found the pipistrelles difficult to catch – even resorting to shooting twenty bats out of the air in 1988 for his study – in 1994, Lumsden found catching them relatively easy. Harp traps – the go-to trap for all bat experts – look exactly how you'd expect: a frame like their musical namesake is strung with numerous nylon strings. These 'harps', which are 3–4m tall, are positioned in a bat's flight path. When a bat collides with a string, it slides down it unharmed into a catch bag underneath. Lumsden needed narrow paths the bats would pass through in which to position her harp traps; and in 1994 there were lots of spaces exactly like that on Christmas Island, thanks to the 1966–9 exploration lines. 'They were starting to grow over, but they created these perfect bat flyways for putting harp traps on,' she says. Before long, Lumsden was catching pipistrelles all over the island. Once caught, she would carefully handle each bat, measuring and weighing it, and noting its sex and other characteristics. This wasn't work exactly – she didn't even plan to create a formal report of her findings, she tells me – she simply relished the opportunity to learn more about this little-understood creature.

'Tiny' is the word Lumsden uses to describe the pipistrelle in the hand. 'It literally just fits in your hand, like that,' she says, inadvertently using present tense, and miming the action of gently balling her fingers around an invisible prune-sized bat. 'They were very gentle little animals that were quite placid,' she adds. 'They'd just sit there calmly, and then they'd go, "Okay, it's time for me to go now."' But just because the bats were forthcoming didn't mean there weren't challenges.

Around the size of a football, and wielding large muscular claws to crack open whole coconuts, robber crabs are the largest land crab on Earth. They're also one of the most brazen and belligerent species for biologists to work alongside. 'They'd just demolish anything,' says Lumsden. 'There was somebody there who had a big, 12-volt battery and they just munched straight through it.' For Lumsden, the chief concern was protecting her bat detector. By lowering the frequency of ultrasonic

*A Christmas Island pipistrelle sits in the hand of Martin Schulz. (Photo by Martin Schulz.)*

echolocation calls into the human range of hearing, bat detectors provide biologists with a window into the world of bats. With them, it's possible to tell not only where a bat is, but what it is doing, based on the distinctive calls it makes when engaged in different activities. With a species like the pipistrelle, about which next to nothing was known in 1994, it was an especially vital piece of kit for Lumsden to have. But to the robber crabs, this strange black cube was just another potential coconut, and so Lumsden had to improvise, using a fold-up camping chair and umbrella to protect it from both crabs and rain.

The study soon expanded, and Lumsden's colleague, ecologist Keith Cherry, joined her to survey the pipistrelle. Between 8 June and 19 July, twenty-two individual bats were caught, studied, and released; and, with the bat detector, 2,476 bat calls were recorded. These calls are complex and percussive, like Morse code but varying in pitch and timbre. Although these calls were never intended for ecologists' ears, they nevertheless communicated a clear and simple message to Lumsden and Cherry: the pipistrelle's distribution was 'patchy', particularly in the north-east of the island, and its numbers seemed worryingly low compared to what had been expected. In The Settlement in the north-east, where pipistrelles had once flown through outdoor restaurants and into people's homes and bowls of soup, they had completely vanished. A nearby area which contained some of the best primary rainforest habitat on the island was empty of them too. 'The data just didn't fit with what Chris [Tidemann] had described,' says Lumsden. They had caught a reasonable number of bats, but Lumsden had a feeling of foreboding. 'There was a thought in the back of my head: *Something isn't right here.*'

It was only when Lumsden returned in 1998, having been given funding to undertake a comprehensive study, that the species' position became clear. Over six weeks, between 10 May and 20 June 1998, Lumsden and co-researchers John Silins and Martin Schulz deployed bat detectors at eighty-four sites. Approximately 2,500km of road was driven (an impressive distance on an island only 15km long), to monitor bats by eye or with a bat detector pointed out of the window. And at sixteen sites along tracks and open exploration lines, the team deployed harp traps, catching 126 individual bats, which they marked by trimming unique patterns into their fur, giving each a recognizable haircut by which it could be identified. Ten of these captured bats were fitted with tiny transmitters, which enabled Lumsden, Silins, and Schulz to track them back to their roosts – the first time this had been done with this species. 'I was really hoping that I'd be proved wrong,' says Lumsden, 'that my fears that they were disappearing from the north-east part of the island would be unfounded, but they were strengthened.'

In fact, the reality of the situation far eclipsed Lumsden's concerns. Between 1994 and 1998, there had been a 33 per cent reduction in overall pipistrelle abundance. But this reduction wasn't spread equally across the island. The decline that Lumsden had witnessed in the north-east in 1994 had now spread to the centre of the island, where pipistrelle numbers had plummeted. Only in the west, where 96 per cent of recorded pipistrelle activity now occurred, did the species seem to be holding ground. But for how long? Just fourteen years after Tidemann's glowing assessment, the species was being wiped off the island, from east to west, like a drawing being slowly erased from an Etch A Sketch.

IN 2002, THE bulldozers fired up again on Christmas Island. This time, it was not in service of phosphate mining, but to extract a different kind of resource from the land. From the late 1970s, scientists, government officials, and even the mining companies themselves had predicted the imminent decline of

Christmas Island's phosphate industry. The highest-quality reserves were soon to be exhausted, and this raised the spectre of what could be next for Christmas Island, and how its residents could support themselves economically. Over the years, a variety of solutions were proposed. In 1984, as Tidemann was counting pipistrelles all over the island, a plan was hatched to turn Christmas Island into a kind of gambling haven, with a state-of-the-art casino catering to the Indonesian ruling class – extraordinarily wealthy people, hamstrung by Indonesia's strict anti-gambling laws. Decked out with faux-marble surfaces and dinosaur sculptures that breathed steam into the air like dragons, the casino opened in 1993, only to close five years later due to low footfall. In 2000, a Korean-backed company proposed Christmas Island as the site of a new satellite-launching facility, an idea that petered out into nothing. Other ideas, among them commercial marijuana plantations, similarly fell flat. But in 2001, a solution presented itself.

The MV *Tampa*, a Norwegian freight vessel, was en route to Singapore on 26 August 2001, when it received a mayday call. A fragile wooden fishing boat called the *Palapa* had become stranded in international waters, its passengers scrawling 'SOS' and 'HELP' onto its roof. When the *Tampa*'s crew located this boat, they discovered 433 asylum seekers, among them forty-three children and three pregnant women, crammed into its 20m-long hull. Bringing these people aboard just minutes before the *Palapa* sank, the *Tampa* made for the nearest land 140km south of their position – Christmas Island – and then waited just outside of Australian waters for permission to land.

Boats like this had been turning up more and more frequently, here and in other Australian territories. The people onboard were mostly fleeing from Afghanistan, where the ruling Taliban had unleashed a wave of brutal violence, and were hoping to find a safe haven in Australia. But attitudes towards the increasing numbers of inbound refugees were fiercely divided in Australia.

Perhaps smelling a political opportunity, the Prime Minister, John Howard, immediately refused permission for the *Tampa* to land the rescued asylum seekers on Christmas Island. A three-day stand-off ensued, during which the government repeatedly insisted – in contravention of international maritime law – that the asylum seekers be returned to Indonesia. Eventually, the captain of the *Tampa* declared a state of emergency due to the deteriorating health of the people under his care, enabling him to enter Australian waters, after which the Royal Australian Navy transported the asylum seekers to the island of Nauru – now infamous for its inhumane immigration detention facilities.

After 'the *Tampa* affair', a plan emerged to construct a $400 million immigration detention centre on Christmas Island, where up to 800 refugees could have their claims for asylum assessed, far from the Australian mainland. This promised to solve the island's economic woes by providing jobs for islanders and tax revenue for the local authorities, but the richest rewards were to be reaped in Canberra, the seat of the Australian government. Howard had embraced an increasingly populist anti-immigrant stance in the run-up to the 2001 general election, with Christmas Island's proposed immigration detention centre the jewel in his rhetorical crown. Invoking 'the national interest', the Howard government sidestepped its own environmental policies, allowing construction of the detention centre to begin in 2002 without any of the usual environmental assessments or controls.

Trees bulldozed, foundations poured, cells assembled – in the remote north-western corner of the island, the immigration detention centre slowly took shape. Meanwhile, in the rainforest, the inverse was happening. As the walls of the detention centre were going up, the roosts of pipistrelles were coming down. It wasn't known why, but pipistrelles were more often than not choosing dead trees for their roosts, rather than living ones, which puzzled researchers. The bats were anything but safe in these trees. A strong gust of wind could rip the bark right off them, exposing the young pups to the elements. Particularly severe storms had even been known to sweep flying foxes – and, it's thought, pipistrelles too – far out to sea, where they likely tired and drowned. And so, as the Australian government prepared to move refugees into the newly built Christmas Island facility, the homes of pipistrelles were becoming more precarious.

This might have gone unnoticed. However, as when the bulldozers had rolled out during the creation of the 1966–9 exploration lines, there was something of a silver lining in the construction of the detention centre. Parks Australia – the government organization responsible for Australia's six Commonwealth national parks, of which Christmas Island National Park is one – received 'compensatory' funding from the Department of Finance and Deregulation to assess the project's environmental impact. This influx of apology money allowed ecologist David James, the newly hired head of the wildlife monitoring programme, to beef up conservation efforts island-wide. Soon, the Christmas Island pipistrelle became, in the words of Lumsden, 'one of the most intensely monitored bat species on the planet'.

Much of what we know about the pipistrelle was learned during this time, including their increasingly puzzling roosting preferences. Lumsden, and now also James and his colleague, Kent Retallick, made an enormous effort to understand the little bat that was quickly slipping away. The speed and consistency with which it was disappearing alarmed them. Between Lumsden's 1994 and 1998 visits to Christmas Island, pipistrelle activity had decreased across the island by one third; but by 2004, the remaining two thirds had halved, with a decline of roughly 10 per cent per year. Plot these figures on a graph, as Lumsden and her colleagues often did, and you'll see a picture-perfect portrait of decline: a straight line dissecting your graph from top left to bottom right, pitched at 45 degrees. But, hidden somewhere amongst these data points, Lumsden knew had to be the signature of something else – of whatever was corralling the pipistrelles into ever-dwindling pockets of habitat and, ultimately, towards extinction.

IN CHRISTMAS ISLAND'S rainforest, there is a strange animal that is rewriting scientific textbooks the world over. Until very recently, it was considered a fundamental law of biological inheritance that every living thing carries only a single set of genomes in its cells – usually a blend of both its parents' DNA, but sometimes derived from just one. However, a recent genetic study has identified the single known exception to this rule. Males of the yellow crazy ant (*Anoplolepis gracilipes*) carry two entirely independent sets of genomes, one inherited from the ant's mother and one from its father. The exact purpose of this chimeric trait is a mystery, although it's thought that it might give the ant useful biological latitude to avoid inbreeding in situations where a new population is founded from just a few individuals. If this is true, it perhaps explains the extraordinary success that this species has enjoyed as a colonizer. Having spread widely across tropical parts of Asia, Africa, and the Americas, and to many oceanic islands – including Christmas Island – it is one of the world's most successful invasive species.

The yellow crazy ant was one of many possible culprits that Lumsden and her colleagues had to consider when trying to ascertain the cause of the pipistrelle's decline. Other threats included the endemic Christmas Island goshawk (*Accipiter fasciatus* ssp. *natalis*) and the introduced nankeen kestrel (*Falco cenchroides*), a fearsome raptor known to predate on bats elsewhere and whose presence had increased significantly on Christmas Island since the 1980s. Cats, adept climbers and famed tormentors of winged things everywhere, were another obvious consideration, as was the black rat, whose extinction scorecard was already packed with strikes, including at least three from Christmas Island. Of these threats, however, the yellow crazy ant had made the most noticeable impact on the island.

It's for their erratic movements – scurrying back and forth, as if they can't quite decide which way they want to go – that yellow crazy ants get their 'crazy'. With bright amber-coloured bodies, they are an extraordinary-looking insect. When the males mature and sprout wings, their unusually long legs make them look like mosquitoes that have been dipped in golden syrup. Fittingly, the lives of these ants are structured around another kind of syrup, a carbohydrate-rich 'honeydew' secreted by sap-drinking bugs called scale insects. In return for this honeydew, the ants clear away the scale insects' predators, and transport the juveniles to new, untapped plants. Via this mutualism, yellow crazy ant colonies grow exponentially, by as much as 3m per day, joining with other colonies to form so-called 'supercolonies'. These gigantic ant-megapolises contain multiple queens, have population densities of over 2,200 individuals per square metre, and cover areas as large as 8km$^2$. Yellow crazy ants will attack anything that dares enter these territories, whether invertebrate, reptile, bird, mammal, or even biologist. Trespassers are immediately hosed with corrosive formic acid, sprayed from the ants' abdomens, which blinds and chokes animals 'like a mustard gas attack', as Parks Australia researcher David Slip puts it. Then, the ants swarm and, wherever possible, eat their prey.

In 2002, it was estimated that between 10 million and 15 million Christmas Island red crabs (a quarter to a third of the population) had been killed by ants in just a few years, triggering what biologists termed an 'invasional meltdown' on Christmas Island. A keystone species, whose foraging habits maintain the distinctive ecology of the island's rainforest, the crabs had simply been 'deleted' from around 30 per cent of that forest, with these areas then undergoing a radical transformation. Now, all the detritus of forest life – fallen seeds, leaves, and fruit – was left to pile up and rot. Seedlings that would normally be pruned away by the crabs had sprung up en masse, clogging the usually airy understorey with overgrowth. Invasive species like the giant centipede and giant African land snail (the same species as in Chapter Three), which had been predated on by Christmas Island red crabs, now thrived.*

How did all of this affect the pipistrelle? The yellow crazy ants farmed their scale insects in the forest canopy, marching up and down the trunks of trees almost continuously. This could have brought them into contact with pipistrelles roosting in those trees. Much like anything else they came across, the ants would have made short work of a sleeping pipistrelle. The researchers now had a possible explanation for the pipistrelles' changing roosting preference: it was thought that this new traffic on living trees could explain them moving from live to dead trees.

But the ants could also have had other less direct effects. For instance, giant centipedes, fearsome hunters which the crazy ants had unburdened of their usual predators, the crabs, had spread. A pipistrelle would be killed and eaten within minutes of encountering a giant centipede, and the invertebrate-rich space under the bark of dead trees, where the pipistrelle now roosted, was one of their favourite stomping grounds. Measuring up to 30cm long, a giant centipede can snatch mice and lizards out of trees, carrying them off with only the front three pairs of its pincer-like legs. Pipistrelles, with their diminutive size, would

---

*   Another devastating side effect of yellow crazy ant invasions is sooty mould outbreaks. Caused when excess scale insect honeydew drips onto the leaves and stems of plants, sooty mould can cause vegetation dieback. On some Great Barrier Reef islands, there have even been instances of sooty mould outbreaks leading to island-wide tree extinctions.

have been easy pickings. Given that the first recorded ant super-colony had appeared in 1989, right in between Tidemann's report of pipistrelle abundance and the decline that Lumsden had discovered in 1994, the timing seems to line up.

'The yellow crazy ant was such a big issue on the island, roughly at the same time the pipistrelle was declining, that I think everybody just assumed they were the main cause,' says Lumsden. But she was unconvinced. The pipistrelle had already been in decline before the explosion of super-colonies in 1990. Also, the majority of super-colonies were in the west of Christmas Island, exactly where the pipistrelle had retreated to, while the section of primary forest in the north-east, where Lumsden had been most shocked to find the pipistrelle absent in 1994, had never housed a super-colony. Then, there was the fact that a 2002 eradication programme, which used insecticide to successfully drive down ant numbers by 99 per cent, prompted no bounce-back from the pipistrelle.

Most of the other predator theories also fell short in some way or another. Dietary studies hadn't found pipistrelle remains in the stomachs of Christmas Island's feral cats, and black rats had been on the island since it was settled, making it unlikely that, after a century living side by side with the pipistrelle, the rodents had suddenly risen up against them. Likewise, despite the suspicious timing of their 1980s population heyday, the nankeen kestrel had long appeared in both areas where the pipistrelle was common and areas from which it had vanished.* Nonetheless, ants, kestrels, goshawks, cats, rats, and centipedes were thought to have contributed to the decline of the

---

\* The nankeen kestrel may have been responsible for a significant change in the foraging behaviour of the pipistrelle, however. By 2004, pipistrelles were leaving their roosts around thirty minutes after dark – rather than an hour and a half before sunset, as had previously been observed. It's speculated that this might have been to avoid the kestrel, which hunts in daylight.

pipistrelle; poisoning by insecticides and mining by-products, habitat destruction, introduced diseases, and even vehicular strikes were all considerations too. But Lumsden and others believed that there was likely something else – a primary driver that could explain the pattern of the pipistrelle's decline.

As it turned out, there was one suspect that ticked a lot of the boxes left empty by the rest. The common wolf snake (*Lycodon capucinus*) was first recorded on Christmas Island in 1987, likely introduced from Southeast Asia by a supply ship. There was plenty of scepticism about this hypothesis. Wolf snakes weren't thought to be good tree climbers, and they are a specialized hunter of lizards. However, the timing of the wolf snake's arrival, and the pattern of its expansion, matched the decline of the pipistrelle like a bloody boot print. Tidemann had found bats everywhere in 1984, including in The Settlement. A decade later, this same area supported extraordinary numbers of wolf snakes – up to 500 per hectare – but no pipistrelles. By 1998, wolf snakes were found in the centre of Christmas Island and then, in 2003, they were found in its south-west. Records of the wolf snake's westward advance and the pipistrelle's westward retreat fit together like puzzle pieces. Perhaps the most damning piece of evidence came in 2007, when James and Retallick captured footage of a wolf snake climbing the trunk of a known pipistrelle roost tree, demonstrating that they were in fact far better climbers than anyone had thought. However, since one hadn't been caught red-handed with a bat in its mouth, it proved impossible for Lumsden or anyone else to convince the authorities to focus squarely on the wolf snake, no matter how guilty it appeared.

Soon enough, however, none of this mattered. In January 2009, Lumsden could only locate four pipistrelles on the island. Wolf snake or no, whatever was causing this decline had almost finished the job. Lumsden and Schulz issued the following warning to the Environment Department: 'There is an extremely high risk that this species will go extinct in the near future, without urgent intervention. While it cannot be precisely predicted when this would occur . . . it is highly likely that it will be within the next six months.'

In the rainforest, as the last few remaining pipistrelles went about their lives, perhaps they were in some way cognisant of the disappearance of their kind. For a species so dependent on sound – who are often said to 'see' with it, even – the silence must have been blinding.

OVER A DECADE later, Lumsden still finds it difficult to talk about the Christmas Island pipistrelle. She agrees to speak with me only on the provision that she can have a box of tissues on standby. 'To predict what was going to happen, and then have it all come true . . .' she says. 'I just wish I had been wrong, but unfortunately, I was right.'

By 2009, Lumsden, James, and others had been making the case for captive breeding for years; but the puzzle of the pipistrelle's decline had hindered decision-making. Without a clear picture of what was happening to the species, there was little support offered. Lumsden suggested building a captive-breeding facility on Christmas Island, rather than transporting the bats elsewhere, estimating that this would cost $800,000, with a yearly running cost of $300,000. This wasn't a cheap solution, by any means, but she reiterated the desperate nature of the situation, telling officials: 'If we don't get an insurance population into captive breeding I really think they could be extinct by June.'

The Australian Environment Department's response was to set up an 'Expert Working Group' (essentially a board of experts that decides what the next best steps are for an endangered species). Their remit was to take a holistic look at the plight of the pipistrelle and other conservation issues on Christmas

Island. However, the group contained no pipistrelle experts. Due to this and their overly broad brief, they moved at a snail's pace, completely out of step with the rapid response necessary to save the species. 'There were really good people in the group but, really, it needed input from people who understood the ecology of the island and understood the pipistrelle, but that just wasn't how it was set up,' says Lumsden. Perhaps indicative of this lack of direct experience, the working group stipulated that, first, a captive-breeding trial be conducted in Darwin, using another pipistrelle species to see if it would work. Lumsden, however, knew that this wasn't necessary. 'I've kept bats in captivity for thirty years. I had two bats of a different species that I'd kept alive for twenty-two years. They're not hard to keep, and we knew that other species of pipistrelle had been born in captivity.'

With help from the Australasian Bat Society, Lumsden and the society's president, Michael Pennay, successfully lobbied for a meeting with the Federal Environment Minister. 'I pleaded with him: "Just let us go and try,"' she tells me. '"I can't guarantee we're going to be successful, but if we just don't do anything . . ."' The sentence finished itself. Despite Lumsden's pleas, however, the minister declined. 'I really believe that he thought if he let us try and we failed, then he'd get blamed for the extinction,' says Lumsden. She tells me that, if the pipistrelle had been in her home state of Victoria, in that moment, she'd have been tempted to go and catch them herself, permission or not: 'To hell with the consequences,' she says, recalling her frustration during this time. Whatever the reason for the minister's decision, action on the pipistrelle would have to wait until the working group had concluded its report.

On 8 August 2009, in a quiet room at The Pink House, Lumsden leapt up and let out a yelp of joy. She'd flown to the

island earlier that day, along with 250kg of specialist equipment, having finally been given the go-ahead to initiate captive breeding. However, until this moment, she'd wondered whether they would even find the bat at all. In January 2009, Lumsden had advised that due to the extremely low number of pipistrelles remaining, captive breeding had to commence within three months. 'In reality, January was probably too late anyway, but at least it would have given us a fighting chance,' Lumsden says, 'but [the working group] had stalled for a further five months.' For a species in such a precarious position, every wasted day diminished the likelihood that a bat would be caught, let alone that it would be enough to establish a new, healthy population. Nonetheless, once the working group had approved the project, the Australasian Bat Society, along with representatives from Melbourne Zoo, senior vets, conservationists, and local workers, had all rallied to the pipistrelle's side. Lumsden led a team on the ground, and while they all hoped that they were not too late, after half a year of waiting, the odds were dismal. And yet, here was Lumsden, yelping at the telltale trill of a pipistrelle passing through the night.

The call that Lumsden had heard had been recorded by a bat detector in the west of the island. And soon enough, Lumsden was hearing more pipistrelle calls. 'We picked up a call in the roosting area . . . then in the main foraging area. We had so many bat detectors out, and we'd pick up a call there, then there, there, there,' says Lumsden, gesturing as if pointing at a map. But any excitement was quickly dashed; they were only ever hearing one call at a time, never two, three or four together, as hoped. 'And that's when we came to the devastating realization,' says Lumsden, 'that there was probably only one pipistrelle left.'

Even though the bat they were following looked increasingly like the last of the species, it had to be captured. Another pipistrelle might show up later and, if the team were lucky, the two might form a breeding pair. However, unlike the pipistrelles

*Ross Meggs demonstrates an improvised, self-supporting ladder built for Lindy Lumsden to access fragile pipistrelle roosts high up in trees. (Photo by Lindy Lumsden.)*

Lumsden had trapped with ease on earlier visits to Christmas Island, this bat seemed wise to her act. Harp traps, for instance, were completely avoided, as if the instincts of this one bat had been sharpened over years spent watching its brethren tumble into capture bags. New types of traps were invented on the spot. Ross Meggs, a wildlife equipment specialist, built a 9m-tall free-standing ladder that would enable access to a pipistrelle roost without damaging the brittle tree housing it. To trap the bat itself, Meggs built a cylindrical Perspex case that was made to fit perfectly around a specific roost tree that the bats had used for years. However, when it was time to deploy this equipment, the bat was nowhere to be found, and didn't visit the tree at all. Not one to give up, Lumsden devised a new kind of trap she dubbed the 'mist net tunnel trap'. A mammoth 240 metres of fine sewing later, she and her team had created a 15m-long, 5m-tall, and 7m-wide tunnel, out of sixteen monofilament mist nets, that was enclosed at one end. Slotted into an exploration line that the bat had been frequently flying, this tunnel trap looked like the discarded sock of a giant.

In the pitch dark, team members stood either side of the trap's opening, ready to quickly draw its cloth doors shut the moment the pipistrelle flew in. The stage was set, the trap's doors hanging in the middle of the rainforest like theatre curtains, two of the team seated either side poised and ready like stagehands. But, like everything else they had tried, the pipistrelle evaded this trap, flying over it once and never looking back.

Few people on Earth know what it's like to witness an extinction. Even fewer can tell you what one sounds like. At 23:29 (precisely 38 seconds past the minute) on 26 August 2009, the last Christmas Island pipistrelle fell silent. The following night, the bat detectors picked up nothing. 'You know you asked me whether there was a time when I realized what was happening?' says Lumsden. 'It was that night. I just knew that it had gone.' Her colleagues remained positive, waiting for the bat's calls to be picked up again, but Lumsden, who had known the species most intimately, and had watched its decline for the past fifteen years, knew otherwise. 'I was inconsolable,' she says. 'We were there for another week, so we continued doing what we were doing, setting the detectors . . . but I just knew there was no point.'

The last call of the pipistrelle can be listened to online. After a few bursts of percussive clicks, and squeaks – a polyphony of sounds that is hard not to perceive as coming from many sources – there is silence. In that moment, one individual bat vanished, and took its entire species with it.

What silenced the last pipistrelle? Had a wolf snake or an army of yellow crazy ants caught it resting on a branch, or had a feral cat snatched it out of the air as it foraged near the ground? Perhaps it had simply reached the end of a long and fruitful life, and succumbed to old age, free at last from whatever had driven the rest of its kind off the island. We'll never know.

Lumsden arrived back in Melbourne the same day the extinction was publicly announced in a press release by the Federal Environment Minister. The next morning, at 6 a.m., she was woken up by a phone call from a radio station asking to interview her. 'It was a day of lots of people wanting to talk . . . I think I did five radio and a couple of newspaper interviews,' Lumsden says, 'and then the next day, the world moved on to something else.'

THERE HAS BEEN much debate about the cause of the Christmas Island pipistrelle's extinction. Now, with the species gone, it's unlikely we'll ever know for sure. However, what's clear is that the scientists who worked with this species did their jobs; they monitored the pipistrelle's decline, translating the disappearance of the swooping and meandering shapes made through Christmas Island's forests by the bats into charts and graphs, and then did everything they could to save it. But there is only so much scientists and conservationists can do. It's an unfortunate and bewildering fact that the people most qualified to make decisions about how best to help an imperilled species have to take their lead from political institutions. In this case, the Australian government didn't do the job they were meant to do. Instead, when the pipistrelle needed the government to act, it essentially ignored the problem. In 2006, the Environment Department listed the pipistrelle as critically endangered; but no funding was set aside for its conservation. At the same time, $3.2 million was allocated to the conservation of a much-loved mainland species, the orange-bellied parrot (*Neophema chrysogaster*).

In his 2018 book, *A Bat's End: The Christmas Island Pipistrelle and Extinction in Australia*, John Woinarski explains that underlying some of the indifference towards the Christmas Island pipistrelle might have been a bizarre taxonomic decision, made decades earlier. Described in a 1997 obituary in *The New York Times* as 'an authority on every kind of bat, all over the world', the American zoologist Karl Koopman had, nevertheless, once written the Christmas Island pipistrelle out of existence. In 1973, Koopman declared that it was not a species in its own right. His argument, since described by experts as 'truly facile', hinged on the comparison of three cranial measurements and conflated the pipistrelle with a handful of other bats, all under the comically appropriate species designation *tenuis* – a Latin word meaning 'thin, fine, or delicate' from which we get our English 'tenuous'.

Australian bodies ignored Koopman's reclassification of the pipistrelle, continuing to view it as a species. However, in much of the wider scientific community, including at the IUCN, there simply wasn't a Christmas Island pipistrelle to lose in the first place. Because of this controversy, the working group had wanted to confirm the species' taxonomic status before agreeing to captive breeding. The resulting 2009 study solidified the pipistrelle's status as a species, but massively delayed efforts to save it – wasted time, during which the population dropped from at least four bats to one.*

---

* In 2023, Koopman's classification came back to haunt the Christmas Island pipistrelle. The 2009 study that had resolved its taxonomy was not externally published, so when a US-based taxonomic group looked at the classification of the species, Koopman's 1973 decision was all they had to go on. The group planned to fold the pipistrelle back into *Pipistrellus tenuis*, and the IUCN would have followed this lead, until Lumsden and others lobbied to have the 2009 paper released, and the Christmas Island pipistrelle was – in a manner, at least – saved.

Beyond Koopman, the underpinnings of this extinction run much deeper. Since settlement, life on Christmas Island has mostly revolved around phosphate. It was annexed by the British for phosphate, invaded by the Japanese for phosphate and, in 1958, when Australia took possession of the island, the goal was clear – the government granting itself 'full licence and authority to cut timber and to get all phosphates . . . in, on and from [Christmas Island]'. Perhaps it was this singular focus around which island life was arranged that led to other priorities being easily set aside.*

In *A Bat's End*, Woinarski details a string of 'major management and policy failings' that contributed to the pipistrelle's extinction. Among them are insufficient resourcing of conservation efforts, a paucity of adequately enforced environmental protections, and '[an] apparent lack of empathy or responsiveness . . . to clearly expressed warnings that emergency measures were needed to prevent extinction.' When it came to Christmas Island, something was always more economically important or politically expedient than the island's biodiversity, which is ironic considering that ecotourism has been touted, since the 1970s, as a potential solution to the twin challenges of long-term economic stability and biodiversity conservation. Positive steps to protect the island's wildlife have included the establishment of a national park in 1980, which today covers 63 per cent of the island, and there have been some major conservation wins for critically endangered endemic species such as the Abbott's booby. However, despite the great work of many conservationists on the island, it's clear that the scale of its environmental

---

\* One example of this came to light in 1966: the Deputy Prime Minister of Singapore, Toh Chin Chye, described his visit to Christmas Island – then plagued by racial segregation, extreme gender inequality, and inferior pay and facilities for workers of Asian origin versus white workers – as 'like walking into the colonial past of fifty years ago.'

problems has far outweighed the will of politicians and government institutions to resolve them.

In 2002, when the rapid decline of the pipistrelle was already well documented, Parks Australia included the following line in a management plan for the national park: 'Christmas Island provides an invaluable opportunity to observe the long-term processes of immigration, colonisation and extinction.' It's a statement that, in content and tone, hews closely to the indifference John Murray had shown the island a hundred years earlier, in the prologue to Charles Andrews' monograph.

DESPITE EVERYTHING, WHEN it comes to the pipistrelle, Lindy Lumsden feels responsible. 'When you do this sort of work, trying to protect a species, and you're there on the day it actually goes extinct and you can't stop it . . . there's just no worse feeling in the world,' she says. 'I should have been able to stop it.'

It's difficult to hear this from someone like Lumsden, who I know, from speaking with her and others involved, gave more than anyone to try and save the pipistrelle. But as I listen to her list the things she feels she should have done – lobbying the government harder, speaking up more, writing reports more quickly, recommending captive breeding earlier – it's impossible not to think about the people who were in power, who could have heeded her warnings but, for whatever reason, chose not to. Those in Lumsden's position often blame themselves for the extinctions they witness, even when it is clear to everybody else that they deserve this burden the least.

'I thought about giving up working in conservation after [the extinction],' says Lumsden, 'but I made the decision not to, and it really strengthened my resolve to try and help other species, so that it doesn't happen again.'

A few days after we speak, Lumsden is back in the field, tracking southern bent-wing bats, a critically endangered Australian microbat. Recently, she and a team of other bat experts and enthusiasts succeeded in having this species designated 'Australian Mammal of the Year', meaning it has received increased press coverage and that there is more awareness of its conservation. It is a moment that reveals how ecologists often have to be chameleons, dipping into other professions in order to save species. In this instance Lumsden is, in a sense, a publicist, all in the name of bat conservation. Later, she shares some footage from a recent expedition of a bent-wing bat soaring through the night, barrelling after moths and other flying insects. Momentarily, it gets away from her, disappearing from view, and the air is still and silent; and then it comes back into the frame.

# 7 Run, Forest, Run!

Christmas Island Forest Skink
*Emoia nativitatis*

A FEW MONTHS after the ill-fated attempt to save the last Christmas Island pipistrelle, The Pink House was changing. In late 2009, this building, nestled in the rainforest at the centre of the island – which for decades had been a hub from which scientists from all over the world conducted research on Christmas Island – was to become a home. In preparation, one of The Pink House's large, spartan dormitories was emptied of old desks, cabinets, and bunk beds, along with the tattered garden chairs that in years past had propped up bat detectors. Soon, a whole new range of bespoke furnishings and accoutrements was being carefully fitted, to suit the tastes and needs of The Pink House's new residents.

Wheelbarrows full of rich, red-brown soil, freshly dug from the rainforest floor, were turned out into large metal trays. These trays of soil were baked in an oven to remove bacteria and other stowaway organisms, left to cool, and then tipped into custom-built tanks that now filled the old dorm room. Next – after being meticulously checked for centipedes – twigs, branches, leaves, rocks, and other bits and bobs found their way into the enclosures. The result, a delicately assembled lattice of rainforest detritus, looked something like a giant game of pick-up-sticks. A makeshift 'hide', a small cardboard box with holes cut into it, completed the scene in each tank. Nearby, in one of The Pink House's other rooms, a different kind of enclosure took shape. The air buzzed with the sound of tens of thousands of crickets, all rattling their hind legs, housed in plastic containers stacked up on shelves and cabinets. This room was a farm of sorts: these insects were to be food.

What was all this for? Lizards. Three of Christmas Island's four endemic lizard species, to be precise: Lister's gecko (*Lepidodactylus listeri*), the smallest of the group, growing to about 5cm in length, not including its tail; the Christmas Island blue-tailed skink (*Cryptoblepharus egeriae*), named for its iridescent blue tail; and, lastly, the Christmas Island forest skink (*Emoia nativitatis*).

An intricate mosaic of black and brown scales came together in the 'metallic brown' skin of a forest skink. At 20cm long – most of which was tail – the majority of its time was spent burrowing through leaf litter or scampering over the exposed buttress roots of trees, in search of tasty invertebrates. Any break in the forest canopy, however, and the forest skink was there; a fallen tree, lying flat in the bright clearing created by its own toppling, would quickly fill up with as many as eighty individuals, all basking in the warm glow of the sun. The Christmas Island forest skink was the most abundant of the island's diurnal reptiles, inhabiting the rainforests of its plateau and terraces, as well as the low forest at its coastal fringes.

Endemic to the same island as the Christmas Island pipistrelle, these two species would almost certainly never have met. While the pipistrelle's world was the sky and trees of the island, the skink's was the rainforest floor. The pipistrelle occupied the night, while the skink occupied the day. Yet introduced threats to Christmas Island's ecosystem, from ants to centipedes to snakes, penetrated both worlds. In the 1980s, the forest skink, blue-tailed skink, and Lister's gecko began declining rapidly, with their remaining populations moving eastwards, in synchrony with the Christmas Island pipistrelle. Now, in 2009, these three lizard species pitched their last stand against extinction at the island's most westerly point: a peninsula called Egeria Point. With nowhere left to run except literally off a cliff, these species were hanging off a figurative one too, unless they could somehow be rescued. In response, a plan was put into action to pull the lizards out of the wild and place them into the safety of a captive-breeding programme, for which The Pink House was hurriedly repurposed.

ON CERTAIN DAYS throughout 2009 and 2010, the low forest and grassland along the clifftops of Egeria Point resembled a crime scene. Like a police search team looking for signs of a missing person, scientists and workers from the parks service walked slowly forwards, eyes locked on the ground ahead of them. By August 2010, they had managed to capture an impressive sixty-four blue-tailed skinks and forty-three Lister's geckos, which were immediately secured within the captive-breeding facility at The Pink House. The third species, however, had proven far more elusive.* A pathside sign, erected years earlier, when the species was two a penny, seemed like a taunt to the struggling search team: 'Look around at the forest skinks active nearby,' it read. But after months of intensive searching only four forest skinks were caught, one of which immediately escaped.

Under the watchful eyes of tanks-full of blue-tailed skinks and Lister's geckos, the three remaining new arrivals to The Pink House were measured, weighed, and checked for signs of illness or injury. In a disappointing twist, it was discovered that all three were female. This meant that, as of this moment,

---

* Christmas Island's three imperilled endemic lizards each had to be caught using a unique method. The nocturnal Lister's geckos were easiest to locate at night, when their retinas reflected back torchlight and glowed orange, like cat eyes. Christmas Island blue-tailed skinks were plucked from the ground using so-called 'sticky poles': 2–3m-long sticks with masking tape secured to the end, sticky side up. The larger forest skinks could wriggle free of these contraptions, however, so the search team had to anticipate exactly where they would flee to and catch them by hand.

there could be no captive-breeding population established for the species. The team would need to keep searching for forest skinks, and hope that the next they stumbled upon was male – if, that is, they found another one at all.

Anyone who has owned a small lizard like a skink or gecko will be aware of its Houdini-like aptitude for escape. Given the smallest window, lizards can vanish, as if into thin air. Unfortunately, this universal aspect of lizard nature coincided with a moment of human error at The Pink House, to disastrous effect. Shortly after the forest skinks had settled into their tank, members of the team discovered that the lid had been left unsecured. Two had made a bid for freedom, scampering up the Perspex walls of their enclosure and down the other side – eager, perhaps, to return to the forest. This dream ended quickly, however; the next morning, they were found dead. For some reason, the third forest skink hadn't joined this escape attempt. Now, the continued existence of the Christmas Island forest skink depended on whether the team could find a mate for this last lizard. Workers returned to Egeria Point to scour the forest floor and porous limestone, desperate to find a male. A helicopter dropped supplies and equipment to remote areas so they could camp overnight; but, still, they came up empty.

Meanwhile, the last forest skink adjusted to a solitary life inside her 130cm x 65cm x 65cm tank. Being the sole representative of her species, she quickly became a favourite of her keepers, who nicknamed her Gump – a reference to both the 'forest' in 'forest skink', and Tom Hanks' character in the 1994 film *Forrest Gump*, with whom she shared a passion for running. Feasting on the invertebrates that fell from the sky every three days, and basking under the strange new sun that hung stationary in the air above her, Gump silently awaited her fate.

The story of the Christmas Island forest skink began between 9 and 18 million years ago. Somewhere in the Indonesian archipelago, an unidentified species of skink diverged into at least two species, one of which was the ancient ancestor of Gump. It lived in Indonesia's rainforests until, millions of years later, a castaway population of the species made its way across the ocean.

Once an undersea volcano, Christmas Island began to form during the Late Cretaceous period (100.5–66 million years ago). At some point the island surfaced, but was plunged back underwater by a global rise in sea levels. Eighteen million years ago, Christmas Island lay sunk beneath the waves – a 'drowned coral atoll', in geologists' parlance. Despite the morbid connotations of this label, however, it was just as alive as it is today. A vast coral reef covered its surface, teeming with marine life, including calcium secretors like foraminifera, whose fossils make up the sand on some islands and coastlines (including on Bramble Cay) and which can still be found living in marine and brackish waters. These tiny creatures contributed to the building of the cap of limestone rock the island wears today.

Five million years ago, the drowned coral atoll of Christmas Island was lifted clear of the water's surface once again. As the Australian tectonic plate subducts beneath the Sunda and Java oceanic trenches, the seabed underneath Christmas Island bulges, pushing the island upwards at a rate of 1.4cm every 1,000 years. In the distant future, this same process in reverse will once more drown Christmas Island; but, five million years ago, it was thanks to this geological uplift that Gump's ancestor could make the journey across from Indonesia, which had – evidence suggests – become too hostile, perhaps due to a predator, for the species to survive.

It isn't known exactly when the forest skink arrived on Christmas Island. Around 630,000 years ago, global sea levels were much lower, and most of Southeast Asia formed one giant landmass, enabling many species to freely migrate across this vast area. Christmas Island, however, despite its proximity to this landmass, was not connected to it, and was instead surrounded by extraordinarily deep water. Just off the coast, the seafloor plummets to a depth of 5km (roughly equivalent to the height of Mount Everest). No shift in sea levels, barring the complete emptying of the world's oceans, would have provided pedestrian access to Christmas Island. Instead, Gump's ancestors most likely floated there aboard a piece of driftwood or some other debris, riding westward currents from somewhere in the Indonesian archipelago. This is all we know of the origin of this species. However, its journey also tells us something about Christmas Island itself.

The Wallace Line is an invisible border that runs for thousands of kilometres, cutting through the Indonesian archipelago in a vague north–south direction. Unlike the geopolitical borders that dissect our world into countries and states, however, this one was drawn by nature itself. It was discovered in 1859 by Alfred Russel Wallace, an English biologist who collected 125,660 natural history specimens over the eight years he spent travelling the Indonesian archipelago. Through a combination of deep water and strong currents, the Wallace Line does an extraordinary job of keeping apart the biogeographical realms of Asia and Australia. On either side of the line are entirely different types of species. It is the reason you find no monkeys, sun bears, or slow lorises in Australia, and no wallabies, kangaroos, or wombats in Malaysia. South and east of the line are marsupials and monotremes (mammals that lay eggs); north and west of it are only placentalia (mammals that give birth to live young). With few exceptions, the Wallace Line keeps these radically different biogeographical realms separate, even on Bali and Lombok – two islands situated just 35km apart.*

---

\* Charles Darwin is often described as the father of evolution, although it's much more accurate to say that this groundbreaking scientific theory

Christmas Island, 1,100km to the west of the Wallace Line, ought to be firmly rooted in the biogeographical realm of Asia, with species originating in the west, but this is only half true. Many of its native species, including the Christmas Island pipistrelle and Christmas Island flying fox, originate from west of the border, but many others, including the Christmas Island forest skink, come from the east. Christmas Island is, biogeographically speaking, a melting pot – a meeting point between east and west. The island's proximity to Java, allowing the migration of birds and bats from west of the Wallace Line, and the strong surface currents that carried skink-laden driftwood from the east, are thought to explain this ecological smorgasbord.

The forest skink enjoyed a mostly carefree existence after it arrived on Christmas Island. 'Benign' is how John Woinarski sums up the environment at the time. Invertebrates were plentiful, and the forest skink soon developed the broadest palate of any lizard species on the island: snails, scorpions, spiders, slaters, springtails, damselflies, cockroaches, termites, earwigs, crickets, true bugs, beetles, flies, moths, caterpillars, wasps, ants, centipedes – virtually everything that crawled was on the menu. Lounging about after a foraging session, they blended back into the surrounding leaf litter, in perfect camouflage. Not that they needed to worry, however; at this time, there was next to nothing for which the forest skink was itself a delicacy. Protected by the surrounding deep ocean, the island was a haven from mainland predators. Christmas Island goshawks or hawk-owls might have snatched up a sunbathing forest skink here and there, but – as the thousands basking peacefully in the island's forest clearings attested – this was of little consequence overall.

---

actually has two dads. Alfred Russel Wallace arrived at the concept of evolution by natural selection independently from Darwin, and wrote on it first, prompting Darwin to publish his now famous 1859 abstract *On the Origin of Species* . . . and the rest is his story.

Their communal sun-lounging does paint a slightly misleading portrait of the species, however. Forest skinks lived solitary lives, and were drawn together to bask in much the same way that holidaymakers will fill every square inch of a beach, getting uncomfortably close to each other only out of necessity. It was only occasionally that a more sociable side to the forest skink emerged. Positioned side by side, 10cm apart, and facing in opposite directions, pairs of forest skinks sometimes chased each other's tails, forming ouroboros-like circles that spun across the forest floor. Whether this was territorial or some kind of prelude to romance isn't known, but this behaviour was observed for up to ten minutes at a time. After mating, a female forest skink would lay a pair of white eggs (one from each ovary), out of which tiny skinklets hatched to begin their own solitary lives.

For perhaps hundreds of thousands or even millions of years, depending on when the skink arrived on the island, this cycle repeated. With nothing much to worry about, the Christmas Island forest skink enjoyed a kind of cloistered existence. But then humans arrived, bringing the natural borders that had protected this idyll crashing down. And though the forest skink and Christmas Island pipistrelle had almost nothing in common – one originating from east of the Wallace Line, the other from the west – their fates were about to become inextricably tied.

IN 1887, A strange new skink arrived at London's Natural History Museum, then a part of the British Museum. She had been plucked from the Christmas Island rainforest by the crew of the *Egeria*, the first of two British ships sent to investigate the island for signs of phosphate, at the behest of oceanographer John Murray. Perhaps it was during this experience that she'd lost her tail, shaking it off in a desperate attempt to distract the terrifying creatures pursuing her; or maybe this dismemberment was evidence of an earlier, more successful, fight for her life. Either way, her tailless corpse now sat submerged in an ethanol-filled specimen jar at the museum, under the curious gaze of the zoologist George Albert Boulenger. Lifting her small, supple body out of the specimen jar, Boulenger carefully laid the skink flat, then set about measuring each aspect of her physiology, slicing open her abdomen to inspect her organs. Satisfied that she represented a new species, he returned her to the ethanol and interred her in the museum's sprawling zoological collection. This was the Christmas Island forest skink, which Boulenger ascribed to the genus *Lygosoma* (since revised to *Emoia*) and gave the species name *nativitatis* (from the Latin 'nativitas', meaning 'birth', and alluding to the Nativity, and the day on which Christmas Island was discovered).

Shortly after, Boulenger sat down to study a very different kind of reptile – the common wolf snake, whose jaws and teeth he had been tasked with drawing for a catalogue of the museum's snake specimens. Described by the legendary nineteenth-century zoologist Albert Günther as 'one of the most formidable enemies of the skinks', the wolf snake had earned this reputation because of its dentition. The first thing a skink would see, staring into the looming mouth of a wolf snake, would be its large anterior fangs. Curving sharply back towards the snake's throat, they are perfectly adapted to slide smoothly under the overlapping scales of a skink, without breaking its dermal armour. Prey pinned in place, the wolf snake would then bring its posteriormost teeth into play – a pair of long fangs, the backs of which taper into blade-like edges, ideal for crunching through the scales of a skink and into the soft flesh beneath. These two sets of teeth are what make the wolf snake a highly successful 'durophagous predator', or hard-prey consumer, and a nightmare for small lizards like skinks and geckos, which form the majority of its diet.

In Boulenger's drawing, the upper and lower mandibles of a wolf snake are drawn in profile, gently curving upwards into a sinister smile. As he traced the line of this snake's angular teeth, Boulenger probably never imagined that they would one day meet with the 'metallic brown' skin of the new species of skink from Christmas Island that he'd described. At this point, the extent of the forest skink's experience with snakes was limited to just one species – the Christmas Island blind snake (*Ramphotyphlops exocoeti*). The blind snake is about as harmless a creature as you can possibly imagine, spending most of its time underground, its behaviour and thin pink body more closely resembling an oversized earthworm. If this is what snakes are like, a forest skink might have thought, what harm is one more?

On 3 November 1987, a package arrived at the Western Australian Museum that sparked panic. Inside were the preserved remains of a wolf snake – the first to be found on Christmas Island. Back then, virtually everything travelling to the island did so via a supply ship called the *East Crystal*, which ran to and from Singapore once or twice a month. Delivering most of the supplies needed to feed the now roughly 2,000-strong human population, the *East Crystal* was the island's lifeline. But, sometime between April and October 1987, the ship had brought death to the island instead. Coiled in bags of potatoes, pallets of lumber, or some other piece of cargo, were wolf snakes.

'Should it become established,' wrote L. A. Smith, one of the museum's herpetologists, '... it could be as catastrophic as the introduction to Guam of the [brown tree snake].' That invasive species had wiped out multiple bird, reptile, and mammal species on Guam, and caused island-wide electrical outages by tripping power cables. At the peak of its powers, snakebites accounted for one in every 1,200 emergency room visits on the island. Similar to the brown tree snake, the wolf snake is a highly successful predator and colonizer, with a track record of devastating native species populations. While it hadn't yet

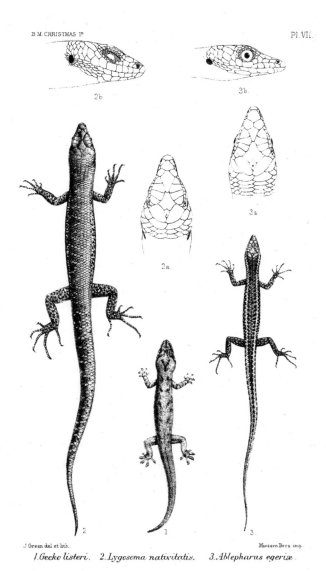

*An illustration of (from left to right) the forest skink, Lister's gecko, and the blue-tailed skink from Charles Andrews' A Monograph of Christmas Island, 1900.*

brought the electrical infrastructure of an entire island offline, the species could pose just as potent an ecological threat as the brown tree snake.

The first wolf snakes to disembark from the *East Crystal* on Christmas Island found themselves in a kind of durophagist's paradise. Flowerpot blind snakes (*Indotyphlops braminus*), a species known for violently beheading its termite prey, house geckos (*Hemidactylus frenatus*), stump-tailed geckos (*Gehyra mutilata*), and Bowring's writhing skinks (*Subdoluseps bowringii*) had already been introduced to Christmas Island by 1987, each a familiar and welcome sight to newly arrived wolf snakes. However, perhaps far more intriguing were the faces they didn't recognize. Christmas Island blue-tailed skinks, commonly found lounging on garden walls and the sides of houses in The Settlement, were one of the first exotic, new items on the menu for wolf snakes. But as their numbers grew, the snakes pushed out into a new frontier: Christmas Island's rainforests.

Going about their usual business of hoovering up the island's invertebrates and basking in the sun, the forest skinks were blissfully unaware of what was coming their way. As a diurnal species, sunset was a signal to the skinks to seek shelter in the leaf litter and to rest. Conversely, for the nocturnal wolf snake, dusk set off the night's hunt. Coiling around the body of a resting forest skink, the snake would clamp its jaws around the lizard's head. When the forest skink became immobile, the wolf snake would then slowly engulf it. At dawn, it was the snake's turn to rest, slinking off into the shadows to digest the contents of its now misshapen belly, as the next night's specials fattened themselves in the sunlight.

Not only was the forest skink utterly unprepared in terms of defensive adaptations to deal with a durophagous snake, but this particular snake was deadliest precisely when the skink was least able to defend itself or flee. As matchups go, it's difficult to imagine how the skink could have been any more disadvantaged. Of course, this was Christmas Island, home to twelve of the world's hundred deadliest invasive species, many of which were only too happy to place a finger on the scale. Any Christmas Island forest skinks living within the vicinity of a yellow crazy ant super-colony would probably end up feeding that colony. Likewise, the boom in the giant centipede population unleashed by the ants also impacted the forest skink, bits of which have been found in the stomachs of the centipedes. And then there were feral cats, perhaps the most feature-complete of the island's predators, whose razor-sharp claws could unzip the belly of a forest skink just as adeptly as Boulenger's scalpel. All of these species likely played a part in the decline of the forest skink. However, the scientific consensus is that it was most likely the wolf snake that led the charge.

In 1998, a reptile survey turned up fewer forest skinks than usual. Although this low turnout was blamed on a spell of unusually overcast and wet weather, it was perhaps not a coincidence that wolf snakes were also found at the centre of the island for the first time. Six years later, the situation came more abruptly into focus. David James, the ecologist then leading the island's biodiversity monitoring programme, concluded his 2004 survey of Christmas Island's reptiles without seeing a single forest skink anywhere on the island. Coupled with this, James found that numbers of Christmas Island blue-tailed skinks and Lister's geckos were also dramatically reduced. Wolf snakes, on the other hand, were everywhere. L. A. Smith's warning about the dangers of the snake establishing itself on Christmas Island now seemed prescient.

In 2005, James sent an urgent memo to the Environment Department, requesting that the forest skink be listed as an endangered species, but this recommendation was ignored. Eventually, however, there was a glimmer of hope. In 2008, a reptile monitoring survey, sampling 900 sites across Christmas Island, finally located forest skinks at a single secluded site on the far west of the island. This was Egeria Point, where Gump and her two ill-fated tankmates had clung on as the last of their species, before being scooped up and deposited into the newly transformed Pink House. In 2012, a last-ditch attempt to locate more forest skinks, involving experienced Parks Australia staff and reptile experts flown in from the Australian mainland, ended in failure. Other than Gump, the forest skink was never seen or heard from again.

'SLUGGISH, MOTTLED BROWN, with few redeeming features' is how Woinarski describes Gump, the last Christmas Island forest skink. This frank and unsentimental appraisal wasn't at all what I was expecting to hear from someone like Woinarski, a herpetologist directly involved in the effort to conserve the island's endemic reptiles. 'No, that's too harsh . . .' Woinarski says – and I wonder, in the pause that follows, whether he has also surprised himself. 'Its significance as the last individual of the species far exceeded its individual virtues as an interesting creature in its own right,' he continues, before summarizing, with a smile, '[Forest skinks] wouldn't sell in pet shops, particularly.'

Woinarski's description of Gump, while on its face unflattering, cuts to the heart of an issue facing most threatened species. It is the reason the World Wildlife Fund fills our ad breaks with footage of rhinos and whales, rather than rats, bats, and skinks – and why they chose a panda for their logo, rather than a snail. Whether we call it 'charisma' or 'personality', only a select few animals possess the wide appeal of something like an elephant or a koala. Nevertheless, as I've heard from many people, being in the presence of the last of a species is a unique experience; an individual animal that perhaps usually wouldn't turn any heads has, in these circumstances, become something unlike anything else in the world. 'It was haunting,' says Woinarski, 'looking at

Gump and knowing this was the last individual of that species, that if she died the species would become extinct.'

Gump, meanwhile, would have been oblivious to the strange status she now held as the sole survivor of her species. Life had been completely transformed following her capture from Egeria Point, even if certain things remained the same. There were the same leaves, twigs, rocks, and logs, and, underneath them, the same red-brown earth. Bounding from log to log were the same tasty crickets as before. Caught in webs spun by familiar spiders were the same delicious moths and beetles. Much was as it should be in Gump's world. Now though, this world ended a foot or so in each direction, in a clear and impassable barrier. Although she found many of the same invertebrates as she rifled through the leaf litter or crawled over logs, Gump couldn't have failed to notice how, every few days, they inexplicably showered down from above, like a biblical plague. Rain worked differently now, too. Whereas experience had taught her that water fell from the sky, sporadically and in droplets, here it arrived at the same time every day, like clockwork, in a huge torrent that came pouring from a plastic hose, and that drenched the soil and everything sitting on top of it, leaving her enclosure pleasantly muggy for the rest of the day.

Though she was likely unaware of it, Gump was receiving an extraordinary level of care and dedication at The Pink House. The humans keeping her fed, watered, clean, and healthy might have terrified and bewildered her, but they were working hard to make her life comfortable, stimulating, and as contiguous as possible with the one she'd led before capture. The term 'keeper' doesn't quite cover the role these people played in Gump's life. It was their job to control the weather, to reassemble a fragment of her world in a Perspex box and, ultimately, to take on the aspects of forest skink life that, from her small enclosure, Gump could not. While the crickets bred on-site were a great bulk feed for Gump, she needed much more

*Gump, the last Christmas Island forest skink, in captivity at The Pink House. (Courtesy of Parks Australia.)*

nutritional variety. Since breeding dozens of species of bugs at The Pink House would be hugely impractical, her keepers turned instead to the natural larder beyond its doors, swishing butterfly nets through the grass outside The Pink House to catch crickets, moths, spiders, grasshoppers, and other bugs. Called 'sweet feed', this broad mezze of organic invertebrates was one of three meal varieties, alongside termites gathered from nearby termite mounds and the crickets farmed on-site, which Gump received three days per week.

'I remember her being a very voracious eater,' says Brendan Tiernan, Senior Threatened Species Program Coordinator at Christmas Island National Park, when we speak. 'Whenever we went in to check on her, she would always acknowledge our presence and look up . . . When it was dinner time, though, she wouldn't hold back. Let's just say, you wouldn't have wanted to be a cricket at that time.'

Once a month, keepers would gently remove Gump from her enclosure, empty it of soil, logs, sticks, water bowls, and unfinished invertebrates, and give it a deep clean. Then, they built her world anew, with freshly baked soil and a carefully arranged selection of rocks, logs, branches, and leaves, before topping it all off with a cardboard hide, a resting or hiding place for whenever she felt threatened. While keepers diligently continued to replicate the outside world for her, the search for a mate was ongoing. But when it concluded without success, the lavish care she received became essentially palliative in nature. 'Comfortable, if lonely,' is how Woinarski describes Gump's life at the end.

When staff arrived at the captive-breeding facility on 31 May 2014, Gump was nowhere to be found. Inside her enclosure, her hide hid nothing at all. Like the fallen tree trunks that researchers had seen slowly empty over the years, the logs on which Gump usually basked under her own private sun now lay bare. It was as if the extirpation of her species had played out in microcosm, within this tank; as if a pint-sized supply ship had docked in the night and unwittingly unloaded a wolf snake, which now lay bloated and hidden, somewhere in the enclosure. Of course, something had lain hidden, just not a wolf snake. Curled up amongst leaf litter that had been collected from the forest floor just a few days earlier, and carefully combed through for centipedes, was Gump's lifeless body.

'It was extremely sad,' Tiernan tells me. 'But we knew that this was the last one and that we'd already lost the species. So, in a way, we'd already come to terms with its passing before it had even passed.'

Samantha Flakus, who was then the manager of the captive-breeding programme, tells me of the moment she learned of Gump's death. On 31 May 2014, she had been at a science forum in Canberra. There, staff from across Parks Australia's science team had converged to discuss work and projects operating throughout the organization's estate, which includes Christmas Island. 'Our director at the time was talking about extinctions,' says Flakus, 'and I remember her saying: "There's not going to be any extinctions on my watch."' Later that same day, Flakus received the call from Christmas Island confirming the opposite. Gump was dead, and her species was extinct.

A 2017 PAPER by Woinarski and his colleagues from the Threatened Species Recovery Hub brings the Christmas Island forest skink together with the Bramble Cay melomys and the Christmas Island pipistrelle. The paper is described as a 'coronial investigation' and follows the structure of a coroner's report. Coroners – independent judicial officials who investigate violent, unnatural, or unexplained human deaths – scrutinize not only the direct causes of deaths, but any wider circumstances that may have contributed, such as systemic or organizational failings. Their investigations are particularly important in helping society come to terms with, and learn lessons from, tragic events. In the paper, coronial methods are applied to an investigation into the extinctions of the three species that all occurred within Australian territory.

Perhaps inadvertently, the authors of this paper take on the role of translators; the language of ecology and of the natural world, from which we humans often see ourselves as separate, is exchanged for language from our world – language of the mortician, or of the medical professional. 'Extinction' becomes 'death'; and 'species' is replaced with 'the deceased'. This simple reframing of extinctions with language we so closely relate to our own experiences of loss and grief adds weight to the paper's findings: that all three extinctions were 'predictable and probably preventable'.

Australia's environmental legislation and policies at the time made 'no commitment to attempting to prevent avoidable extinctions', the authors explain. The paper cites instances where key conservation actions weren't taken, including that no captive breeding was suggested in the melomys recovery plan. For the skink, official 'threatened' status was given only four months before Gump's death, and fifteen years after a significant decline in the species was demonstrated (and nearly a decade after David James' recommendation).

A coroner's report will often expose inequalities between different groups of people – between rich and poor, different races, genders, communities, ages. In the same way, this paper demonstrates the inequalities between species. In the case of the three Australian species, their lack of 'charisma, utility, cultural significance [and] ecological role' all worked against their favour. They were all, also, 'remote from major human population centers. Hence, they were largely unfamiliar to a potentially supportive public.' The 'deceased' were, in essence, unpopular and anonymous and, in part, allowed to be lost due to that unpopularity and anonymity. 'These are sorry tales,' the report concludes.

When I speak with Woinarski, he tells me that 'trying to balance despair and hope, problems and successes, that's part of the high-wire act of doing ecology and conservation.' There is a little hope in his and his colleagues' coronial report. 'Our primary interest', the authors write, 'was not to apportion blame.' Rather, it was to identify 'remedies' so that 'future extinctions may be less likely.' A table of fifteen recommendations is provided, and it includes: making sure there is a 'clear chain of accountability . . . for the prevention of extinction', ensuring that 'the process for listing species as threatened is timely and comprehensive', that 'the process for recovery planning for threatened species is timely and effective', and that the public is 'fully informed and involved in the governance of the recovery process'. The relatively short list is, the authors write, 'likely applicable to efforts to prevent extinction in most parts of the world'. They hope it will 'achieve a positive legacy' for the three Australian species.

AT LONDON'S NATURAL History Museum, I'd been shown specimens of the Bramble Cay melomys and of the Christmas Island pipistrelle, whose wings the curator spread out for me on the palm of my hand. Now I was peering into a jar of Christmas Island forest skinks, attempting to thread my gaze through the tangle of tails and limbs, towards the first – that unfortunate lizard with no tail, which Boulenger had used to describe the species in 1887. The term 'type specimen' is what biologists call the individual specimen, or collection of specimens, on which the scientific definition of an organism rests. They are, essentially, a perfect example of whatever they happen to be, which is ironic in the case of this incomplete originator of the Christmas Island forest skink. In my excitement, I'd forgotten to ask museum staff whether I'd be seeing this particular lizard – and yet, there she was, resting at the bottom of the jar, beneath the others, on account of her missing tail.

Lifting Boulenger's forest skink carefully out of the jar with a pair of forceps, I laid her flat in my palm. Although she had been captured in 1887, this lizard didn't seem to have aged a day in the 136 years since. Most striking of all was the elasticity of her limbs, which, with a bit of gentle manipulation, could fully articulate – her knees bending, her long, thin fingers flexing freely. This didn't feel like a museum specimen collected over a century ago; it felt as though I was holding a skink that had just been caught, which might suddenly start struggling in my hand. The bronze-coloured scales on her back and sides looked as vibrant as in any photo of the species, and her skin was smooth and unbroken, aside from in one area. I turned her over, and right across her belly was what I can only think to describe as a wound. A clean scalpel-slit through which her organs had been removed, appearing like an incision made in a coroner's post-mortem: a physical reminder of her species' legacy in death.

Where might the Christmas Island forest skink be today, if things had gone differently? Surprisingly enough, we almost certainly know the answer.

Alternating dark- and light-brown stripes run the length of a Christmas Island blue-tailed skink and, on some individuals, the contrast between the two is so extreme that they look like lizard-shaped humbug sweets. At the base of their tails, this coloration gives way abruptly to electric blue. In the wild, this detachable bright-blue tail is thought to serve as a distraction, drawing the eyes of predators as its owner breaks free and scampers to safety. Today, these vibrant streaks are what you see most of at the captive-breeding facility where Gump was kept. In tank after tank, they whip back and forth like little sparks of electricity; here, blue-tailed skinks forage, fight, and, crucially, mate.

While Gump and her two companions were the only three forest skinks to make it out of the wild alive, they were joined at The Pink House by forty-three Lister's geckos and sixty-four blue-tailed skinks. Years later, as Gump sat alone in her tank – artificially extending, by virtue of her hopeless existence, the lifespan of her species – these other lizards were busily making a spectacular recovery. 'At one point we had nearly two thousand Lister's geckos and about the same number of blue-tailed skinks,' Samantha Flakus tells me. With so many lizards, and an insurance population at Toronga Zoo on the Australian mainland, a tank of twenty blue-tailed skinks was able to be kept at a local school, to involve the next generation in caring for the island's wildlife.

As with any project, there were bumps in the road. In 2014, an outbreak of a new species of bacterium caused facial deformities, lethargy, and death in some blue-tailed skinks and Lister's geckos at the facility, but this was dealt with by isolating infected individuals and administering antibiotics. Outside of this, the success of both species' recovery eventually presented a problem. Without funding to continually expand the facility, something needed to be done with any excess lizards once the population grew too large. 'Pretty much anywhere on Christmas Island where there are wolf snakes, you will have 100 per cent mortality of reptiles,' says Flakus, 'and we wouldn't release animals into an area where we knew they would face certain death.' This led to a scenario where staff at the captive-breeding facility had to care for blue-tailed skinks and Lister's geckos with one hand, and humanely euthanize them, at the egg stage, with the other.* 'That was a really challenging time to work through,' says Flakus.

Today, a change of address is in order for some of these lizards: the island of Pulu Blan, in the Cocos (Keeling) archipelago, 980km west of Christmas Island. Just 2.08 hectares, this flat island doesn't look much like Christmas Island, but it is densely forested and its climate is similar. Uninhabited by humans, Pulu Blan has managed to avoid the wolf snakes and yellow crazy ants that cruise the world's shipping lanes. On 7 September 2019 – coincidentally National Threatened Species Day in Australia – 150 captive-bred Christmas Island blue-tailed skinks were released into the wild on Pulu Blan, many of which had never seen sunlight before.

---

* The eggs were euthanized within two weeks of being laid. This is considered the most humane way of managing population growth within captive lizard populations and is a method used commonly by zoos.

'There are lizards everywhere on that island now. I can't think of anywhere I've been in the world with that kind of density. There are just lizards on everything!' says Brendan Tiernan, who leads reintroduction efforts. I ask him if he means *literally* everything. 'I mean that on every tree that's in the sun, on every branch, there are lizards,' he replies, triumphantly. The abandoned shack that Tiernan and his team use as a makeshift field office is often 'covered in blue-tailed skinks,' he says. 'You'll be working and a wild skink will shoot across the desk or between your feet.' Tiernan soon expects the blue-tailed skink population to reach what is called 'carrying capacity' – the maximum number of a species that a habitat can naturally support. 'Right now, the population is probably close to five thousand. That's a higher density than would have been on Christmas Island in their ideal habitat.'

Another batch of blue-tailed skinks released on nearby Pulu Blan Madar are also thriving, and now Tiernan and his team are eyeing up much larger islands on behalf of their one-time captives. 'We're looking at islands that are over a thousand hectares. One day, if we're successful, there could be hundreds of thousands or even millions of them living back in the wild.'

Flakus and I speak about these reintroductions, and when I ask if the forest skink could have made a similar recovery she replies in an instant: 'It would have been successful, if there had been animals to breed. I have no doubt at all.' Skinks, Flakus explains, are generally easy to keep in captivity, and forest skinks and blue-tailed skinks had very similar husbandry requirements. Had things gone differently, then perhaps Pulu Blan would be home to one more species today.

# 8 The Gritador

Alagoas Foliage-Gleaner and Cryptic Treehunter
*Philydor novaesi* and *Cichlocolaptes mazarbarnetti*

It's early morning one day in November 2008, at the Museu Nacional, the National Museum of Brazil. Overlooking the city of Rio de Janeiro, this grand, colonial manor house could almost be a museum artefact itself. Built in 1803, it has witnessed revolutions, housed both Brazilian and Portuguese royal families, and is nearly two decades older than the nation whose history it safeguards. Now, it is home to one of the most important museum collections in South America: 20 million scientific and cultural artefacts spread across more than a hundred palatial halls and rooms, some still adorned with gilded stucco reliefs, chandeliers, and royal portraits. In just a few hours, the museum will be overtaken by a bustle of activity as families, tourists, and school groups jostle for glimpses of Ancient Egyptian mummies, dinosaurs, meteorites, and 'Luzia', the oldest known human inhabitant of South America, whose 11,500-year-old skull is its star attraction. For now, though, as the first hint of dawn suggests itself on the horizon, the museum is quiet and still.

In a room away from the public exhibits, half a dozen small bird specimens are laid out neatly on a table. Freshly exhumed from a drawer in the ornithological collection, they are between 19cm and 22cm long from beak to tail. They have amber-coloured plumage, with buff and dark-brown highlights around their blank white eyes – little windows into their cotton-wool-stuffed souls. Their beaks are tied shut with string, and labels attached to their feet denote that they are *Philydor novaesi* – a species of rainforest-dwelling, insectivorous bird from the north-east of Brazil, commonly known as the Alagoas foliage-gleaner. One of them is the holotype, the original Alagoas foliage-gleaner upon which the entire concept of this species rests. However, in the minds of Juan Mazar Barnett and Dante Renato Corrêa Buzzetti – the two ornithologists now hunched over the table – there is, amongst these specimens, an imposter.

A 'cryptic species' is a species that has been erroneously classified as something else. Their discovery is often a by-product of scientific progress, as advances in the technology we use to scrutinize the natural world have brought the once soft boundaries between organisms into ever-sharper focus. In the past, we drew boundaries between species based on things like appearance, habitat, diet, and behaviour; now, DNA analysis helps us recognize genetic differences between two similar-looking species. In some instances, one species will open up to reveal a cascade of further species, each hidden inside the last like Russian nesting dolls. And with the fact that climate change will bring with it an unprecedented deluge of extinctions, there has been a renewed effort to identify cryptic species so that they can be conserved; more and more biologists are charting expeditions away from the jungles and deserts of the world, and instead into the dusty collections of natural history museums.

Barnett is no stranger to cryptic species. An Argentinian ornithologist, his interest in birds began in 1984, when he was nine years old, shortly after which he became renowned within Aves Argentinas (the Argentinian ornithological society) as a boy wonder. Known affectionately as Juancito, even at that tender age he knew the scientific name, range, patterns of migration, diagnostic characters, and abundance of every single bird in the country. Later, as a professional ornithologist, he was especially skilled at locating birds in the field, an intuition that led him to become something of a ghost-hunter. Extraordinarily rare birds, some that hadn't been seen for decades, seemed to present themselves before him – such as the austral rail (*Rallus antarcticus*), which had vanished forty years before Barnett and a group of other ornithologists rediscovered it in a Patagonian marsh.

Some of his most important discoveries came in the form of cryptic species, however, such as the pink-footed shearwater (*Puffinus creatopus*), an undescribed species whose holotype Barnett found as he rifled through drawers at Buenos Aires' Museo Argentino de Ciencias Naturales, Argentina's museum of natural science.

Now, Barnett and Buzzetti scrutinize every part of the Museu Nacional's six Alagoas foliage-gleaner specimens. Using a dial calliper, the pair measure the beaks, wings, tails, and other aspects of the specimens to the nearest 0.1mm. They also note details such as the mass, length, coloration, wingspan, and gonad condition of the birds as recorded at the time of collection, and even the precise make and model of binoculars used in the field to identify them. Within this wealth of data, the outline of a cryptic species soon emerges – a hidden bird that has managed to fly under the radar of the ornithological community for decades.

The Alagoas foliage-gleaner was once itself a cryptic species. In 1979, the type specimen was captured by Ph.D. students Dante Martins Teixeira and Luiz Pedreira Gonzaga in the north-eastern state of Alagoas. At the time, neither Teixeira nor Gonzaga realized the significance of what had tumbled into their mist net. After all, the little amber bird they'd caught looked a lot like a black-capped foliage-gleaner (*Philydor atricapillus*), which is how they labelled it. It was not until four years later, in 1983, that Teixeira and Gonzaga realized that they in fact had discovered a new species. They formally described the Alagoas foliage-gleaner, giving it the species name *novaesi* in honour of the Brazilian lawyer and ornithologist Fernando da Costa Novaes. In their 1983 paper announcing this discovery, they write of a species that – noisy rummaging through dead foliage aside – is notably quiet, vocalizing only occasionally. These rare vocalizations, the researchers write, take one of two forms: one is a 'thürr', 'thoor', or 'theer', and the other, a descending 'uüarrr, uüarrr'.

It was birdsong that had roused the suspicions of Barnett and Buzzetti. On 12 October 2002, the pair had been in the north-east Atlantic Forest – dense, humid rainforest, where every tree is sheathed from trunk to crown in epiphytic plants (plants that live on other plants, often on the branches of trees). There, Barnett and Buzzetti saw a bird that looked just like an Alagoas foliage-gleaner; but its behaviour was odd. It was foraging inside a bromeliad, something foliage-gleaners weren't known to do. Then it sang, and its song didn't end as the ornithologists expected. Instead of petering out into silence, the bird followed up its 'uüarrr, uüarrr' (which they describe as a high-pitched 'rattle') with a pair of short, sharp screams, the second louder and pitched slightly higher than the first. When further sightings and recordings only deepened their suspicions, the two men requested and were eventually granted an audience with the Museu Nacional's specimens.

'Cryptic treehunter' (*Cichlocolaptes mazarbarnetti*) is the name Barnett and Buzzetti gave the new species they discovered amidst the Museu Nacional's Alagoas foliage-gleaner specimens in 2008. During their examination, they noted that two of the six specimens showed significant morphological departures from the others of the same age and sex. A longer and heavier body was the most obvious difference, but their bills were also larger. They also had flatter heads, compared with the more rounded heads of the Alagoas foliage-gleaner, and a variety of subtle variations in plumage. Along with their distinct vocalizations and behaviours, these physical differences justified assigning the two specimens not only to a new species designation, but also to an entirely different genus. This meant that the two species, that had for so long been labelled identically, were in fact only very distantly related. They were in separate genera, but in the same family, which is a much larger group of classification: the

Furnariidae family of birds. This is the equivalent of our own taxonomic family, Hominidae, which includes species as diverse as orangutans, gorillas, chimpanzees, and us, humans.

Barnett and Buzzetti's findings were published in 2014. Overnight, two of the specimens the ornithologists had inspected had moved species, genus, and drawers at the Museu Nacional; and, like that, the cryptic treehunter was cryptic no more.

This chapter tells the story of these two species which, for most of the time we've known them, were thought to be the same. Although differences were eventually discerned between them, they ultimately shared the same fate. In 2014, as the cryptic treehunter was revealed to the world, both this species and the Alagoas foliage-gleaner had likely already vanished from it.

COVERING 6.7 MILLION square kilometres, roughly 40 per cent of South America, and containing half the world's remaining tropical forest, the Amazon is, by a long shot, the largest rainforest on Earth. It's also the densest in terms of biodiversity, home to more than half of all plant and animal species, and 10 per cent of all known species. It is often referred to as 'the lungs of the Earth'. Although this nickname sprang from the erroneous belief that the rainforest produces 20 per cent of the planet's breathable oxygen (in reality, its net contribution is close to zero), it speaks to how synonymous the Amazon has become in our collective imagination with nature, and even life itself. As a result, the ongoing destruction of these 'lungs' by wildfires, slash-and-burn agriculture, and logging, is understandably perhaps the single most talked-about environmental issue of our time. However, withering in its long shadow is another South American forest, whose destruction is rarely, if ever, mentioned.

Running south along a 4,000km stretch of Brazil's Atlantic

coast before reaching inland into parts of Paraguay, Uruguay, and Argentina, the Atlantic Forest cuts a slender figure next to its larger neighbour. Patchy and highly fragmented, centuries of unrestrained deforestation have left less than 15 per cent of the original forest intact – most of this is isolated scraps of woodland under fifty hectares in size. The situation is especially dire in the north-east Atlantic Forest, such as in the state of Pernambuco, where as little as 2 per cent of the original forest remains. However, despite the unbridled destruction with which the Atlantic Forest has been afflicted, it remains one of the most biologically rich forests on earth and a global biodiversity hotspot.

Roughly 2,200 vertebrate species are found here – 5 per cent of the world's total. Plant diversity is so high that one study was able to count twice as many tree species in a hectare as exist across the entire eastern seaboard of the United States. In total, around 20,000 plant species grow here, including over 2,000 species of epiphytes. Species diversity in the Atlantic Forest actually exceeds that found in most parts of the Amazon rainforest; and, crucially, many of these species are endemic. With 85 per cent of the Atlantic Forest gone, these figures represent just a sliver of the ecosystem's original biological wealth.

This forest was home to the Alagoas foliage-gleaner and the cryptic treehunter. Just over 500 years ago, when both birds are thought to have been abundant, the Atlantic Forest spanned an area the size of Germany, France, Spain, and Portugal combined, and was the second largest rainforest in the world, after the Amazon. Go back even further, however, and the boundary between these two great forests disappears completely. If we could rewind time and watch the story in reverse, we'd see, at around 10,000 BCE, the tail end of the last glacial period (or ice age), the dry savannah between the Atlantic Forest and the Amazon rainforest flood with trees in the cooling climate. Back in the last interglacial period, another 105,000 years earlier than that, they'd separate again. Over hundreds of thousands of years, as the climate swung repeatedly between freeze and thaw, this dance between the two huge forests would continue, and they'd separate and come together and separate, over and over. From above, it would appear as if the two forests were expanding and contracting like a pair of gigantic green lungs, the ice racing back and forth from the Earth's poles, their breath. At some unknown point, in the last few million years, we'd see the Alagoas foliage-gleaner and cryptic treehunter diverge genetically from their most recent ancestors, and the stories of these two species begin.

To 'GLEAN' IS to gather something, often with great difficulty. It is a word with Celtic roots that makes its way to us via Latin, Anglo-French, and Middle English, and was originally used to describe the act of scouring fields for leftover grain after harvest. It's also the perfect way to describe the foraging behaviour of the Alagoas foliage-gleaner. In the humid montane rainforests of the Atlantic Forest, it would flit between orchids, bromeliads, and other plants on an agitated hunt for invertebrates. Hanging upside down to snatch spiders from the underside of a branch and to pluck insects from a tangle of dead vegetation, then hammering a beetle larva out of a rotten branch with its beak – this was laborious work. As this bird darted around the midstorey and canopy like a circus acrobat, a near-identical-looking bird would sometimes cross its path. Halfway along a branch, a clump of bromeliad leaves might suddenly shake like a pom-pom before a cryptic treehunter shot out.

The Alagoas foliage-gleaner and cryptic treehunter went about their busy work as part of large, noisy, mixed-species flocks. It's not known which specific species comprised these flocks; but they would have been a diverse mix of woodcreepers, woodpeckers, ovenbirds, antbirds, and tyrant flycatchers, all with different jobs. Some members of the flock quietly pottered about the understorey looking for millipedes, centipedes, and other arthropods, while at the other end of the spectrum in terms of personality would have been a 'flock-leading' bird whose constant vocalizations cohered the group or triggered its dispersal whenever danger approached.* Together, these flocks formed noisy rabbles that thrummed branches and rattled bromeliads, turfed up bark, and terrorized the insect community wherever they went. And occasionally, in amongst the chirps, tweets, and squawks made by these birds, the high-pitched rattle of an Alagoas foliage-gleaner rang out through the trees, followed by the frantic screams of a cryptic treehunter.

The world that these two birds emerged from was in many ways much like it is today. Jaguars stalked their prey – giant anteaters and tapirs – through the understorey and savannah at the periphery of the forest, and herds of capybara wallowed in streams and rivers, under the greedy gaze of green anacondas. A kaleidoscopic array of colourful birds squawked, shrieked, and gleaned their way through the forest, which teemed with the same dizzying array of amphibian, insect, and plant life. In amongst all these familiar sights and sounds, however, there were also giants. Armadillos the size of bears, one-tonne rodents, and sloths the size of Asian elephants; nearby in the surrounding savannah were *Stegomastodons* – gigantic woolly elephants, closely related to mammoths – stalked by sabre-toothed cats.

---

* We only know which species the Alagoas foliage-gleaner and cryptic treehunter flocked with recently, when their habitat was highly fragmented. Birds in these modern flocks included chestnut-coloured lesser woodcreepers (*Xiphorhynchus fuscus*), a small ashy-grey bird called the Alagoas antwren (*Myrmotherula snowi*), and the 'flock-leading' cinereous antshrike (*Thamnomanes caesius*). Many of these species are critically endangered.

The Alagoas foliage-gleaner and cryptic treehunter may have shared their forest home with megafauna. Perhaps the task of gleaning invertebrates was occasionally complicated by giant sloths reaching into the canopy and bending entire branches towards their gaping mouths. Or maybe something so large and lumbering might have seemed like just another piece of the environment to a pair of comparatively tiny birds.

Around 14,000 years ago, however, things began to change. At night, flickering lights started to appear in the forest and, suddenly, the gigantic sloths, rodents, and other large animals began to disappear. 'The Pleistocene overkill' is the theory that the spread of our species, *Homo sapiens*, was the primary cause of the near total extinction of global megafauna. A growing body of evidence supports this; however, some scientists have argued that our role as hunters was probably just one of many causes, including climate change. Whatever level of culpability we bear for this particular moment of ecocide, the arrival of people – like 'Luzia', whose ancient remains lay nearby as Barnett and Buzzetti separated the cryptic treehunter from the Alagoas foliage-gleaner at the Museu Nacional – would have undoubtedly changed the balance of life within the Atlantic Forest. Estimates for the date of human colonization of the Americas vary widely. For the two birds at the heart of this chapter, however, this first wave of human migration was largely insignificant. Only in the lights, the fires we brought to the Atlantic Forest, was there an omen of what was to come.

'HUMANS CAN'T HELP but change the environment, simply by existing in it,' Phil Riris tells me. An archaeologist and Senior Lecturer in Archaeological and Paleoenvironmental Modelling at Bournemouth University, Riris specializes in the archaeology and historical ecology of tropical South America. 'There's a lot of controversy about the peopling of the Americas, particularly now it's suggested that humans may have entered the Western Hemisphere as early as 25,000–24,000 years ago,' he says. 'If that proves accurate, it's plausible that people were in eastern Brazil 14,000 years ago.'

Charcoal and the remains of stone tools evidence our early presence in the Atlantic Forest. Then there is pollen, which spread with us as we moved about the forest, and that remains today in fossilized form and can be read like a travel journal. These tiny fragments of ancient flowers show that early humans in the Atlantic Forest were transient hunter-gatherers. Humans then transitioned to a more settled existence around 6,000 years ago, as shown – Riris tells me – by the remains of *Cucurbita*, a genus of flowering plants that produce berries with a hard rind, commonly known today as the squash. This was the earliest domesticated plant in the region, and was widely cultivated for food. The Brazilian pine tree (*Araucaria angustifolia*), whose edible nut-like seeds have been an important, highly nutritious, and culturally significant food source for indigenous peoples across the country, is particularly elucidative of the movement of humans around 1000 CE. '*Araucaria* is like a cultural keystone species for some indigenous groups in southern Brazil,' says Riris. What this means is that the tree has played such a significant part in traditional dietary, spiritual, symbolic, and medicinal practices that it has shaped the cultural identities of these peoples. As they moved, they were often accompanied by its starchy seeds, and over a thousand years later the remains of these seeds are now being unearthed all over the Atlantic Forest. Beginning in the sixteenth century, however, the forest starts to tell a very different story.

On 23 April 1500, a Portuguese navigator called Pedro Álvares Cabral was blown off course and accidentally came upon the north-eastern tip of a new, 'undiscovered' land. As was the European custom at the time, Cabral immediately claimed it for his monarch, King Manuel I of Portugal, and a new possession was added to the already bulging property portfolio of the Portuguese crown. The name 'Brazil' took a while to attach itself to this new land. Believing it to be an island, Cabral had called it Ilha de Vera Cruz (Island of the True Cross) – probably referencing the same legend after which St Helena was named – while sailors returning to Lisbon gave it the less lofty title of Terra dos Papagaios (Land of Parrots). In a secret 1502 'master map' of the world drawn up for Manuel I, the country makes its first appearance on paper – this time as the 'Land', rather than the 'Island', of the True Cross. It is represented here as a disembodied chunk of coastline, the only part of the country that the Portuguese had seen at this time, suspended in empty space. Nevertheless, it paints a vivid picture of exactly what this new land meant to Lisbon: behind a trio of colourful parrots are huge, oversized trees painted in rich blues and greens.

It can be said that the country that we now call Brazil has its roots in trees. When Cabral and his crew came ashore in 1500, they soon discovered a new species of spiny hardwood tree that seemed to bleed when cut, as if flesh rather than wood lay beneath its bark. When the sapwood was ground up and soaked in water, the resulting crimson pigment produced a similar dyeing effect to that of the sappanwood tree (*Biancaea sappan*), a species from tropical Asia whose supply into Europe was exclusively controlled by Arab merchants. With this new tree, Portuguese agents stood to profit enormously by providing the nobility of Europe, particularly those of the French court, a more affordable route to their beloved royal reds. The tree was given the name 'Pau-brasil', from the Portuguese 'pau', meaning wood, and 'brasil', meaning reddish or ember-like – also the root of both its scientific name, *Paubrasilia echinata*, and common name, brazilwood. Then, like so many European colonial territories – among them the Gold Coast, the Ivory Coast, the Pepper Coast, and the Slave Coast – this new country was

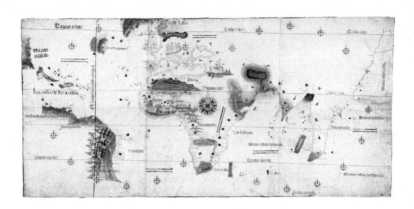

*The Cantino World Map, named after Alberto Cantino – the Venetian secret agent who smuggled it from Portugal to Italy in 1502. The coast of Brazil is depicted in the bottom left.*

stamped with what was essentially a product label, denoting the objective of all extractive efforts to follow. 'Terra de Brasil', the land of brazilwood, was born, and with it the destruction of the Atlantic Forest began.

Soon, fire cleared the way, up and down the coast, for fields of sugarcane. These plantations didn't just gobble up the land they grew upon; the mills that processed sugarcane required an enormous amount of fuel, in the form of wood or charcoal, which the Atlantic Forest had no choice but to provide. Coffee was perhaps the most destructive new plant, thanks to the prevailing attitude that the crop required untouched 'virgin forest soils'. Plantations raced inland, overtaking undisturbed parts of the Atlantic Forest and leaving bare land in their wake, which became cattle pasture. Even the design of the plantations was wasteful, with individual coffee plants spaced as much as three times further apart than in other grain-producing colonies, to enable foremen to better oversee enslaved workers. Underpinning all of this was the belief at the time that Brazil possessed an inexhaustible supply of land, ripe for exploitation.

Over the centuries, the Atlantic Forest shrank back from the coast and started to fragment. Pockets separated to form even smaller pockets, the space between them overtaken by monocrops and pastureland, until even these isolated fragments were absorbed. To a very limited extent you can see the tail end of this destruction for yourself via Google Earth's 'timelapse' feature, which collects together a slideshow of global satellite images going back nearly four decades. In this sequence, the Atlantic Forest fades from green to brown, as if someone has turned down the colour saturation on your screen. In front of your eyes, the forest disappears, like a puddle evaporating on a hot day, leaving behind only a smattering of small droplets. And it was in one of these droplets that the Alagoas foliage-gleaner and cryptic treehunter were left stranded.

It isn't every day that an entire forest changes hands for the sake of a bird. But this is exactly what happened in 2004, when a Brazilian environmental NGO called SAVE Brasil (with help from the Aage V. Jensen Charity Foundation) paid a landowner $120,000 for a small forest on a hill, in north-east Brazil. It was not exactly large – at 390 hectares, it was roughly double the size of London's Regent's Park. However, what made the deal worth it was not so much the land itself, but what lived there. 'We purchased this patch of forest because of the Alagoas foliage-gleaner; at the time, we didn't realize that there was also another species there [the cryptic treehunter had yet to be discovered],' says Pedro Develey, a biologist and Director of SAVE Brasil. 'We bought the area, because otherwise the forest would have been completely destroyed.'

A few months earlier, Develey had visited the north-east Atlantic Forest and learned first-hand how severely fragmented it had become. 'I remember returning to São Paolo and I just thought, "Wow!"' he says. 'Even today, in the south Atlantic Forest, we still have continuous blocks of Atlantic Forest covering 200,000 hectares – if you look on a map, you'll see a green spot,' says Develey. 'But when you go to the north-east, it's just small tracts.' The largest of these tracts is 6,000 hectares, Develey explains, but this is an outlier; 80 per cent of the remaining fragments in the north-east Atlantic Forest are less than 50 hectares. Paradoxically, while these scraps of habitat are smaller than those in the south, they're often far more biodiverse. A 100,000-hectare protected area in the south of the forest will contain less than ten globally threatened bird species, whereas, in the north-east, nearly twice that number are known to exist in a fragment of just 400 hectares. This is one of the reasons that The Nature Conservancy (an environmental non-profit) considers the north-east Atlantic Forest one of the key global conservation priorities of our time.

The forest that SAVE bought was in Pernambuco. This state had been one of the fifteen hereditary fiefdoms created by Manuel I in the 1530s, then doled out to the Portuguese aristocracy and its associated lackies. From that point on, it was spared virtually no destructive act. Home to the richest reserves of brazilwood, Pernambuco became the state most severely impacted by its trade; later, it was the locus of the sugarcane industry, but was also a major producer of coffee and other crops. The destruction of the Atlantic Forest is often talked about as if it belongs to the distant past; however, in truth, it has never really let up. Agriculture, logging, charcoal production, and other industries have continued to eat away at the edges of forest fragments, leaving just 2 per cent of Pernambuco's original woodland intact.

The Alagoas foliage-gleaner – and, it turned out, the cryptic treehunter – were, Develey explains, 'indicators of good forest. They needed a well-preserved forest, with large trees and lots of bromeliads.' If the habitat was too degraded, the birds would be unable to survive, since both species' feeding behaviour revolved around bromeliads and the insects that flock to them. And so it was hoped that, by purchasing this rare surviving piece of their original habitat, both the forest and birds could be saved. After merging it with another nearby forest reserve, they christened it 'Serra do Urubu', but it also became known by another name: 'forest of hope'.

There was little to be hopeful about, however. It was thought that there might be twenty or so Alagoas foliage-gleaners in the newly purchased Serra do Urubu. But the results of the first bird census suggested that just three birds remained. 'When you have a really small population, protecting the habitat is not enough. You need to start direct management, in terms of breeding biology, food supplementation, or changing the structure of the habitat,' says Develey. However, to do any of this required a deep understanding of the species and how it lived, something that simply did not exist.

'Bird conservationists often rely on the information gleaned by people who maintain birds in captivity,' says Develey. 'It's an unfortunate situation, but here many of these people are responsible for bird-trafficking.' One example is the Spix's macaw (*Cyanopsitta spixii*), a parrot and popular exotic pet that was until recently extinct in the wild but is now being bred in captivity and reintroduced, thanks in part to the knowledge base developed by the same bird-traffickers who once drove its decline. Ironically, this means that although this species was subjected to rampant exploitation by the illegal exotic pet trade, it was easier for conservationists to mount its recovery as a result. But, as Develey explains, this paradoxical phenomenon cuts both ways. 'With small passerine birds like the Alagoas foliage-gleaner, there has never been any interest in keeping them as pets, because they are tiny little birds, and not colourful,' he says. This meant that Develey and his team needed first to study the bird, and hope that they learned enough to save it before the species slipped through their fingers.

'I never imagined that I would be one of the last people in the world to see that bird,' says Develey. This sighting was not a very good one, he tells me. And he admits he doesn't remember what year it was – '2007 or 2009,' he says. At the time, he didn't realize how significant it was. Perched on a branch, surrounded by the long, frond-like leaves of bromeliads and the delicate flowers of orchids, it was difficult to make out the small amber shape above him. 'But the bird was there,' he says. By this point, despite all efforts, little had been learned about the Alagoas foliage-gleaner, and the population showed no signs of rebounding. Then, after a final sighting in September 2011 at Serra do Urubu, the species vanished for ever. 'I'm not trying to deny the fact that we failed as conservationists, because we *failed*. We lost a species,' says Develey. 'But we failed because we arrived too late.'

Not long before the final sighting of the Alagoas foliage-gleaner, Barnett and Buzzetti uncovered the cryptic treehunter at the Museu Nacional. It was a fascinating if bittersweet discovery. A species that was on the very brink of extinction had suddenly spawned another. The holotype of the cryptic treehunter was a large female bird, caught in 1986 by Teixeira and Gonzaga; the label identifying it as an Alagoas foliage-gleaner was carefully slipped off its leg and replaced. Along with the reassignment of this specimen and a smaller juvenile female, the discovery necessitated a certain degree of retroactive detective work to tease free the ecological record of this new species from that of its progenitor. This meant combing through the notes, photographs, videos, sound recordings, and other evidence of the cryptic treehunter that had been unwittingly misattributed over the previous three decades. Attempts to add to this record, by painstakingly searching for the new species at Serra do Urubu and other forest fragments, turned up nothing. In all likelihood, Barnett and Buzzetti knew, it had gone before they or anyone else had noticed its existence.

In 2012, as Barnett and Buzzetti prepared their paper announcing the discovery of the cryptic treehunter, there was another loss. Barnett, having long struggled with an autoimmune disorder, contracted meningitis and died at the age of thirty-seven. 'Juan was a great friend with a big heart. He always wanted to help everybody without expecting anything in return,' Buzzetti tells me when we speak. Barnett's obituary, published in the ornithological journal *The Condor*, describes him as 'the pride of a generation'.

After Barnett's death, Buzzetti struggled to finish the paper describing the cryptic treehunter. Barnett had been the lead author and had written the introduction and most of the first chapter already. He'd also done a lot of work on the ground, speaking with people in Pernambuco about the Alagoas foliage-gleaner and the mysterious new bird the two ornithologists had found. After encouragement from friends and colleagues, Buzzetti finally finished and published the paper in 2014, keeping Barnett as the lead author. On the naming of the cryptic treehunter, Buzzetti writes: 'the second author dedicates the name of the new species to the first author, a good friend and colleague who suddenly passed away before this manuscript was finished, in recognition of his important contributions to the conservation of the Atlantic Forest of northeastern Brazil and its declining avifauna.'

*Mazarbarnetti* was the species name chosen for the bird, alongside the English common name, the cryptic treehunter. But Barnett and Buzzetti's paper also proposes another name for the species: 'Gritador-do-nordeste', meaning 'screamer of the north-east'. In Brazilian folklore, the Gritador is a figure who commits suicide in anguish after accidentally killing his own brother. 'Now his soul sometimes can be heard as it wanders

*Specimens of the Alagoas foliage-gleaner (left) and cryptic treehunter (right) side by side at the Museu Nacional. (Photo by Dante Buzzetti.)*

through the forest in the top of the hills, screaming in pain while searching for his brother,' write Barnett and Buzzetti.

'This name was Juan's idea,' Buzzetti tells me. 'He learned about the legend of the Gritador while speaking with the people of Pernambuco.' This legend was the perfect analogy for the species. 'The impression of the cryptic treehunter was of a bird screaming for another of the species that it could not find,' he tells me. 'That was my impression when I listened to this bird. That he was screaming for another, but there were no more.'

ON THE EVENING of Sunday 2 September 2018, a fire took hold at the Museu Nacional. A spark from a malfunctioning air conditioning unit started it off, and the museum – with its many wooden interiors – went up like a tinderbox. By 9 p.m. the entire building was alight. In the auditorium, which housed the five-tonne Bendegó meteorite – the oldest ever found in Brazil – the temperature reached 1,000 °C. Across Rio de Janeiro, the city's residents looked on in horror at the flames visible in the night sky, and at the scenes of violent destruction playing out on their television screens.

'A lobotomy of the Brazilian memory' is how the politician Marina Silva, now Brazil's Minister of the Environment and Climate Change, described the fire the following morning. Inside the roofless shell of the Museu Nacional, 95 per cent of the museum's collection had been turned to rubble – including 'Luzia', fragments of whose skull were found in the cooling ashes. 'It was a loss so significant to Brazil that, to give it an equivalent, it would be as if both the UK's Natural History Museum and Buckingham Palace had gone up in flames,' Constance Witham of the British Council wrote. In truth, however, this was no equivalent. In the fire, masks, weapons, tools, clothing, and other artefacts belonging to the indigenous peoples of Brazil, including a huge archive of recordings and other records of indigenous languages and music – sounds that had long ago vanished from the world – were destroyed. These objects, many of which dated from long before colonization, told a much broader, further-reaching story of Brazilian history than the building that had consumed them. The collection of flora and fauna specimens at the Museu Nacional was the largest in Latin America, and a lot was lost from here, too. The entomology collection was an early casualty in the fire. In room after room, cabinets stocked with roughly 5 million specimens, everything from butterflies to bedbugs, were incinerated. This included a globally unique collection of lacebugs and an estimated 800 type specimens of arachnid.

So it was that in November 2018, when a paper recommended that the IUCN list the Alagoas foliage-gleaner and cryptic treehunter as extinct, the museum in which the latter was discovered was in ruins. However, amidst the devastation, the bird specimens, which had been housed in a separate building, survived. Along with thousands of items that have been salvaged since and are being restored, they now form the foundations of a new archive of Brazil's history. In 2019, the IUCN declared the Alagoas foliage-gleaner and cryptic treehunter extinct.

A FEW MONTHS after having finished this chapter, I received a mysterious email. The sender was someone I'd hoped to interview, but who in the end couldn't speak. At the end of the email they signed off with this: 'I can tell you that *Cichlocolaptes mazarbarnetti* [the cryptic treehunter] simply doesn't exist.'

Later, I learned that the basis for this claim was the work of a postgraduate student at the Federal University of Rio de Janeiro, who had conducted genetic analysis of the specimens of the cryptic treehunter and Alagoas foliage-gleaner in 2023. According to my correspondent, this currently unpublished study proved that these two species were in fact one all along. It was a remarkable twist in the tale. I tracked down the student in question and sent them an email asking to speak about their work; then, I waited.

'I'm here to take away your doubts,' says Marcos Raposo at the top of our call in March 2024. Curator of Ornithology at the Museu Nacional, Raposo co-supervised the postgraduate's thesis that I'd been told had undone the species-hood of the cryptic treehunter. 'Actually, the results of this study were not conclusive,' he tells me. 'But this doesn't matter, in my opinion, because this is a species that should never have been described to begin with.' Raposo and Barnett had been friends, which I sense makes the topic of our conversation today a little awkward. 'Juan and I talked about all the evidence he had collected in support of the new species. I also had in-depth conversations with Dante [Buzzetti] about it, but in the end none of this convinced me of the species' validity,' he says. 'When a species is heading towards extinction, it isn't unusual to see morphological variation. In fact, due to many different processes, you almost expect to see it.' By this logic, it's possible that the larger birds amongst the Alagoas foliage-gleaner specimens were simply larger Alagoas foliage-gleaners, he argues. Another sticking point for Raposo are the vocalizations attributed to the cryptic treehunter. 'These could simply have been new vocalizations or a different part of the vocal repertoire of the Alagoas foliage-gleaner that hadn't been heard before.' Raposo also questions why a species from the *Cichlocolaptes* genus would adapt to look so much like a species from the *Philydor* genus. 'That would require a much more complex evolutionary process than the alternative,' he says. 'It's far more parsimonious to imagine that these two sets of specimens are the same species.'

Raposo readily admits that none of this proves that the

cryptic treehunter didn't exist – rather that, in his opinion, the current evidence doesn't warrant its status as a species. 'I can't tell you whether this species is valid. But we will test this in the next few years,' he says. The high-quality genetic material needed to answer this question sits in the freezer at the Museu Nacional, which at the time of writing is still being rebuilt. For now, the mystery remains unsolved. Complicating the matter, however, is that Barnett and Buzzetti did request this material to perform the tests themselves. In their 2014 paper describing the species, they write: 'our requests for permission to X-ray and take samples for molecular analysis from the specimens ... were denied in September 2004, November 2008, and June 2013.' It is unclear why their requests were denied.

Later, doubts still firmly intact, I came across the transcript of a 2016 discussion between members of the American Ornithological Society's South American Classification Group. In this document, scientists argue back and forth over the merits of Barnett and Buzzetti's description of the cryptic treehunter and vote as to whether they as a body should accept the new species. Some of the statements echo those made to me by Raposo – 'I imagine that a single population of *novaesi* [Alagoas foliage-gleaner] going extinct could exhibit abnormalities due to genetic problems, including gigantism,' says one member. Others argue in favour of the species' validity, and remind me of my conversation with Buzzetti; most are people who have spent time out in the field in Serra do Urubu and other forest fragments, searching for these birds. The meeting inevitably circles back to the same issue, however. 'The obvious remedy is there, begging to be applied,' says one member. 'I would join all of the other voices in asking that the MNRJ [the Museu Nacional] revisit its position regarding allowing tissue from across the type series of *P. novaesi* . . . to be applied to a genetic analysis.' In the vote, the 'no's' carried the day, and the group declined to recognize the species-hood of the cryptic treehunter. However, the International Ornithological Committee and the IUCN both accepted the species.

Taxonomy is an ever-changing discipline, with new discoveries and technology forcing near-constant reappraisal of designations. Genetic analysis may well undo the cryptic treehunter in the next few years. If this does happen, the knowledge that has been accrued by Barnett, Buzzetti, and others will still be important. Instead of marking out a new species, however, it will serve to deepen our understanding of the Alagoas foliage-gleaner. It will also tell us something about extinction: about what can happen to a species in the final moments of its existence – how a bird might lose the shape of itself right before the end, and scream.

# 9 Ghost of the Galápagos

Pinta Island Tortoise
*Chelonoidis abingdonii*

On 1 December 1971, József and Maria Vágvölgyi came face to face with a ghost. The couple had travelled to Pinta Island in Ecuador's Galápagos archipelago so that József, a malacologist, could study endemic snails. It was as they rummaged through the brush-like vegetation, scanning for the shells of potential specimens, that they saw it, just a little way off in front of them. Pigeon-toed feet and green-grey skin that bunched up in folds around four huge knees, a prune of a head at the end of a metre-long neck, snaking out from underneath a metre-wide shell. This was a Pinta Island tortoise (*Chelonoidis abingdonii*) – a species of Galápagos giant tortoise unique to the island, which had been extinct for almost seventy years.

To the Vágvölgyis, this giant tortoise was, although certainly a pleasant surprise, just another giant tortoise. The Galápagos archipelago was full of them, after all – fifteen species in total. Not being tortoise experts, the couple were oblivious to the significance of the moment they'd stumbled into. Their chance meeting with this enormous beast represented a kind of rising from the dead, an extinct species becoming extant once more. And so, unaware that this was the beginning of a decades-long story that would grip millions of people around the world, they snapped a quick photo and returned to their snails.

Disbelief greeted József Vágvölgyi when he later described his and Maria's tortoise encounter on Pinta Island. Over dinner, he told the sea turtle expert Peter Pritchard, who reacted first with scepticism, then with giddy excitement. That night, Pritchard didn't sleep; 'I practically lost my teeth,' was how he described his reaction. When the Charles Darwin Foundation, the organization responsible for conserving the archipelago's giant tortoises, caught wind of the sighting, they too were sceptical. Not so long ago they had searched the island for this species and found nothing but sun-bleached tortoise bones. Four months later, however, they instructed an expedition of park wardens, who were travelling to Pinta Island to cull goats, to keep an eye out for a tortoise. At the same time, Pritchard joined another expedition bound for Pinta Island, under the auspices of conducting sea turtle research. However, what he really wanted to see, if it did indeed exist, was this tortoise that had seemingly come back from the dead.

Manuel Cruz had been on Pinta Island for several weeks by the time Pritchard's expedition set sail. As a member of the goat eradication team tasked with studying what the invasive goats on the island were eating, the trainee zoologist had spent these days stalking the island with a rifle slung over his shoulder and a teenage park warden named Francisco in tow. Each day proceeded much as the last: find goat, shoot goat, open goat's stomach, note down the contents, rinse and repeat. But 20 March 1972 was very different. While exploring the western flank of the island's volcano, Cruz and Francisco found themselves pointing their rifles at what they thought was a goat nibbling a tree some sixty metres ahead of them. This could easily have been where this story ended, had it not been for a last-minute realization: the creature, happily munching away at the other end of their barrels, was not a goat at all, but a tortoise. Pritchard was still at sea when he heard the news over the captain's radio. It was true, the Pinta Island tortoise lived!

THERE AREN'T MANY stories like that of the Pinta Island tortoise. Over the past 122 years, 351 species have been rediscovered. Some of these species had been considered extinct for over a century, such as the Kandyan dwarf toad, a 3–4cm-long toad endemic to Sri Lanka, thought extinct for 137 years but rediscovered in 2009. Others were rediscovered after only a few decades; the tiny bird, the Cebu flowerpecker, endemic to Cebu Island in the Philippines, was declared extinct in the mid-twentieth century but rediscovered in 1992 in a small limestone forest.* Where the story of the Pinta Island tortoise differs from these, other than the size of the animal thought lost, is the number of people who cared that we'd found it again.

The announcement of the tortoise's discovery and transfer to a captive-breeding programme at the Charles Darwin Research Station on the Galápagos island of Santa Cruz became an international news story. TV networks ran it during prime time, with American comedian Johnny Carson even riffing on it during his blockbusting *The Tonight Show* in February 1976.

Somewhere in all this media attention, a nickname for the tortoise emerged: 'Lonesome George'. The real Lonesome George was George Gobel, a famous comedian who had earned the moniker thanks to his world-weary persona as a downtrodden husband. US media personalities noted similarities between Gobel's character and the newly discovered Pinta Island tortoise – chiefly that they shared a profound misfortune when it came to the opposite sex. The tortoise, a large male, was very likely to be the last of his species, unless a mate could be found, and searches continued to be fruitless. The nickname 'Lonesome George' stuck, and a star was born.

Tens of thousands of people were soon visiting the Galápagos Islands every year, hoping to catch a glimpse of the archipelago's newly famous resident. The last Pinta Island tortoise now spent his days at the Charles Darwin Research Station

---

* The short period the bird was considered extinct for, however, was still long enough for most of its habitat to be cleared, under the assumption there was no longer any need to conserve it; in conservation, this is called the 'Romeo error'.

in a secure enclosure complete with cacti, a shelter, a large pool, and a sea view. While he often hid, tourists lucky enough to visit on feeding days had the chance of witnessing the tortoise emerging from the undergrowth and making a beeline for the food brought by his keeper. His fans included some of the most famous humans: Leonardo DiCaprio, Jane Fonda, King Charles, and David Attenborough. Admirers who couldn't visit in person wrote to him instead; fan mail arrived at the station regularly. In time, he became a kind of unofficial mascot for the Galápagos Islands, with local shops stocking postcards, T-shirts, keyrings, and other trinkets bearing his image or silhouette.

But this tortoise, whom we had come to call George, was not a natural celebrity. He was known to be shy, especially when around other tortoises, and he lacked the natural curiosity that typifies all species of Galápagos giant tortoises. In his enclosure, he spent a lot of his time hunkered down in the cool shade of a bush. When female tortoises of a similar species, with which he should have been capable of producing hybrid offspring, were released in his enclosure, he snuck around as if hiding and tried his best to avoid contact. The last Pinta Island tortoise was, by all accounts, an unusual individual; but this is unsurprising, when you consider what his life had been like.

This tortoise had lived out his days wandering Pinta Island – a place once described by Herman Melville as 'a distant dusky ridge . . . so solitary, remote, and blank, it looks like No-Man's Land'. It's possible that the tortoise – born, it is estimated, around 1910 – had spent the last few decades prior to his capture entirely alone, without even seeing another tortoise. Giant tortoises, though independent, come together to share waterholes; on other islands in the Galápagos where populations are healthy, they can appear like boulders, peeping out of muddy ponds where they bathe together to regulate their body temperature in the heat. Male tortoises fight over females, butting heads and shells, and emit bellowing calls during mating that can be heard for miles around. Then, the females travel down to the shorelines of their islands and bury their fertilized eggs in the sand. Yet this Pinta Island tortoise likely bathed in mud alone, a solitary boulder; he probably spent his adult life never butting heads with another tortoise, and the island would have been silent of the calls of mating males, the beaches empty of eggs. The few tortoises he had grown up with would have disappeared one by one when he was still young, taken by humans to eat, to collect, and to study. This tortoise's life, as his name now reflected, had been a lonely one.

Then, all of a sudden, two strange bipedal creatures appeared in front of him. Later, more came and lashed him to branches that they carried on their shoulders, like pall-bearers at a funeral, as they stumbled over jagged lava fields to their camp. There, the tortoise was tied by his ankle to a cactus – what was usually a tasty snack, now a post for him to be shackled to – then later flipped onto his back and laid on the floor of a dinghy, and rowed out to a larger boat, which carried him away from his island home for ever.

The life that awaited him in captivity was, in many ways, arguably preferable. Having lived exclusively on the flesh of prickly pear cactuses, he was now fed exotic foods that were beyond his wildest dreams (bananas became a favourite). The health-care package was also comprehensive and state of the art, and he would have the opportunity to meet female tortoises. All of this was, in a sense, the beginning of the Lonesome George story: the beginning of our attempt to save his species. However, something also ended here. This tortoise's life in the wild was over. When we pulled this final specimen from his home, the species became extinct in the wild. The images of the tortoise tied up and being carried to the boat are bittersweet. There is also a kind of poetic symmetry here; the way this tortoise left the life he knew – lying helplessly on his back, rocked by the motion of the waves as he was ferried over the ocean – was strikingly reminiscent of how the story of his entire species had first begun.

IF YOU THREW a tortoise into the sea, would it sink or float? This might sound like the start of a parable, but this question was once at the heart of a scientific debate. And, in 1923, an American naturalist named William Beebe travelled to the Galápagos Islands in search of an answer.

Ever since its discovery in 1535, the Galápagos archipelago – a 280km-long chain of islands, each comprising a volcano and lava flows that have cooled on the surface of the sea – has been a source of wonder and mystery. Early sailors referred to it as 'Islas Encantadas' (Enchanted Islands) and for good reason: the Galápagos Islands exist at the confluence of great natural forces. Positioned directly on the equator, 906km west of the Ecuadorian coast, the sun rises at around 6 a.m. and sets at 6 p.m., year-round. Cold water that flows on the Humboldt Current from Antarctica, warm water carried on the South Equatorial, and fresh water from the mouth of the Rio Guayas all flow into the sea around the Galápagos Islands. This divergent watersmeet creates unpredictable currents and winds that surge chaotically in all directions, but it also delivers the vital waterborne nutrients that underpin an ecosystem packed with bizarre and beguiling creatures.

Marine iguanas make their living not on the land, like all other iguanas, but underwater, grazing algae from rocks on the seabed. Meanwhile, Galápagos fur seals have traded the sea for dry land, spending 70 per cent of their time out of the water. The wings of flightless cormorants have grown stumpy, and their bellies fat, as they've given up on flying altogether. Red-lipped batfish strut about coral reefs on their fins, which are better adapted to this purpose than they are for swimming, with bright red 'lips' that always appear freshly painted. The Galápagos archipelago is a land of living, breathing contradictions.

To many in Beebe's time and earlier, however, the strangest of all the Galápagos Islands' residents were the giant tortoises. These huge, lumbering reptiles – which sometimes weighed in excess of 400kg, could grow up to 1.8m long, and could live for 180 years – were a source of fascination to all who visited. It was after these goliaths that the archipelago had first been named, 'galapago' being an old Spanish word for 'tortoise'. Charles Darwin was especially fascinated by them, writing about them in 1839 in *The Voyage of the Beagle* as being so huge that they seemed 'to my fancy like some antediluvian animals'. What was strange about the tortoises to Beebe and his contemporaries, however, was that they were in the Galápagos Islands at all.

It's easy to imagine how a bird or a seal might find its way to a remote archipelago. Plants have numerous routes they can exploit, even the most challenged in this arena like the St Helena olive in Chapter Two. Storms and birds can carry snails and other small creatures. But a large terrestrial reptile, like a giant tortoise, is less easy to place in this image of vast ocean migration. One early theory was that they had been ferried by humans from the Seychelles and Mascarene Islands in the Indian Ocean, where the only other extant species of giant tortoises live. Another was that they had waddled from continental South America along a hypothetical land bridge, which would have been the entry point for most Galápagos wildlife – lizards, snakes, and tortoises with the seeds of cacti, flowers, and trees as their luggage – before it disappeared into the sea.

Both theories were eventually shown to be flawed. In 1837, Albert Günther, Director of the Department of Zoology at the British Museum, confirmed striking physiological differences between Galápagos giant tortoises and those from the Seychelles and Mascarene Islands, ruling out the notion that the two were closely related. Then, experiments in 1891 using sound waves to measure the depth of the Pacific Ocean revealed that there had been no land bridge between the Galápagos Islands and the South American mainland.

The only possibility that remained was nevertheless difficult to believe. The ancestors of all Galápagos giant tortoises, so the theory went, had floated to the archipelago from the South or Central American mainland. In the most conservative scenario, this would mean an open-sea crossing of 1,000km; but, depending on where this journey had begun, it could have been much longer. A floating tortoise would have had to survive for weeks, keeping its head above the waves, as the Humboldt Current swept it along. Then it had to beach on an island where it could quickly and easily find food and water. Its descendants would then have undertaken a series of smaller sea crossings, radiating out among the other islands of the archipelago and evolving into the fifteen different species of Galápagos giant tortoises. To buy into this theory, you needed to believe that tortoises could float well enough to spend long stretches of time at sea, without food or fresh water – and, in 1923, William Beebe wanted proof.

With a splash, the giant tortoise disappeared beneath the waves. Following a gruelling four-day search, Beebe and his crew had captured this tortoise on the Galápagos island of Pinzón. They had performed a series of experiments on land first, to test how well the tortoise traversed various types of terrain. Then, they had loaded it into a boat, rowed out to sea, and dunked it overboard. The tortoise's head was first to emerge, rising straight up out of the water on the end of its long neck, like a periscope. Next, its shell resurfaced, then it bobbed about in the chop like a buoy.

From the boat, Beebe watched as the tortoise floated with ease and even swam against a one-knot current. 'I could see the throat vibrate in breathing, without any detectable lowering or elevation of the body,' he wrote in his 1924 book *Galápagos: World's End*. 'So for a time at least these creatures have perfect control over themselves in the water.' The tortoise swam towards the crew's yacht and then the rowing boat, trying and failing to haul its gigantic body up onto either, before Beebe and his crew retrieved it. They then repeated the experiment numerous times, tipping the tortoise overboard and watching as it struggled back towards them. It floated. More, it swam. However, one week later it died, likely due to congestion in its lungs caused by inhaling seawater. Beebe believed that the death of his test subject negated the possibility that tortoises had floated to the Galápagos. Today, however, we now know that he was wrong to dismiss his theory so quickly.

In 1999, Adalgisa Caccone, a geneticist at Yale University, led a study that analysed the DNA of Galápagos giant tortoises and other tortoises from around the world, on a search for the former's closest living relative. Blood was taken from twenty-four live tortoises so that their DNA could be compared to the Pinta Island tortoises. Lonesome George's blood contributed to this study, as did the DNA taken from the skin of three dead specimens collected in 1906. Surprisingly, the closest living relative turned out to be not one of the giants of the Indian Ocean, but a tortoise species around a hundred times lighter. Chaco tortoises (*Chelonoidis chilensis*) measure at most 43cm in length, but they have in common with their giant cousins in the Galápagos a ravenous appetite for cacti. They share more than this, however: they share an ancestor. The geneticists worked out that this ancestor species split into two branches while still on mainland South America (more recent genetic analyses have suggested that this occured either 11.75 million years ago or as far back as 25 million years ago). The Chaco tortoise remained small, setting up home in dry areas of Bolivia, Argentina, and Paraguay. Meanwhile, the ancestor of all giant Galápagos tortoises grew big, and set its sights on the sea, its shell a ready-made raft.* In this part of the world, floods were common along the coast. Once washed out to sea, the Humboldt Current was ready to carry an adrift giant tortoise to the Galápagos Islands.

The giant tortoise specialist Linda Cayot observed giant tortoises' propensity to move uphill during heavy rain, which, she asserts, is probably a precaution against being caught up in flash floods. It's reasonable, then, to assume that the mainland predecessors of giant Galápagos tortoises would have fled uphill whenever rain came. However, female tortoises travel to lower ground to lay their eggs, often all the way to the shore, no matter the weather. Knowing this, it's not difficult to picture a female tortoise being caught out by a wave while digging a nest for her eggs on a beach somewhere along the western coast of South America, and then being washed out to sea. The El Niño Southern Oscillation is a climate pattern that occurs sporadically, at two- to seven-year intervals, and causes surface waters in the eastern tropical Pacific Ocean to warm, leading to more intense weather and currents. During El Niño years, the waves lapping up against the western coast of South America would have been bigger, rain would have been heavier, floods deeper, visibility poor, and winds extremely strong, even against the heft of a giant tortoise. In 1982 during El Niño, Cayot witnessed a male tortoise on Santa Cruz Island deliberately floating himself down a river: 'he floated, bounced against rocks, walked a bit, then floated again,' she writes. 'I followed, crawling with the current . . . giving me a new sense of empathy for my study subjects.'

It is now suspected that female giant tortoises were carried out to sea, as they performed their dangerous duties as expectant mothers, fairly regularly during El Niño years. Most

---

* Giant tortoise fossils have been found on the mainland, leading Adalgisa and her team to the theory that Galápagos giant tortoises developed their giantism before drifting to their island homes.

of them would have perished in the open ocean or soon after landing, like Beebe's specimen, but only one had to make landfall and survive for the story of Galápagos giant tortoises to begin. Some time between 2 and 3 million years ago, a single female tortoise is believed to have landed somewhere in the Galápagos carrying a clutch of fertilized eggs, or a packet of sperm stored safely in her oviducts. And, like that, the Galápagos' strangest resident had arrived.

As more islands emerged from the volcanic hotspot under the ocean, tortoises descended from the first foundling made journeys of their own. Swept into the sea and ferried on the currents, they soon found their way to every island in the archipelago, establishing new populations, which eventually became the fifteen Galápagos giant tortoise species.

Around 250,000 years ago, one of those seafaring tortoises made its way to Pinta Island. This individual made the longest journey of all, from Española or San Cristobel in the south-eastern corner of the archipelago all the way to the northernmost island of Pinta, roughly the distance between London and Paris. Caccone has hypothesized that this foundling tortoise rode a strong current from the northern coast of San Cristobel that leads directly to Pinta Island. Carried swiftly along, this tortoise would have eventually arrived on the usually barren-looking island. There, it was greeted by a beautiful sight. The palo santo tree (*Bursera graveolens*) grows just inland and, with its bare branches and silvery bark, appears devoid of life. The tree has no leaves or flowers for most of its life, except during rare moments of rainfall. When the tortoise arrived, there was likely rain due to the effects of El Niño, and the small white flowers of this tree – nicknamed the holy tree and seen as a metaphor for resurrection – would have been in bloom. Under these flowers, the tortoise found a cactus to quench its thirst, and then proceeded to bury its eggs. A few months later, the first of the Pinta dynasty, that would evolve to become the species now known as the Pinta Island tortoise, was born.

THERE ARE CERTAIN species around the world that ecologists refer to as 'ecological engineers', thanks to the significant roles they play in their ecosystems. Elephants, which uproot and push over trees, creating habitats for smaller animals, and whose dung creates micro-habitats for seed germination in the soil, amongst much else, are a famous example. Galápagos giant tortoises are another. Tortoises have shaped the look and ecology of the Galápagos Islands more than any other animal, earning them the nickname 'gardeners of the Galápagos'. And just like their cousins on other islands, the tortoises that emerged on Pinta Island remade the island in surprising and remarkable ways.

Not long after tortoises arrived on Pinta, paths appeared, criss-crossing the island. These were tortoise highways, connecting tortoise-points-of-interest to minimize the energy spent getting between them. Some led down to the coast, where the females dug nests in soft soil and sand to lay their eggs. Others led inland, towards places where the tastiest fruits grew from trees. Some gave out to clearings, in which, when conditions were wet enough, a dozen or so tortoises wallowed in muddy water. These mud wallows, dug by tortoises to catch rainwater, served a dual purpose. In hot weather, when the heat of the sun could kill a careless tortoise, it could slip into the water and cool off. Then as the air got colder during the night, the muddy water would reduce the amount of heat the tortoise lost.

More impressive than these landscaping projects were the sculptures the tortoises created from stone. On exposed rocks, moisture settled and pooled in tiny puddles before dawn. Made thirsty by Pinta Island's arid climate, tortoises would seek them out, in the process dragging their clumsy feet and plastrons (the almost flat undersides of tortoises' shells) over the rock. Over thousands of years of this erosion, smooth craters formed on the surface of these rocks, creating rows and rows of uniform holes that collected more and more water as they grew.

Extraordinary things soon happened to the plants of Pinta Island, too: they started to move. Plants that only grew in specific areas started cropping up all over the island, as though they were using the trails blazed by the tortoises. And, in a way, they were. As tortoises moved about, grazing and trampling new paths where they pleased, seeds got caught in the folds of their leathery skin or under their shells and were carried off, to work themselves free somewhere else. In this way, tortoises cleared dense vegetation from an area and a meadow would spring up in its place, sprouting from the seeds the tortoises had inadvertently carried and then sown. And it wasn't just on the bodies of tortoises that seeds would be transported, but inside them too. For seeds that were ingested by Pinta Island tortoises, the slow crawl through the stomach and intestines allowed them to travel further across the island. This detour via tortoise guts actually enhanced the germination of some seeds: it is thought that prickly pear cactus, the favourite foodstuff of the tortoises across the archipelago, is specifically adapted to being ingested and dispersed like this.

This species of cactus and the Pinta Island tortoise interacted in other ways, too. When the first tortoise arrived, the island's prickly pear cacti looked very different. Shorter and thinner, their pads grew much closer to the ground. But then they started to change. Their trunks thickened and grew significantly, and their pads, in which they stored valuable water, climbed higher and higher off the ground, giving them more of a tree-like shape. These changes were an evolutionary response; taller, thicker-trunked cacti, with pads that grew ever higher, became much better at surviving than slender, stunted ones. The reason for this was simple: the newly arrived giant tortoises.

As the name suggests, the prickly pear cactus is not without its defences. Sharp, stiff spines protrude from its flesh in all directions, but in the mouth of a Pinta Island tortoise, these were just a garnish. The tortoises snapped at the pads of the cacti, crushing them using the tough bony plates and sharp edges of their beaks. Unable to fend off the ravenous tortoises with its usual weapon of choice, the only recourse left for the prickly pear cactus was to raise its pads off the ground and toughen its trunk. And this is where the relationship between these two species gets really interesting. Not only were the Pinta Island tortoises forcing the cacti to change shape but, as the prickly pear cactus obliged, it placed a selection pressure back onto the tortoise – one that would radically alter the way this tortoise looked.

On some Galápagos Islands, like Pinta, rainfall is scarce and the climate is arid. In this environment, the moist pads of the prickly pear cactus were essential to the giant tortoises' survival. To make sure it could continue to reach this essential resource, the Pinta Island tortoise underwent a redesign. Its neck lengthened to keep up with the cactus pads that were moving ever further out of reach and, to make the most of this extending neck, its shell bent upwards at the front, forming a saddle-like shape, allowing the tortoise to angle its neck up higher. The changes that these two species prompted in each other took place over tens of thousands of years – a delicately balanced evolutionary arms race that hugely benefited both combatants.

Other changes to the tortoise's design were made by the island itself. For example, the hot, dry climate of Pinta Island made it advantageous to expose more skin, so that thermoregulation was easier; so the Pinta Island tortoise's legs grew longer and its skin area increased. Similar stories of co-evolution have played out around the archipelago, with tortoise, cactus, climate, and landscape all in deep conversation with each other.* The islands of the Galápagos also play intimate roles in individual tortoises' lives. The position of trees along the coast, the thickness or absence of undergrowth, cloud cover, and climate all contribute to determining the sex of hatchling Galápagos giant tortoises. If a clutch of eggs is buried in a nest that is warmer, the hatchlings that emerge will all be female; if the nest is cooler, the hatchlings will be male. On Pinta Island, the soil played a role too. Pinta Island soil is unusually rich in a volcanic mineral called plagioclase, something which geochemist William M. White explains leads to a cooler soil than found elsewhere in the Galápagos archipelago. Indeed, twentieth-century records have shown that Pinta Island seems to have had an unusually large male tortoise population.

The Pinta Island tortoise was well adapted to life on its island home. While temperatures created an abundance of males, the population still seems to have thrived for the best part of 250,000 years. The only danger for adult Pinta Island tortoises was their own clumsiness; with their thick-skinned, long, lumbering legs, and their heavy shells, they often lost their footing. Thanks to the shape of their saddleback shells and their low centre of gravity, they could right themselves without too much bother, most of the time. But slip in the wrong place – say, near one of the many crevices in the island's rocky ground – and a tortoise could plummet to its death. Trees were a much less

---

* Right across the Galápagos Islands, we can see how individual tortoise species have been moulded by their environments in similar ways. Those that evolved on other dry islands share characteristics with the Pinta Island tortoise, such as a saddleback shell and longer limbs; meanwhile, giant tortoise species on cooler, more humid islands, where retaining heat is the priority, have dome-shaped shells and short limbs – for example, the Eastern Santa Cruz giant tortoise appears virtually neckless.

dramatic, although equally dangerous, hazard. The clumsy hind legs of a Pinta Island tortoise could easily become ensnared in the thick branches of a tree stump – the bear trap of the giant tortoise world – leaving it to starve or cook to death in the open sun, unable to seek cover or cacti.

When the tortoises were still small, Galápagos hawks were a threat. While they eventually grew into big boulders whose mouths even cactus spikes couldn't penetrate, as hatchlings they were fragile. A baby giant tortoise weighs no more than 80g, the weight of a tangerine, and has a length of just 6cm. They undergo a thousand-fold increase in body weight before becoming fully grown adults at around five years old. The tiny Pinta Island hatchlings' only weapon was the small, sharp protuberance that grew between their upper jaw and nares – the nostril-like holes above their mouths – called an 'egg tooth', which they used to break out of their eggs. But these egg teeth fell off soon after birth. After feeding on their yolk sacs underground for up to a month, the hatchlings would emerge defenceless, and scurry desperately towards nearby undergrowth to hide. Above them, Galápagos hawks would be circling, waiting to swoop down on any lagging behind. However, once a Pinta Island tortoise grew out of this treacherous early phase of its life, it would no longer be troubled by hawks or any other predators.

Cloistered away in the safety of their remote archipelago, Galápagos giant tortoises were blissfully unaware of the changes underway elsewhere. The world outside was becoming increasingly dangerous for giant tortoises. Soon the Galápagos Islands would be one of their last safe havens, and the tortoises themselves one of the world's rarest treasures.

GIANT TORTOISES ONCE inhabited all the continents with the exception of Antarctica and Australia. Prehistoric giant tortoises could grow to be even larger than the giant tortoises we have today. *Megalochelys atlas*, which once roamed huge swathes of Asia, had an estimated length of 2.5–2.7m, a height of 1.8m, and a weight more than double our modern giants, averaging at around a tonne. Between 50,000 and 10,000 years ago, these tortoises began to disappear, along with other megafauna. By far the most compelling and generally accepted explanation for this is, quite simply, that we hunted them to extinction. The only giant tortoises that survived this time were those which had managed to hide themselves away somewhere far out of our reach. The Galápagos giant tortoises were one such group. We would eventually catch up with them, however.

It was completely by accident that the Bishop of Panama, Fray Tomás de Berlanga, discovered the Galápagos Islands. On 10 March 1535, Berlanga set sail from Panama, bound for Peru, but a strong current carried him off course. After six days, Berlanga and his crew were running dangerously low on water when a black shape appeared on the horizon: a volcano, emerging from the waves. The first island of the archipelago came into view. After landing and searching for a water source, to no avail, Berlanga took a leaf out of the giant tortoise survival manual, squeezing water from the pads of cacti. He and his crew saw tortoises during their stopover, describing them as being in a sense like islands themselves – 'tortoises so great, that they [could carry] a man on top' – and so the archipelago was named after them.

With little fresh water, treacherous seas, and often arid conditions, the Galápagos archipelago wasn't seen as particularly hot property. For centuries after its discovery, it was most useful to British pirates, who used the islands as a hideout. In 1798, however, the archipelago's stock skyrocketed when the British Royal Navy officer, explorer, and fur trader James Colnett announced that the Galápagos Islands could serve as a base of operations for sailors hunting whales in the Pacific Ocean. At the time, whale oil was the fuel on which Western countries depended to light homes, to lubricate machinery, to make soap, varnish, and paint. Spermaceti, the curious oil found in the heads of sperm whales, was used to make candles and to light street lamps, and a waxy substance formed in their bowels, called ambergris, was the de facto fixative in perfume manufacture (a utility it retains to this day). 'Whalebone' – technically not bone, but baleen (the teeth-like bristles that act as a food filter in a whale's mouth) – was the plastic of the nineteenth century, used as the bones of corsets, as typewriter springs, collar stays, buggy whips, the ribs of umbrellas, chess pieces, piano keys, and walking-stick handles. Whale meat was widely consumed in England and France, while whale tongues were a delicacy reserved for royalty and the clergy. All of this meant that when Colnett published the first accurate navigational charts of the Galápagos, whalers flocked to the archipelago. Quickly, the waters around the islands filled with European whaling boats loaded with spears. Steam rose into the sky from big vats that melted down whale blubber to be transported back to England. And the sea, for miles, turned red with blood. A huge industry, which fuelled nineteenth-century Western society, took over the archipelago. And the fuel that powered these many whalers was the flesh of giant tortoises.

Large, docile, and by all accounts delicious, the giant tortoises of the Galápagos Islands were an invaluable resource for whalers and other visitors. The tortoises could survive for months without food and water, meaning they were cheap livestock to keep; and so, the same quality of physical endurance that had brought their ancestor to the Galápagos archipelago now led them on countless further voyages. On many ships, they were left to roam the deck until their number was up; on others, as they were so heavy and sturdy, they served as ballasts. They were also taken en masse to California, during the Gold Rush, to cheaply feed labourers. By 1870, an estimated

200,000 tortoises – roughly half the population prior to human contact – had been removed from the archipelago.

Darwin also developed a taste for Galápagos giant tortoises when he visited in 1835. He saw the tortoises of the archipelago as both a perfect example of natural selection at work, for how each 'race' adapted to conditions on its separate island, and a great snack. Describing a part of his trip in which he lived entirely off tortoise meat, he writes: 'the breastplate roasted . . . with the flesh attached to it, is very good; and the young tortoises make excellent soup'.

As attested in *The Voyage of the Beagle*, Darwin frequently amused himself by leaping onto the backs of unsuspecting giant tortoises, as if they were donkeys, taking advantage of some species' saddle-shaped shells. 'Upon giving a few raps on the hinder part of the shell, they would rise up and walk away; – but I found it difficult to keep my balance,' Darwin writes of his rodeo days. On 20 October 1835, he left the

DARWIN TESTING THE SPEED OF AN ELEPHANT TORTOISE (GALAPAGOS ISLANDS).

*Sketch depicting Darwin measuring the walking speed of a Galápagos giant tortoise by Meredith Nugent.*

Galápagos Islands after spending just over a month there, feasting on their giant residents. The *Beagle* was stocked with forty-eight giant tortoises plus two small tortoises – sustenance for the crew as they sailed to Polynesia. With this information, we are forced to acknowledge the absurd irony of the father of evolution – the man who would later proclaim the biological uniqueness of the Galápagos Islands to the world – partaking in the feeding frenzy that eradicated half of its giant tortoises. As the *Beagle* sailed north, it passed Pinta Island. Mercifully for the tortoises here, the ship didn't stop. Although many before it had done.

It's estimated that tens of thousands of tortoises once lived on Pinta Island. However, after the whalers arrived, it became a kind of island-pantry for passing ships, whose logs paint a harrowing picture of the degree to which sailors exploited Pinta Island tortoises, eating their flesh to stave off hunger, and drinking from their bladders to quench their thirst. This all started with Colnett, who was the first to make a written account of the species. Colnett initially sent his hungry crew to fish, but on Pinta Island they found a much more abundant food source – the island's tortoises and, almost certainly among them, the ancestors of Lonesome George.

What had been initiated by hungry whalers was then continued by collectors. From the late nineteenth to the early twentieth century, museums, scientific establishments, and private individuals all clamoured for specimens of Galápagos giant tortoises, and came to Pinta, seeing it as the perfect location to shop around for a particularly big or impressive specimen. Numerous stories from the time describe tortoises being injured on the journey down the volcano, slung to poles and lowered off cliffs by collectors, or dying on board ships, their remains cut up, dried, or pickled.

The Pinta Island tortoise was, by the turn of the twentieth century, in its twilight years. This new scarceness only increased its value to collectors, however, and a series of expeditions aimed at collecting the few that remained, in the name of science, thinned the population further. Perhaps most egregiously, in 1901 the only female Pinta Island tortoise found on the island was killed and skinned after a shipping company refused to guarantee that if captured it would reach England alive – where it was intended to join the collection of living giant tortoises assembled at the country estate of Walter Rothschild, the banker, politician, and naturalist. A giant tortoise had become the ultimate status symbol for the European upper classes over the last half a century. To have one roaming around your manor house gardens, or in your living room, stuffed and mounted, was considered the peak of elegance.

In 1906, what were thought to be the last tortoises on Pinta Island were collected by the California Academy of Sciences to be brought back as specimens (i.e. not alive). One, 'a very fat male', was drinking from a pool and turned to look at the collectors as they approached. They killed it, skinned it, and carried it back to the boat. This was repeated with two other males. After this, the Pinta Island tortoise was believed to be extinct. That is, until George turned up sixty-six years later, and presented us with a second chance to ensure that his species would live on.

'¡Muerte al Solitario Jorge!' came the screams of the protesters – 'Death to Lonesome George!' On 3 January 1995, just over two decades after Lonesome George's capture and removal to the Charles Darwin Research Station (CDRS), a group of around thirty fishermen, wearing masks and wielding clubs and machetes, blockaded the facility's entrance. The fishermen, who referred to themselves as 'pepineros' (meaning 'cucumbers' in Spanish), had gathered to protest the forced closure of the local sea cucumber fishing industry, which had been instituted to protect local wildlife. In response to the ban, the pepineros threatened to storm the CDRS, burn down facilities, and butcher the animals kept there. Lonesome George, as the world-famous figurehead of wildlife conservation in the Galápagos Islands, quickly became their prime target.

Unfortunately for sea cucumbers, they contain something extremely valuable to humans. Fucosylated glycosaminoglycan, a chemical that occurs in their skin in high concentrations, is something of a miracle drug. Since the late twentieth century, it has been used as a treatment for blood clots and some cancers, and is currently in review as an anticoagulant. It has been used in Traditional Chinese Medicine since the Ming Dynasty and as a folk medicine in the Middle East, treating cancer, arthritis, fatigue, impotence, and other ailments. In many Asian countries, sea cucumbers are also a delicacy consumed for their health benefits.

By the early 1990s, the growing demand for sea cucumbers made them an increasingly valuable commodity. Fishermen in the Galápagos Islands, where the average income was less than $1,600 per year, found that they could make hundreds of dollars per day catching sea cucumbers rather than fish. Unsurprisingly, international fishermen wanted a piece of this pie too, and the archipelago became crowded with fishing vessels once more. To much outcry, in 1992 the area around the archipelago was protected as a marine reserve and commercial fishing was banned to protect sea cucumber numbers. However, the authorities charged with policing it were under-resourced, and fishermen ignored the new designation.

It wasn't just sea cucumbers at risk. Gangs of poachers processed their catch on the islands' beaches, boiling the sea cucumbers in large vats and then draping them over rocks, or on wire racks, to dry in the sun. They chopped down local vegetation to fuel the fires underneath these vats and left behind human faeces and rubbish. Most shocking of all, however, was the discovery that they were killing giant tortoises – in 1994 alone, eighty tortoises were slaughtered just on Isabela Island. It was as if a portal had opened to several centuries earlier, and bands of hungry whalers had leapt through. Only this time, the killing of tortoises was more likely a protest rather than a means to stave off hunger at sea; bodies were hacked apart with machetes while the invasive goats that authorities encouraged people to hunt were ignored.

Soon the poachers' behaviour led to disaster. On 12 April 1994, a huge fire broke out on Isabela Island, started by hunters who had forgotten to extinguish their cooking fires. Black smoke rose up in a pillar, visible for miles around for two whole months; between 3,500 and 4,500 hectares of vegetation was burned. Although Isabela's tortoises were not at risk from the fire directly, the danger posed by poachers could not be ignored. Donkeys were immediately enlisted on a tortoise-rescue mission, ferrying them to safety on their backs. Helicopters initially deployed to fight the fire also airlifted tortoises to a nearby breeding centre. These events, which were beamed to news stations around the world, prompted a more aggressive stance from the authorities. In December, after it was discovered that the three-month quota of 550,000 sea cucumbers had been exceeded nearly twenty times over, the archipelago's designated sea cucumber fishery was finally closed. One month later, the blockade began.

Ominous chants calling for Lonesome George's death filled the air. The voices belonged to men who had come to depend on making hundreds of dollars per day from sea cucumber catches. It was not fair, so the narrative went, that the conservation of animals was prioritized over the livelihoods of thousands and thousands of people. What did a sea cucumber do in its ocean home, anyway, that was more valuable than what it provided for humans – the fishermen and the many people for whom it was a medicine? And, more pressingly, what value did a big boulder of a tortoise have that meant that its black and barren island home had to be preserved?

Three years earlier, a similar protest by the pepineros had prompted Linda Cayot, then Head of Protection at the CDRS, to smuggle George out of his enclosure. She was so concerned, she substituted a body double – a large female from another species – while the protesters surrounded the centre. This time George stayed put inside his enclosure, obliviously munching through his banana quota for the day. The team around him, however, braced themselves for the worst.

Lonesome George had lived at the CDRS for twenty-three years by this point. Efforts to get George to procreate were made all the more nerve-wracking by his various accidents and ailments. He had become troublingly fat from his luxuriously sedentary lifestyle, had multiple bouts of constipation after eating entire cacti all in one go, and he often fell and injured himself. He received around-the-clock care but there was concern that one day, before he had produced offspring, his clumsiness might be fatal. George, however, continued to show no interest in female tortoises from other islands who were introduced into his enclosure. Other ideas to get him mating included rehousing him with a group of mixed-sex tortoises, and showing him videos of giant tortoises mating. Neither idea panned out. When placed with others, George simply hid and kept to himself, and the tortoise pornography was decided to be too off-putting for the constant stream of visitors that came to see him.

In 1993, Sveva Grigioni, a Swiss geologist who volunteered at the research station, was assigned the task of getting Lonesome George to produce semen so that artificial insemination could be attempted on two female tortoises. This was not an easy task. Before entering his enclosure, Grigioni washed off her own scent with an unscented soap, then coated her hands in genital secretions collected from female giant tortoises kept at the centre. Every day, she spent an hour or two with George, slowly moving closer and closer to him each visit, until he grew so comfortable with her he would happily let her stroke his shell. Grigioni then gently rubbed his back end, legs, and tail, and George would lift his body off the floor to expose his penis, which was usually tucked inside his body. These efforts were unsuccessful in producing semen, but they did awaken in George some interest in the female tortoises. However, this waned once Grigioni left to complete her doctorate back in Europe and, before long, George returned to a life of celibacy.

And so, as the pepineros pressed in around the Charles Darwin Research Station on 3 January 1995, the story of the Pinta Island tortoise seemed, one way or another, to be reaching its conclusion. After a few days of protest, the military intervened; soldiers arrived on Santa Cruz, and the pepineros sailed away. On 1 September, however, the Ecuadorian president had to personally step in and veto a law devised by Eduardo Veliz, congressman for the Galápagos, which weighed heavily in favour of the pepineros by permitting the use of natural resources within the national park. Three days later, protests erupted again, this time at the airports on the islands of Baltra and San Cristobal, as well as at the CDRS and the compound of the Galápagos National Park Service (GNPS). Veliz, now the pepineros' de facto leader, wrote a letter to the Ecuadorian president: 'We are prepared to take extreme steps such as taking tourists hostage and burning any parts of the National Park,' it read. On public radio, he instructed the pepineros to sack and burn the CDRS and GNPS buildings. Soon after, the National Park Service was stormed by pepineros armed with machetes and bludgeons, who searched the compound for the institution's director Arturo Izurieta – reportedly to kidnap and kill him. Izurieta survived by barricading himself and his young children inside his house, before escaping. Throughout the protest at the CDRS, with no one allowed to enter or leave the facility, students and visiting scientists took turns nervously guarding Lonesome George. A plan was hatched by the pepineros to enter the CDRS and take animals and staff hostage, and Molotov cocktails were prepared.

The unrest only subsided after tragedy struck. On 13 September, a research station associate named Arnaldo Tupiza had to use a motorbike to drive up to his research site rather than his usual jeep, which the protesters had impounded. As he was driving back, Tupiza crashed head-on into an approaching truck and was killed. This sobering moment was a turning point. A few days later, the pepineros let staff leave the buildings to attend Tupiza's funeral. Hundreds of people accompanied his casket through the streets. Meanwhile the authorities brokered a deal with the pepineros and by the time the mourners returned from the funeral, the protesters had left.

Calm returned to the Galápagos Islands. For Lonesome George, life continued much as it always had since he'd arrived at the CDRS.

As the new millennium arrived, an ambitious project began on Pinta Island. Three goats introduced to the island by fishermen in 1959 had, by the time of George's capture, multiplied to over 40,000. As herbivores, goats occupy a very similar niche to giant tortoises – only they do so much more effectively. Nimble and sure-footed, a goat can easily out-graze a Pinta Island tortoise and, being mammals, they don't have to plan their days around the sun. Their horns and muscular strength mean they can also frighten tortoises off the best grazing spots. Culling expeditions, like the one on which George had been captured, ran regularly, but did little to dent the overall population. And, by 2000 their numbers were over 100,000. For George or any of his potential offspring to stand the best chance of returning and thriving on Pinta Island, the 100,000 feral goats had to be reduced to zero.

There was another reason to eradicate them. Somewhere on Pinta Island, it was hoped there would be a female tortoise for George to mate with, or at least another male to take some of the pressure off George. With the goats gone, any survivors would be much easier to locate.

A new hunting technique made this goal feasible. Rather than set out with rifles and search for the goats, the rangers turned the goats' own herding instinct against them. After capturing a single goat and fitting it with a GPS collar, all they had to do was follow it home. Using this method, entire herds were killed all at once, and those lone goats who had inadvertently betrayed their kin became aptly known as 'Judas goats'. The tactic was so successful that by 2003 there were no goats left on Pinta Island. This success was bittersweet, however. The home of the Pinta Island tortoise had been tidied, but no Pinta Island tortoises were found.

In 2010, thirty-nine sterilized hybrid tortoises were released on Pinta Island – the plan being to eventually replace these with the offspring of Lonesome George. In the meantime, the tortoises were like house sitters, maintaining the island's ecosystem, holding the space for the Pinta Island tortoise to return.

BACK AT THE CDRS, at around the same time each morning, Lonesome George heard the gate to his enclosure creak open, and poking his head out from the bushes, he saw the same friendly, familiar face. Fausto Llerena had been a member of the goat-culling team that had found George and had been George's keeper for as long as he'd been in captivity. The two had a special bond. Llerena was, in a sense, a tortoise-whisperer – or more specifically, a George-whisperer. Each time George suffered indigestion, it was Llerena who noticed, asking him, 'What's wrong?' and curing him with papaya. Sometimes, Llerena would arrive in the morning carrying bunches of bananas or the leaves of porotillo and otoy plants. Even when he arrived empty-handed, the tortoise seemed pleased to see him. He would potter up to meet Llerena, stretch out his neck, and open his mouth. 'And there he stood for a while, with his mouth open, staring at me without blinking, as if he wanted to say something,' Llerena described. George, he was certain, wanted to communicate. When he stretched his neck out, it was 'as if he was asking, "How are you?"' he explained. After greeting George, Llerena would often simply sit and spend time with him. He did this every day of the week, even on Sundays; and it was a Sunday like any other when he entered George's enclosure to find him sleeping.

The date was 24 June 2012 – forty years since the two had first met. The pair had aged, together. Llerena was now seventy-two, and George was presumed to be at least a century old. Llerena reached out a hand towards the tortoise and immediately realized that something was wrong. George was still as a stone. A wave of light-headedness overcame Llerena. He left to get Wacho Tapia, Director of Research for the Park Service. They returned together, and Tapia confirmed it. George wasn't sleeping, but had died. The pair, beside George's body, embraced.

Linda Cayot wrote two years later, 'When Lonesome George died . . . the world lost the last of a species. Many who knew George well lost a friend.' A few days after George's death, Llerena described the loss to numerous media outlets. He said, 'I feel like I've lost a best friend. There is a void and there is sorrow, especially when I see the photos. In my heart I'm not convinced he's dead.' Adalgisa Caccone said, 'I grieved and I thought of the things I could have done that I did not do. I thought I had more time.'

The news of George's death spread throughout the CDRS, then rippled out into the local community and further afield. Hundreds of news articles announced George's death to the world, and the world responded in letters, emails, and tweets. Messages reached the CDRS from people who had known George for years, others who had seen him only once on a visit to the islands, and many who simply knew of him from news reports or books. On an A-board positioned by the roadside in Puerto Ayora, Santa Cruz, a few minutes from the CDRS, was written in chalk: 'Today we have witnessed extinction. Hopefully we will learn from it!' A small cross was drawn at the bottom of this message, like a headstone, next to the last Pinta Island tortoise's name.

'I'd love to tell you a nice story about Lonesome George,'

*Lonesome George in his enclosure at the CDRS, Santa Cruz. (© Fotos593 / Shutterstock.)*

conservation biologist James Gibbs tells me. 'But the truth is I don't think he really liked me. I was the scientist assigned to take his blood samples.' While Llerena would sit quietly with George and feed him bananas, Gibbs was 'the guy who came with the needles and turned him upside down and jammed them into his leg. That was not the role I wanted. But it was the one I had,' he says.

With the death of Lonesome George, Gibbs's role changed. It was decided that, after tissues were extracted for San Diego Zoo's Frozen Zoo, George's frozen remains would be flown to New York so a taxidermy could be made, and that Gibbs would accompany the tortoise on this next leg of his journey. He describes the box George was being transported in – the 'casket' – being driven down the street towards the port. 'The response along the way was incredible,' he says. Like during a funeral procession, people came out of their houses and places of work as the truck passed to touch the box in a final gesture of thanks or goodbye; some cried as they did so. 'It really blew me away. There was something transcendent about this one individual tortoise.' For Gibbs himself, however, the enormity of the moment was too much to think about at the time. 'I'll be honest with you, I still have difficulty thinking about it now: that frozen tortoise that was the terminus of this branch that had peeled off on Pinta Island hundreds of thousands of years ago,' says Gibbs. 'I think about all the tortoises [before it], all those generations of tortoises adapting and changing and how, in the blink of an eye, the whalers and sailors came, and then the scientists, and finally the fisherman. And then there was George.'

It's August 2022 when I speak with George Dante Jr., just over ten years after Lonesome George's death. Like most people involved in this story, the first thing he talks about is the sense of honour he feels to have worked with the tortoise. Unlike the others, however, Dante Jr. knew the last Pinta Island tortoise only in death. A master taxidermist, he was commissioned to create a taxidermy of Lonesome George in collaboration with scientists from the American Museum of Natural History (AMNH) in New York.

On 11 March 2013, the frozen remains of Lonesome George arrived at the AMNH, after a twenty-eight-hour journey by land, air, and sea. Almost all the museum staff gathered to witness the moment his enormous wooden casket was unloaded. His journey had been difficult. 'I remember the freezer breaking down in the Galápagos, the power grid being unstable, and I remember worrying about his condition – is he okay?' Dante Jr. recalls. Then, when George arrived, 'It was like royalty was visiting the museum that day. I mean, the only thing missing was a hundred cop cars escorting his truck.' Opening the casket, he felt enormous relief. Still dusted with tiny ice crystals and frost, the body had not thawed at all. It was the ideal starting point for what was to come.

Taxidermy occupies a unique space within conservation. Derived from the Greek words 'taxis' and 'derma', and translating literally as 'arrangement of skin', taxidermy initially developed as a way for hunters to preserve and present the skin and horns of animals they were particularly proud of killing. Although the discipline has taken some interesting turns over the years – for instance, in the anthropomorphic creations of Victorian taxidermist Walter Potter, who dressed his taxidermies of small animals, such as squirrels and kittens, in clothes and posed them like humans within lifelike miniature sets – it was most often utilized to create hunting trophies.

In the late nineteenth century, soon after the British publisher Thomas Greenwood announced that museums were as indispensable to society as drainage, the police, and 'lunatic asylums', public museums sprang up all over Europe and America, and the discipline of taxidermy found a new expression. Carl Akeley, an American taxidermist, writer, and biologist, began making magnificent re-creations of wild scenes from around the world and populating them with taxidermies of animals found in each location. Akeley's 'wildlife dioramas' inspired a wave of similar mock-ups in museums everywhere, and 'the Akeley method', a series of techniques he devised to create taxidermies that feel genuinely alive, is still used to this day. For this reason, Akeley is now known as the 'father of modern taxidermy'.

It's not surprising that the creation of Akeley's work necessitated the killing of scores of lions, tigers, rhinos, elephants, and many other animals, including tortoises. However, alongside his wife, Delia Akeley – the notable hunter most celebrated for gunning down the bull elephant in 1906 that set the size record – Carl Akeley was championed as a conservationist in his time. The Akeleys believed that the animals they were killing in droves were already destined for extinction, so by hunting them and creating beautiful taxidermies of the few that remained, they could conserve the memories of these animals.

Dante Jr. first saw Carl Akeley's work at the AMNH as a boy, and it was this experience that inspired him to become a taxidermist. Now, in 2013, he was the one responsible for conserving the memory of an animal that was truly the last of its kind.

Casts and moulds were taken of Lonesome George's limbs and head, for reference, then his skin was carefully removed and tanned in a chemical bath. Dante Jr. created a detailed, life-sized clay sculpture of the tortoise's head, neck, and legs, labouring for hours over every wrinkle and fold; this would be the underlying structure. Then, the still-wet skin was slid onto this strange doppelganger, and extra clay added wherever it hung loose – a process Dante Jr. describes as 'like tailoring a suit in reverse; you're building the body to fit the suit'. Wrinkles were pressed into the skin in all the right places, and then Dante Jr. and his team waited for the skin to dry. They watched, anxiously.

'As skin dries, it shrinks,' he explains. 'We'd put all these beautiful wrinkles in . . . and now we're watching it every day just worrying, is it going to shrink? Are these wrinkles going to move?' George's skin dried perfectly, however, and was affixed to the clay sculpture with glue before being carefully painted to exactly match its coloration and patterning when George had been alive. The real tortoise's shell was then placed on top.

At crucial junctures, scientists visited the studio to advise Dante Jr. on whether the angle of a knee or the musculature of the neck was technically correct. But while it was important that Lonesome George looked accurate, he also had to feel alive. This wouldn't come from a reference photograph or research paper, but somewhere else. Dante Jr. started imagining Lonesome George, still alive and moving – observing the tortoise in his mind, in all his different moods. On the floor of his studio, one scientist got down on his hands and knees to demonstrate how Lonesome George walked.

A year and a half after Lonesome George's remains had arrived in New York, the taxidermy was finished. The tortoise now perpetually stands as though he is slowly walking forwards. His head is held high in the air, his long neck is outstretched, and in between his hind legs his small tail is tucked up and to one side. This pose was chosen by the people who knew Lonesome George best, the staff and scientists of the CDRS. Before the taxidermy went on display to the general public at the AMNH and other museums, a private view was held so that they could be the first to see him. A documentary entitled *Preserving Lonesome George* captures the moment the taxidermy was revealed.

'When I walked into that exhibit, I was really very touched,' says Galápagos Conservancy President Johannah Barry. 'He really captured the spirit of George.'

In 2015, Lonesome George finally returned to the Galápagos Islands. His taxidermy was transported back to the CDRS and

*George Dante Jr. working on the taxidermy of Lonesome George in his New Jersey studio. (Photo by Roderick Mickens; © AMNH.)*

installed in a climate-controlled room. Under the watchful eye of staff members for almost twenty-four hours a day, the last Pinta Island tortoise, Lonesome George, is now almost never alone.

VULCÁN WOLF IS the northernmost of Isabela Island's six volcanos. Remote and extremely difficult to access, it is an area of the Galápagos archipelago that has remained relatively undisturbed – at least, that is, by humans. In 2008, Gibbs and CDRS researchers discovered large numbers of giant tortoises living there: as many as 10,000, it has been estimated. Species from all over the archipelago have ended up there and, in an ironic twist, they have done so thanks to the nineteenth-century whalers.

In a cove on the west side of Vulcán Wolf, whaling ships anchored before heading out into the Pacific to hunt. 'The whalers often gathered tortoises in large numbers,' Gibbs tells me, 'and seemed to have traded them, and bartered with them' – meaning the whaling ships would often have had many different tortoise species stowed on their decks. When whaling ships were attacked, tortoises were thrown overboard to reduce weight and increase speed; or they were thrown overboard when particularly aggressive tortoises attacked others, or when a crew had overstocked for their journey. Just as their founding ancestor had done, these tortoises would make their way as best they could to the nearest land – the bay, where they found a welcoming refuge on the isolated slopes of the volcano. And so, a sanctuary of sorts was created by, as Gibbs describes it, 'the very agents that caused so much destruction'. He continues: 'The great tortoise murderers, the whalers, have created this opportunity now for us to save the species.'

Extraordinary discoveries have already been made on the slopes of Vulcán Wolf. The Floreana giant tortoise, a species last recorded by Darwin himself over a hundred and fifty years ago, has been rediscovered here, along with another tortoise that might be an entirely new species of Galápagos giant tortoise. And recent genetic analysis of a young hybrid found on the volcano has revealed it to be half Pinta Island tortoise.

'This story is far from over,' Gibbs tells me. 'And it's such an extraordinary story. You know, this could be a real de-extinction.'

Perhaps, with this species, we will be given a third chance to be better custodians. Today, in amongst the huge numbers of tortoises living on Vulcán Wolf, it's possible that a pure Pinta Island tortoise lives on.

# 10 The Eye of Potosí

Catarina Pupfish
*Megupsilon aporus*

Rain rarely falls in El Potosí, a small village in the state of Nuevo León, Mexico. To the north-east, the Sierra Madre Oriental Mountains loom, but in every other direction El Potosí is surrounded by the dry and dusty expanse of Valle del Potosí, an arid basin dotted with the occasional cactus or desert shrub, broken up by enormous circular fields of potato, corn, and alfalfa. The communities in the valley make their living from the land but in El Potosí it is apparent that something about the land is off. Not only have the fields that once surrounded the village retreated east, but the earth itself here has turned a pallid grey.

On some mornings, just after sunrise, before El Potosí's residents start the day, clouds of acrid smoke blow in and settle silently over the village streets. A sulphurous smell hangs in the air and seeps under doors and through open windows. The source is in the land surrounding the village. Smoke billows in plumes from the ground where masses of corn once grew. Here and there, withered tree trunks stand upright, blackened and stripped of their leaves and branches. The ground is covered in ash, and parts of it have sunk, forming craters, giving the landscape a lunar appearance. Life and colour have been drained from this land.

A fire has created this scene, but not in the way you might expect. Rather than racing across the land as a wildfire does, or through fields after harvest when farmland is slashed and burned (a common practice in this area), this fire smoulders underground. A thick layer of peat (partially decayed vegetation and other organic matter) stretches for miles and miles beneath Valle del Potosí, formed of the bed of an ancient lake that existed here 10,000 years ago. In the late 1990s, when a field was torched after harvest, the flames set this peat alight. For nearly thirty years now, it has been burning. Today, hazardous cracks in the ground, too small to fall in but large enough for a foot to slip into, are permanent features of this landscape, offering glimpses into the underground fire, which glows red-hot. It is as if the Earth's core has somehow been knocked off-centre and ended up here, beneath Mexico.*

Everyone in El Potosí has a story about the fire. Some have witnessed the ground smoulder and sink just metres away from their front doors. Others have inherited farmland only to find that the fire has left it unable to sustain life. An area of roughly 22km² has been affected by the peat fire. It has consumed fields and pastureland, and even threatens to swallow up Federal Highway 57, which connects El Potosí with the outside world. The fire has worked its way into every aspect of village life, affecting drinking water, respiratory health, public safety, and agriculture. Most residents remember a time before the fire, however, when the sweet smell of fruit wafted into town from nearby orchards and when doors opened onto green fields, rather than barren earth. Many also remember how, in the very same spot where this fire started, there had once been a lagoon.

Crystal-clear water bubbled up from a natural spring on the northern edge of El Potosí to feed a lung-shaped lagoon at the base of a small hill. Spanning a little over a hectare, the lagoon was large enough to be the centre of village life. Its name in Spanish was Ojo de Potosí, meaning 'the eye of Potosí', and

---

* Underground peat fires like this are nicknamed 'zombie fires', due to how difficult (near impossible) they are to extinguish. Equally monstrous is the effect they have on the climate, with a zombie fire releasing as much as a hundred times more carbon into the atmosphere than a wildfire of equivalent size.

people came here to fish, to think, to go for walks, even to propose. Many generations of villagers learned to swim here, cannonballing into the deeper sections and paddling in the shallows. And it was here and only here, in this lagoon, that the species at the centre of this chapter lived. The Catarina pupfish (*Megupsilon aporus*) was small – males and females of the species were just 2.6cm and 3.6cm long, respectively. Males were an iridescent blue on their backs and sides, while the females were golden and olivaceous. In the shallows of the lagoon, within dense mats of aquatic plants like hornwort, these fish lived in their thousands, snapping up the larvae of lake flies and similar insects. The Catarina pupfish lived here for millions of years – but its life in this lagoon ended quickly and suddenly. Like *Plectostoma sciaphilum* and the Bramble Cay melomys, the Catarina pupfish called a small area home, and when the stability of that home was threatened, so was its existence. After its lagoon home disappeared, however, the Catarina pupfish survived for decades in captivity. Finally succumbing to extinction in March 2015, it is the most recent in this book.

'Pumpkin seed' and 'pursy minnow' are two names that nineteenth-century biologists used to describe pupfish, a group of tiny freshwater fish species, the majority of which reside in pools dotted about the deserts of the American south-west and Mexico. Given that most pupfish are no more than 5cm long, these names were perhaps meant to convey a sense of their scale. However, the name that stuck has nothing at all to do with size and everything to do with personality. Chasing each other in circles, wagging their tails energetically, nuzzling, and in captivity begging their keepers for food – these are all behaviours you'd probably expect to see in a dog rather than a fish, and you wouldn't be alone. 'They play just like puppies,' remarked the

American ichthyologist Carl L. Hubbs, after observing pupfish. Hubbs coined the term 'pupfish' in a 1950 research paper, and the group, which is formally known as the Cyprinodontidae subfamily, finally had a common name. Following the same logic, the Mexican common name is 'cachorrito' – literally 'puppy' in Spanish.

In December 2023, I receive a crash course in Pupfish 101 from Chris Martin, Associate Professor and Curator of Ichthyology at UC Berkeley, California. Martin, who signs off his emails with 'best fishes', has been obsessed with fish since around the age of eight, he tells me. Pupfish came into the picture a little later when he was a graduate student, and now his laboratory at UC Berkeley (the Martin Fish Speciation Lab) is the only facility in the world studying pupfish diversification. Flying in the face of the popular convention of biologists naming new species after their colleagues and loved ones, Martin instead named his daughter after a pupfish – a disclosure which, to my mind, only solidifies his credentials further.

'As a group, what makes pupfish unusual is that they have the highest temperature tolerance of any freshwater fish,' says Martin. At one extreme is the Julimes pupfish (*Cyprinodon julimes*), which lives in 46 °C hot springs in Julimes, Mexico. 'It's hotter than hot tub water where these fish live, year-round,' Martin explains, 'and they're actually breeding there and somehow completing their life cycle.' At the other end of the spectrum is the Atlantic pupfish (*Cyprinodon variegatus*), which can be found as far north as Massachusetts, where in winter the mercury hits −5 °C and the sea freezes over. 'So, they have the widest temperature tolerance range of essentially any freshwater fish, possibly any fish at all,' says Martin, 'and they also are extremely euryhaline – in other words, their salinity tolerance is off the charts.' The Atlantic pupfish is the perfect example of this, capable of surviving in water with a salt content anywhere from 0 to 120 per cent. 'With almost any other fish, if you take it from freshwater and throw it into saltwater, it drops dead. But pupfish have no issues at all,' says Martin.

Even amongst this group, however, the Catarina pupfish stood out. 'It was just a little bit different,' Martin tells me. 'With many pupfish, they're hard to tell apart by just looking at them. But the Catarina pupfish was a bit stouter, and when males reached maturity, they developed a conspicuous black dot on their tails.' Most pupfish are not sympatric, meaning that they do not co-occur with other species from the same group. Ecologist and writer Christopher Norment describes this as a sort of 'ecological loneliness' – cut off from other related species, often in disparate and hard-to-access places. But this wasn't the case with the Catarina pupfish. Though its lagoon was small, it shared it with another pupfish neighbour, the larger and more aggressive Potosí pupfish (*Cyprinodon alvarezi*). Its anatomy was unusual too: it didn't have cephalic sensory canal pores, which makes it completely unique amongst all pupfish. These pores are believed to play an important role in detecting the vibratory waves produced by prey and predators. And, yet, despite the absence of these pores, the Catarina pupfish was a carnivore, unlike the majority of pupfish, feeding on not only algae and detritus, but also on the larvae of insects. 'It was just absolutely fascinating, and totally baffling,' says Martin.

How pupfish dispersed and became isolated in their various habitats isn't known in many cases. Some used to live in much larger bodies of water that have, over tens of thousands of years, dried up. Others, perhaps, were spread by birds. A pupfish egg can survive in the gut of a bird; like seeds, it is possible the ancestors of pupfish species were dropped from the air into isolated ponds and streams, where they established populations, and then remained for millions of years, over time adapting to their environments. Some, like *Cyprinodon simus*, which feeds on zooplankton, and *Cyprinodon desquamator*, which feeds on the scales of other fish, developed new, specialized diets. Others adapted to low or high saline, and hot or cold water. Even the flow of water in the pool or stream, spring or lagoon they ended up in changed the course of their evolution. Some species developed pelvic fins while others didn't, or grew longer or shorter caudal peduncles (the base of the caudal fin where a fish's swimming muscles are found).

Diverging from the ancestor it shared with the *Cyprinodon* genus of pupfish 7–9 million years ago, it is believed that the Catarina pupfish is a relict – a surviving remnant of a much larger ancient population of pupfish. The species became isolated from this population over 5 million years ago and somehow (it's not known how) ended up in Ojo de Potosí, where it remained. 'It was probably there for millions of years,' says Martin – and, over this time, it didn't spread to any other body of water and diversify, like its sister genus *Cyprinodon*, which became fifty-five species. It remained in the shallows of Ojo de Potosí 'not really changing at all'. Meanwhile, things changed around it. During the Pluvial period 10,000 years ago, when many dry habitats became wet, the lagoon grew into a lake, and the desert turned into marshland. Then, the lake shrank back and, eventually, humans arrived, building first a hermitage on the hill next to the spring that fed the lagoon, then a hacienda nearby and finally, the village of El Potosí.

In 1948, a Mexican biologist called José Alvarez del Villar travelled to Ojo de Potosí. At the time, the arid lands of southwestern Nuevo León were thought to be devoid of aquatic life. Alvarez del Villar disproved this when he discovered two species endemic to the lagoon, the Potosí pupfish and a cambarellid crayfish (*Cambarellus alvarezi*). The Catarina pupfish, however, proved more elusive. Over a decade passed before it was finally discovered. On 23 February 1961, two American ichthyologists, Robert Rush Miller and H. L. Huddle, set fish traps baited with chicken liver in Ojo de Potosí and found a single specimen amongst the Potosí pupfish they caught. While they didn't then publish a formal description, they recognized that this species was new to science. They named it, informally, the Catarina pupfish, after Catarino Rodríguez, another name for El Potosí.

In the years that followed, not much changed for the Catarina pupfish. Occasionally, scientists who had heard about Miller and Huddle's newly discovered fish travelled to Ojo de Potosí to see it. Crouching on the bank, or standing stock-still in the shallows, they would quietly observe these little lives unfold, watching as the males circled and chased each other, or dispersed abruptly at the looming sight of a Potosí pupfish. When they lifted the dense mat of hornwort and other aquatic plants, slips of iridescent blue and green would scatter in all directions. Otherwise, however, the fishes' lives continued as before. At night and before dawn, males would rapidly circle the females to initiate courtship, as if attempting to hypnotize them. From the bottom of the lagoon silt would rise up, creating murky circles in the water. Then in the day, when children from the village came to the lagoon to swim and paddle, the Catarina pupfish returned to their hiding place amongst the vegetation, and waited for quiet to descend over Ojo de Potosí.

For more than a decade, the Catarina pupfish existed in a sort of limbo. It had been discovered, yet not formally described, and therefore not declared by any official body, like the IUCN, to be a species. Miller and Huddle got on with other, more urgent work; and in the meantime, with few people studying it, the species kept a large part of its life hidden. We knew next to nothing about its behaviour, and nothing of its genetic make-up, its ancestry, or of the millions-of-years-long journey that brought it to the little lagoon it lived in.

Robert K. Liu was a student in 1968 when the strange new fish arrived at the University of California, Los Angeles. One of Liu's professors, Vladimir Walters, had returned from Mexico with a haul of specimens from various desert lagoons and hot springs, including Ojo de Potosí, and now they were being unloaded at the fish laboratory. Still in the water-filled plastic bags they had been transported in, they looked like prizes at a circus shooting gallery. After being carefully acclimatized to their tanks, with the water temperature slowly raised over five hours, the Catarina pupfish, now darting around not in the murk of a lagoon but behind clean walls of glass, were ready to reveal their secrets.

'It wasn't really part of my dissertation, but fortunately I was able to study [the Catarina pupfish] and observe them interacting with other species,' Liu tells me. 'And probably most importantly, I had a single-lens film camera with a flash.' He smiles, perhaps sensing my surprise that a simple camera would rank top on a list of high-tech laboratory equipment. But, as Liu explains to me, it was essential. The intricacies of some of the species' behaviours played out faster than the human eye could perceive, a pair of fish coming together and colliding in a flash of tails and fins that was over so quickly it looked like time had skipped forward. Liu's camera undid all this. By freezing these split-second interactions, it offered him a portal into the hidden world of his subjects, and through this he witnessed the lives of Catarina pupfish play out with unprecedented intimacy.

At either dawn or dusk, male Catarina pupfish did their best to impress females with a dance. With fins folded flat against his body, a male would rapidly trace tight horizontal circles in the water around his desired mate, a behaviour known as 'looping'. Next came what would become known as the species' characteristic jaw-nudging. Sidling up alongside the female, the male protracted and retracted his jaws repeatedly in a puckering motion, as if blowing kisses, as he moved back and forth along the nape of her neck. This seduction, if successful, elicited a downward tilt of the female's head and a ritualized 'nip' of the plant or other surface on which they would soon spawn. Side by side, the two fish then pressed their bodies together to form a single, S-shaped whole – these embraces, common to all pupfish, are what Norment calls 'S-shaped waves of desire'. Bodies interlocked, the two fish vibrated together for a moment, before the female jerked her body straight, releasing a single egg, which was immediately fertilized by the male's sperm.

Love and war bore certain similarities in the world of the Catarina pupfish. Males of the species performed the same 'looping' movements around other males, as if at any moment the two might exchange kisses and tumble into an S-shaped embrace – but for one difference. Rather than tucking his fins against his body, the charging male instead held them erect. 'In the world of fish, it's all about making yourself as big as possible,' Liu tells me. To this end, the males of the species demonstrated another unique behaviour. During what is known as 'opercular rotation', male Catarina pupfish rotated the flaps covering their gills so they stood out at either side of their bodies – a pupfish analogue to the human ritual of puffing out chests and squaring shoulders. This was often followed by a chase, one fish darting after the other at breakneck speed, and then a tussle. Viewing all this with the naked eye, it would be easy to see why pupfish were seen as cute and associated with playing puppies. However, up close, time frozen by Liu's camera, the true nature of this 'play' was apparent. Defending the space immediately around it, a male Catarina pupfish would charge at any trespasser, biting and slamming his tail into its body. Sometimes, two males would lock their jaws together and tumble into a 'mouthfight', chewing at each other's faces ferociously. Puppy-like this was not.

*A Catarina pupfish (left) and a Potosí pupfish (right), demonstrating aggressive behaviour. (Photo by Robert K. Liu.)*

'I was able to catch them in these very unique behaviours that distinguished them,' says Liu. Not occurring in any of the fifty-five closest-related pupfish species, opercular rotation and jaw-nudging were behaviours so unique that Liu's observations formed part of the basis for Miller and Walter's description of the Catarina pupfish as a new species. Eventually published in October 1972, this paper also drew on a 1971 genetic study of the new fish by Teruya Uyeno and Miller, which discovered that the males possessed unusually large Y-chromosomes and fewer sex chromosomes (forty-seven) than the females (forty-eight). This marked the Catarina pupfish out not only as a new species, but as being so genetically distinct that a new genus had to be created, for which Carl L. Hubbs, the nomenclator of the entire pupfish group, provided the name *Megupsilon*, derived from the Greek 'mega', meaning large or many, and 'ipsilon', the Greek spelling of the letter 'Y'. *Aporus*, from the Latin, meaning 'without pores', was chosen as the species name.

In the early morning of 6 July 1972, three months before Miller and Walter's paper was published, Liu finally observed Catarina pupfish in the wild. He had travelled overnight down to Ojo de Potosí with Walters and three fellow students. Liu tells me that he had hoped to spend a few hours quietly watching the Catarina pupfish through the crystal-clear water of the lagoon. However, word of the arrival of a group of strange men, American fish scientists no less, travels fast in a remote agricultural village. Soon, Liu and his cohorts found themselves surrounded by a crowd of excited local children, stomping and splashing through the water.

'We were in their space, really. They were just there, living their lives. We couldn't tell them to stop. But for observation you need undisturbed water, and it was not really possible given the situation.' The water was not clear at all; the children's feet lifted silt up from the bed of the lagoon, and their movements scared the fish away. However, whether the trip had ended up being sabotaged by children or not, one thing was immediately clear to Liu – the water at Ojo de Potosí had been disturbed in a far more significant and likely irreversible way. 'The spring had been modified,' says Liu. Technically, this process had already begun as early as 1924, when part of the spring had been cemented shut to divert water flow. Then, in 1960, a wall had been built across the lagoon to dam it and increase its size. 'Now, there was a pumphouse, drawing water out for crops,' says Liu. A desert is defined by water scarcity and so, as Liu explains, 'Where there is water in the desert, there is conflict.' Here, the endemic species of the lagoon needed it to survive; but the pressures of farming on an industrial scale meant that, to the people of El Potosí, the water was also needed to sustain their agricultural industry. Norment writes, desert pupfish 'depend on the same desert waters that people desire, and so they are rare and mostly threatened.' The alterations to the lagoon likely made perfect sense to the people of El Potosí. However, to Liu, they were bad omens.

BEFORE THE ARRIVAL of humans, the Catarina pupfish had few neighbours. There were, of course, the larger pupfish that lurked in the deep, and the tiny freshwater shrimps that were Ojo de Potosí's other endemic species. But other than a handful of snails, coots, ducks, woodpeckers, and plovers, a Catarina pupfish would come across little else, tucked away in this isolated spring. The first recorded contribution made by humans to Ojo de Potosí's ecosystem was the goldfish (*Carassius auratus*). They were spotted in the lagoon in 1961, before which they'd likely been introduced, perhaps as a pet someone had decided to release. They maintained only a small population, despite consistently breeding. However, the next introduction would be far more disruptive. Largemouth bass (*Micropterus salmoides*) are a predatory fish species highly prized by anglers for their strength and large size (up to 75cm in length). No one knows exactly when these fish found their way into Ojo de Potosí, but it had happened at some point after 1968. They may have been put there in the hope of establishing a food source, or else by anglers tired of the meagre sport offered by the pupfish. Either way, from the moment these bass slipped into the clear water of Ojo de Potosí, they found themselves surrounded by thousands of bite-sized snacks, in the form of the Potosí and Catarina pupfish – species that had no natural predators.

When the Mexican biologist Salvador Contreras-Balderas visited Ojo de Potosí in 1974, he was shocked to discover a huge quantity of dark shapes gliding through the water. He realized that bass had completely overrun it. The deep part of the lagoon was where the adult bass could be found, circling ominously. Juveniles and small adults, meanwhile, swarmed the shallows. Just ten years earlier, this water had held thousands of Potosí pupfish and Catarina pupfish, the two species living together in a state of inharmonious equilibrium. During the rainy season, the lake swelled, creating more shallow-water habitat and a boom in Catarina pupfish. Then, when the dry season returned and the lake contracted around the deeper areas, there would be more Potosí pupfish. Now, however, these endemic species had all but disappeared – the remnants reduced to cowering in the shadows of their terrifying new neighbours. Contreras-Balderas managed to collect just one Catarina pupfish, but he returned later that year with some colleagues, and together they were able to catch more and transplant small numbers of each of the three endemics (the third being the cambarellid crayfish) into nearby hot springs, which acted as lifeboats. In the meantime, they started culling the bass in Ojo de Potosí. Disaster was averted. The bass, which might easily have spelled the end for the Catarina pupfish, along with the other two endemics, were gone from the lagoon by 1976 and normality returned to Ojo de Potosí. However, this brush with extinction was just a warm-up.

In 1984, vast fields of corn, alfalfa, and potatoes were growing in abundance – the latter specifically for a globally popular brand of crisps. Fruit orchards burst into life and large herds of cattle grazed in fields. The valley was rich in produce. Soon, there'd be a harvest; millions of sheaves of corn would be milled and potatoes picked, to be sliced and baked, packaged and distributed around the world. The price of all this agricultural success, however, was paid in water. The valley's groundwater had almost been sucked dry. Wells had been dug and now punctured the earth here and there. And, in El Potosí, water extraction from Ojo de Potosí was intensified. It was as if the fields' thirst – fuelled by crisp consumers the world over – could not be satiated. By summer's end, the isolated lagoon in which the Catarina pupfish lived shrank to just 15 per cent of its original area. The following year, it was 10 per cent. The year after that, 5 per cent. Meanwhile, nearby, the potatoes swelled underground, like nuggets of gold. Fortunately, Contreras-Balderas and other biologists, alarmed at the situation, had begun collecting Catarina pupfish, along with the other two endemic species, to breed in captivity – a prudent decision since, in 1987, the farming season left no main reservoir of water at all. All that was left were 'a handful of small pools and a trickle of water' that very slowly refilled the lagoon, Contreras-Balderas and fellow ichthyologist María de Lourdes Lozano-Vilano later wrote. The remaining wild individuals of the Catarina pupfish survived in these tiny dregs of habitat.

Rescued from a shrinking lagoon in the Mexican desert by biologists, the Catarina pupfish travelled to many far-flung locations around this time. Populations sprang up in distant scientific institutions in Austria, Spain, and Germany thanks to a joint initiative spearheaded by Contreras-Balderas and another Mexican ichthyologist called Arcadio Valdés González. The idea was to get the Catarina pupfish into as many aquariums as possible, with fish regularly traded between these populations to prevent inbreeding. One of these Catarina pupfish communities landed just a stone's throw from Ojo de Potosí, at the Autonomous University of Nuevo León (UANL) in San Nicolás, Nuevo León – although the fifteen fish that founded this population had taken a circuitous route to get there, having hatched in Vienna at the Mexican Fish Ark in the Hause des Meeres. Receiving them back in Nuevo León in 1989 was Maria Elena Angeles Villeda, a laboratory technician and doctoral student of Valdés González.

'They were hard to keep,' Angeles Villeda tells me. At first, she and her colleagues tried placing some of the fish in outdoor ponds, but they didn't survive. 'It took a lot of effort, experimenting with temperatures, ratios, aquarium size,' she says.

Eventually, a balance was struck – the fish were happiest in 70l aquariums, in water that was heated to 24 °C and had a salinity of 2 per cent. Each tank had an air pump and a water filter and housed a population of ten Catarina pupfish – eight females and two males. 'As long as they had this proportion, nothing would happen,' says Angeles Villeda. 'However, if you wanted them to breed, you had to separate one male and one female, once the male was ready.' Fortunately, the fish would let her know when it was time, as conspicuous black spots appeared on the caudal peduncles of male Catarina pupfish the moment they reached sexual maturity. When this happened, a pair would be removed from the population and deposited into another tank, furnished with artificial nests called spawning mops after the fact that many look exactly like miniature mop heads. These simulated the kinds of plants the fish would have deposited their eggs onto in their natural environment.

Once the couple had privacy, courtship would begin at dusk. The Catarina pupfish at UANL laid roughly five eggs per spawning. These were immediately removed by researchers and placed into 4l oxygenated aquariums. At first, the eggs were often contaminated with fungus, something kept at bay in Ojo de Potosí perhaps due to bacteria in the water or another natural process. This was another problem which Angeles Villeda and her colleagues had to troubleshoot. Methylene Blue, a salt commonly used in fish husbandry to dye aquarium water and which, at higher proportions, operates as a disinfectant, proved to be a solution. Once the team started adding it to the spawning tanks, the effect was immediate. After four days, tiny fry emerged from these disinfected eggs. A month later, they were moved to bigger tanks, and a month after that to 10l 'growing aquariums', each holding thirty fry. 'The fry fed on microalgae and ciliates, which is strange because the adults were carnivorous [fed strictly on mosquito larvae],' says Angeles Villeda. 'But it worked because we had a 100 per cent survival rate.'

As the captive Catarina pupfish were settling into their aquariums at scientific institutions around the world, the conditions at Ojo de Potosí were deteriorating. By 1990, Valle del Potosí and the neighbouring valley were dotted with over eighty wells, each at least 100m deep, which had been dug to extract water for crops. Some springs and creeks dried up completely as a result. 'Valdés González and Contreras-Balderas went into the community to talk about the Catarina pupfish to try to convince the local authorities to help, but the situation was impossible,' says Angeles Villeda. 'They didn't really care about a fish that wasn't in their diet or that they could not fish for, and they didn't believe that [the lagoon] would dry up so quickly, so they didn't listen.' The water extraction continued.

In 1992, Charles J. Yancey, Senior Aquarist at the Dallas Children's Aquarium (DCA), travelled to Ojo de Potosí. Along with his DCA colleague Dave Schleser and Paul Loiselle of the New York Aquarium, Yancey had been invited by Contreras-Balderas and Valdés González to collect four of Mexico's most critically endangered pupfish from Ojo de Potosí and other nearby springs: the Charco Palma pupfish (*Cyprinodon veronicae*), the La Palma pupfish (*Cyprinodon longidorsalis*), the Potosí pupfish, and the Catarina pupfish. Yancey returned to Dallas to start a captive colony of Catarina pupfish in a back room at the DCA. He'd collected these specimens just in time – 'a year later [the] springs were dry', according to an article in the aquarist journal *Drum and Croaker*. And by 1994, the Catarina pupfish was extinct in the wild.

With its natural habitat completely desiccated, the Catarina pupfish was now essentially homeless – its surviving members squatting in laboratory aquariums. Although the captive-breeding effort was proving successful, there was no longer an Ojo de Potosí to return to. As if to underscore this point, in 1998 a fire spread from a nearby field into the hollow basin left over by the lagoon, setting Ojo de Potosí and the peat beneath it alight.

Though the Catarina pupfish were found to be sensitive to environmental changes and prone to infections, the species survived reasonably well in captivity. 'The females were able to breed once they were five months old, and typically lived for three years,' says Angeles Villeda, 'and we bred them as much as we could for those three years.' As the population grew, Valdés González and Angeles Villeda continued to ship Catarina pupfish to other aquariums around the world. Over the next two decades, tanks in North America and Europe filled with these little refugees. But once again, in 2011, the species' fortunes turned.

'It didn't look like anything was wrong. The adults were

*Smoke rises from a crack in the ground at dawn in Valle del Potosí. (Photo by Priyadarsi D. Roy.)*

laying the same number of eggs,' says Angeles Villeda, 'but these eggs no longer hatched.' No matter what she tried, substituting females for other females, or switching in younger males, nothing seemed to fix the problem. Different disinfection methods were used, and the conditions of the aquarium water were checked and then rechecked, but nothing worked. 'It happened really fast. Over three or four months they went from 100 per cent fertility to 0 per cent,' she says. When this happened, Valdés González, Angeles Villeda, and their colleagues contacted all the other institutions to which they had sent Catarina pupfish, only to discover that they were all in exactly the same situation. More or less overnight, the entire stock of the species had become infertile, it seemed. A slew of *Mycobacteria* infections then wiped out most of the adult fish, and the populations that had sprung up around the world collapsed in unison – except for one.

Charles Yancey had successfully maintained twenty-seven generations of Catarina pupfish at the DCA, starting with the handful of specimens he had collected from Ojo del Potosí in 1992. And, somehow, his fish had managed to avoid the disastrous fate of all the other Catarina pupfish populations. Robert K. Liu and his colleague Anthony A. Echelle visited Yancey's fish room at the DCA in 2012 to observe what were then the last twenty surviving Catarina pupfish. 'Nobody had had as much success with them as Yancey,' Liu says. 'It was an incredible operation.' Eventually, however, Yancey's Catarina pupfish went the same way as all the others. The population dwindled as fewer and fewer eggs hatched, and the species headed towards extinction.

IT WAS LUCKY that Chris Martin had a fish room at all in 2013. A postgraduate student at UC Berkeley at the time, Martin

## The Eye of Potosí

almost wasn't given the tiny windowless room in the basement of the university's animal care facility. It was more broom closet than laboratory space; he could touch both sides of the room at the same time if he stretched out his arms. This claustrophobic nook was shrunk even further when Martin's equipment arrived, the walls closing in as rows of shelves went up, laden with twenty 40-gallon fish tanks. 'You could fit two people in the room, and not much more than that,' Martin tells me. However, even a cramped, dark, underground chamber like this provided ample room for pupfish.

Immediately, Martin began looking for Catarina pupfish in the hope that he could contribute to captive conservation efforts for the species, and study them, in his fish room. He emailed the Dallas Children's Aquarium, asking if he could be given some of their fish to start his own colony, but received no reply. Next, he tried calling, and after several weeks managed to get Barrett Christie, the aquarium's director, on the phone. The news was not good. The Catarina pupfish population at the DCA had produced just four eggs that year, from which only males had hatched, and all but two of those had died. 'I was just like "Argh!"' Martin says, raising his hands up to the sky. This was the first he had heard of the aquarium's Catarina pupfish population collapsing – the species going from potential study subject to functionally extinct in the span of a short phone call. Had the dire state of the DCA's Catarina pupfish population been communicated earlier, Martin and other pupfish experts might have been able to help. At the very least, there were tricks and techniques that could have been attempted to help revive it.

The persistence of only one sex, either male or female, is the death knell for many species. However, fish have the benefit of a little wiggle room. Sex-reversal is a process where fish are induced to switch from male to female, or vice versa. For example, by introducing juvenile male tilapia to low levels of oestrogen, biologists can convince them, by way of chemical signals, to alter the course of their development so that they become female. 'Fish sex is very plastic,' says Martin, 'sometimes it's even environmentally determined in killifish – where if you raise them at certain temperatures you get all females or all males.' Whether this would have been enough to save the species, he isn't sure. 'Because the male and female Catarina pupfish had different-sized sex chromosomes, you probably would have had a female with a weird complement of twenty-three chromosomes [rather than the usual twenty-four],' he says. Nevertheless, at this stage anything was worth a try, and Martin, having successfully reversed the sex of a similar fish, was the ideal candidate to attempt it. Or, at least, he would have been, but for one snag. Sex-reversal is only possible with young fish, and the two Catarina pupfish that were now the last vestiges of their species were both over a year old.

Martin, however, wanted to try to salvage something of the species, and there was one possibility remaining. The DCA's male Catarina pupfish could be bred with female Potosí pupfish to produce hybrids, retaining 'at least some of the species' genetic ancestry,' says Martin. From there, the female hybrids could be bred with male Potosí pupfish, a process called 'backcrossing'. The odds of success would be low. However, if it worked, the species – or some semblance of it – would remain with us. And so, a few days after their phone call, Christie flew the two remaining fish out to San Francisco International Airport, where Martin was waiting to meet them. Later that evening, the last Catarina pupfish settled into their new tanks in the tiny basement laboratory at UC Berkeley.

Over the years, Martin's research with pupfish has led to fascinating discoveries, including a new craniofacial function for a gene in humans that is contributing to how scientists understand birth defects. But taking on the last Catarina pupfish in 2013 wasn't really about work at all. 'Actually, it took time away from research that would have helped my career,' he says.

'It was just a personal thing. It's so rare in our lives that you have the opportunity to contribute to the existence of a species . . . and I do believe that species deserve to just exist, just because.'

Martin went to great lengths to care for these fish. Alongside his own groceries, he started picking up fresh greens and fish at the supermarket, which he chopped and froze, and then scattered into their tanks with high-grade commercial pellet food and bloodworms – 'a pretty gourmet diet,' he informs me. Newly hatched brine shrimp were the cherry on the cake: live prey to satisfy whatever innate hunting instincts remained after the species' ten or so generations in captivity. Each day, Martin carved out an hour or two from his busy schedule to visit the laboratory and check on the fish, siphoning dirty water out of their tanks into the drain by hand, before topping them up with clean water.

The fish, for their part, responded well to this treatment. Paired up with female Potosí pupfish (specifically, with juveniles of approximately the same size as the adult male Catarina pupfish), it wasn't long before Martin's tanks came alive with the same looping and jaw-nudging that Liu had observed at UCLA forty years earlier. 'S-shaped waves of desire' paved the way for hundreds of eggs, deposited on spawning mats in each tank by the female Potosí pupfish. At least a hundred of these eggs were viable, Martin tells me, and later hatched to reveal tiny fry. This initial part of the process was a success; in fact, it was so successful that Martin ended up with more hybrid females than he had room for and even managed to send spare fish to Valdés González in Mexico. However, the real test of this hybridization effort – the back-cross – was yet to come.

Gaze into certain tanks within Martin's subterranean laboratory around this time and you'd have seen what looked like ordinary, run-of-the-mill pupfish interactions. The hybrid females and the male Potosí pupfish took to each other well, courting and then mating, with the females then depositing their eggs onto spawning mats. However, something else was also taking place. The term 'hybrid breakdown' describes what happens when the offspring of hybrid fish buckle under the weight of their own bizarre genetics and is especially common when the parent-fish belong to two species that are highly genetically distinct. In such cases, 'things like the misregulation of genetic pathways cause a mismatch between genetic pathways during the entire process of development,' says Martin. The resulting offspring are 'just really messed up', he tells me – that is, if they survive the egg stage at all.

Only two or three fry successfully hatched from the eggs Martin meticulously picked from his hybrids' spawning mats. The rest of the eggs were dead. The tiny hatchlings followed suit within a fortnight. Separated by millions of years of evolution, the genetic gap between the Catarina pupfish and Potosí pupfish was just too broad for Martin's back-crossed hybrids to bridge. Despite the tragic events transpiring in surrounding tanks, Martin's hybrid fish seemed to be doing well. True, they were looking a little long in the tooth, but each was still laying eggs on the day it died. One by one, the hybrids went the way of their doomed offspring. Meanwhile, as Martin continued to work to back-cross his hybrids, the species whose ancestry he was attempting to preserve slowly dwindled in number, until, in 2015, only one remained.

Ironically, Martin had been at a de-extinction conference on 6 March, the day that the Catarina pupfish went extinct. Popularized by Beth Shapiro's 2015 book *How to Clone a Mammoth*, de-extinction is a branch of science that aims to resurrect extinct species through the use of cloning and genetic engineering. Talk at the conference turned excitedly to species like the passenger pigeon, a North American bird famously hunted to extinction in the 1800s – at one point, at a rate of 300,000 birds per day – and the thylacine, commonly known as the Tasmanian tiger, and the possibility that both might one day live again. 'It was a wild conference because everybody there was so optimistic,' says Martin.

After the conference, attendees returned to their offices and laboratories across the United States and beyond with visions of woolly mammoths once again sweeping the Siberian steppe, or of one day meeting a dodo in person. Martin returned to his fish lab at UC Berkeley with a journalist in tow. He'd told her the story of the Catarina pupfish – about its uniqueness, the draining of its tiny lagoon, and the solitary fish at his lab. Now she wanted to write about it and to see the last of the species for herself. Winding their way through corridors and down into the basement of UC Berkeley, they finally arrived at the anonymous broom closet where Martin kept his fish. 'But, we get there and the fish is dead,' says Martin, 'and I was just like "Shit!"'

There was something unnatural about this fish, whose tribe was named for their puppy-like energy, now floating limp in the water. Where once it had zipped about the tank, snapping up brine shrimp, its small body bobbed about, eyes glazed over, mouth agape. Martin had watched this fish loop breathlessly around females. He'd seen it press itself against the bodies of other fish, forming S after S after S, fertilizing the eggs that Martin had collected and carefully transplanted into new tanks. Gently scooping its body out of the water, he put this last fish in the deep freeze at UC Berkeley, where it remains today, along with the other male Catarina pupfish and Martin's hybrids.

When we speak, I ask Martin how he feels about the extinction. 'I don't even like to talk about it anymore,' he says. 'But I do still talk about it with my students.' Each year, at the end of a class on extinction, Martin tells the story of the Catarina pupfish, culminating in the death of the last fish in his lab at UC Berkeley. This year, his students asked him if he cried. 'I was thinking about it and it wasn't just this one-day event,' says Martin. 'It's not like you could just cry and be done with it.' The extinction, he tells me, weighed heavily on his mind for a year. 'I almost felt alone,' he says. 'There was just no one to really share it with who saw everything about this species and its story . . . the millions of years of this one lineage, a monotypic genus, of the story that it had to tell that's just completely gone now.' Four years later, in 2019, the Catarina pupfish was declared extinct by the IUCN.

ON WEEKDAY MORNINGS, a yellow school bus trundles along the streets of Madison, New Jersey. One of nearly half a million such buses that ferry children to and from schools across the US, it is presumably the only one to be operated by a pupfish breeder. Fish caught John Brill at an early age. 'When I was ten or twelve, my father was a TV repairman, and he would fix people's TVs in their homes, and his customers would give him food and all sorts of stuff,' Brill tells me. 'One day, he came home with a pint jar with some guppies in it.' Down in the basement, Brill and his father dusted off an old fish tank, filled it with water, and tipped the guppies in. 'I remember just looking into that tank in the room all by myself and it was so quiet – all you could hear were the bubbles from the aerator – and just watching the fish move. It was *mesmerizing*,' he says. 'I was hooked. It was like a drug.' Since then, fish have been a constant in his life. Brill, whose aptronymic surname he shares with a species of flatfish (*Scophthalmus rhombus*), is not a scientist; but, thanks to his encyclopaedic knowledge of and experience with pupfish, he is regularly consulted by academics. Over the past six decades, he has kept and bred tens of thousands of fish from a wide range of species – including, for a brief spell, the Catarina pupfish.

Brill was a beer-truck driver when, in 1981, the Catarina pupfish came into his life. A big part of the fish-breeding scene at the time revolved around conventions, where hundreds of aquarists would bid in auctions for rare fish. It was at one such convention in Rochester, New York, that the Catarina pupfish made its debut in the aquarium hobby – although this was not quite the glorious unveiling it sounds. 'The fish were just spooked to death . . . they were sitting on the bottom of the tank, with their fins clamped close,' says Brill. Being decidedly unflashy, at least by aquarists' standards, most of the people at this convention walked past these sad-looking fish. However, knowing what they were – a new fish that had only been described nine years earlier – Brill didn't care what they looked like; he had to have them. 'To sum it up, [the species] didn't make a big splash,' he says, 'but it made a big splash to me!' And, so, a few hours and $12 later, Brill was the proud owner of a pair of Catarina pupfish.

While we talk over Zoom, Brill takes me down to his basement fish room. Behind him, in two large aquariums, I can see several bronze and pearlescent shapes darting about – Cuatro Ciénegas killifish (*Lucania interioris*), an endangered Mexican species. In 1981, little slips of blue and green glistened in these aquariums instead: the Catarina pupfish Brill had won at auction. 'They were wild fish,' says Brill. In those days, this was not unusual. Pupfish were taken from springs and creeks all over North America and slipped into the tanks of hobbyists, sometimes by the very same scientists who were studying them. This might shock us today, considering the eventual fate of the species; however, with the Catarina pupfish then thriving at Ojo de Potosí, the impact of this then-common practice was negligible. While scientists like Valdés González and Angeles Villeda were spreading the Catarina pupfish between scientific institutions, a parallel process was underway amongst hobbyists like Brill. In fact, for many pupfish species the work of hobbyists has been crucial to their survival.

The late Al Morales was legendary in the American pupfish-breeding community, at one point rumoured to have maintained as many as fifty populations of pupfish in ponds and tanks in his back garden. Morales, by trade a building inspector for the City of Denver, Colorado, was such a successful breeder that the US Fish and Wildlife Service gave him a pair of the critically endangered Devil's Hole pupfish – a species that the USFWS believed was impossible to breed in captivity; however, Morales managed the feat in the 1990s.* James Cokendolpher, a now-retired arachnologist, likewise managed to breed 800 La Palma pupfish, a species extinct in the wild, in a small pond in his back garden.

In 1992, Morales set up the Cyprinodon and Related Genera Study and Maintenance Group, to coordinate the breeding of pupfish within the North American aquarium hobby. In Europe, there were similar initiatives, and soon hobbyist groups either side of the Atlantic were sustaining captive populations of the Catarina pupfish. Even in 2005, Brill tells me, there was a small subgroup of hobbyists working to breed and conserve the species. 'There were probably ten, twenty different captive populations at one time,' says Chris Martin, who has led the group since Morales' death in 2013, 'including hundreds of Catarina pupfish in captive-breeding colonies in the Spanish Killifish Association.' All these fish, however, went the same way as those bred in scientific institutions.

For decades, the two halves of the pupfish-breeding community – the scientists and the hobbyists – remained mostly separate. This meant that two conservation efforts could be taking place for one species, with neither side effectively communicating what they had learned. Martin, however, has broadened membership of the group started by Morales to include more people from scientific institutions like Vienna Zoo and ZSL London Zoo. Communicating via a biannual e-newsletter, each

---

* Unbeknownst to the USFWS, Tom Baugh, another hobbyist, had managed to breed this species in an aquarium even earlier, in 1983.

member reports the population numbers of the species in their care. That way, if one member signals a population drop in their colony of a particular species, there is a global community of breeders ready to respond and send fish from their own tanks. It's a simple and effective solution, creating a throughline between passionate hobbyists like Brill and scientific institutions such as UC Berkeley, London Zoo, San Antonio Zoo, and others. Had it existed in its current, more holistic form earlier, it's possible it could have kept the Catarina pupfish with us a little longer. However, although captive-breeding efforts have become more coordinated, these alone are not enough to save many similar pupfish today.

'Right now, in northern Mexico there is an ultradiverse oasis called Cuatro Ciénegas at risk,' says Martin. 'It's probably the most diverse desert oasis in the world – with two endemic pupfish, one endemic cichlid, and an endemic box turtle that is the only example of an aquatic box turtle in the world – and for the last ten years there has been groundwater pumping in that area.'

Elsewhere there are pupfish habitats at risk for very different reasons. On the Caribbean island of San Salvador there is a series of hypersaline lakes that are home to two endemic pupfish, including the scale-eating pupfish. As climate change drives up sea levels, large predatory fish will be able to access these habitats and the pupfish within them. 'We know that they'll almost certainly be wiped out, because this has already happened elsewhere,' says Martin. In 2022, Martin and some of his colleagues travelled to Lake Chichancanab in Quintana Roo, Mexico, home to an astonishing seven pupfish species, including the endemic Maya pupfish (*Cyprinodon maya*), the largest of the group at 10cm long and the only one known to actively prey on other pupfish. 'Farmed Nile tilapia, a huge fish that'll eat anything that fits in their mouths, got into Chichancanab in the 1990s,' says Martin. And other predatory fish, such as tetras, have also colonized the lake since. All this has had a devastating impact on the pupfish. 'We searched for two days straight . . . all day long,' Martin says. But two of the endemic pupfish species were nowhere to be seen. 'These are not IUCN-listed extinctions yet,' he tells me, the operative word being 'yet'.

The problems are manifold for those fighting to conserve pupfish. While a species may be identified as at risk, convincing a captive-breeding facility to take it in is another story. 'We're at capacity' is a phrase Martin hears often when trying to place a critically endangered pupfish in captive breeding. A lot of the time, the institutional buy-in just isn't there for pupfish, which are small and understated, and unlikely to draw crowds at aquariums. On top of this, according to Martin, accessing certain habitats is sometimes impossible, due to the activities of drug cartels, and getting the necessary permits is difficult. 'We need a UN committee to go and organize this . . . a task force of people who are able to work with international governments, who can check out these habitats,' says Martin.

'I hope this makes it into your book,' he adds, and I think to myself: *I hope it makes it out of my book.*

BACK AT THE Autonomous University of Nuevo León, the challenges are different. 'We're seeing the effects of climate change every day, particularly in this part of the country,' says Ana Laura Lara Rivera, a research professor at the university and colleague of Angeles Villeda. Drought and heatwaves have been recurring themes in Mexico over the past few years. 'We have water one day and then none the next,' she says.

As a consequence, desert springs and lakes have been drying up all over the region – their waters rerouted into the fields and cattle troughs of farmers unable to source water anywhere else.

'By the time we realize there's something wrong it's often too late,' says Angeles Villeda. 'We have to conduct population studies, wait for results, and analyse the results before we can go to the authorities.' It's a lengthy process, during which the tiny springs and lagoons in question all too often vanish. 'Crop or cattle producers don't care about the fish that live in the water; they just want the water,' says Lara Rivera, 'and you can kind of understand them. If they don't have water, they can't feed their family for the year, and their animals die.'

In the excessive heat, air conditioners are constantly on, placing increased pressure on the power grid, leading to outages that can last for days, and this has obvious implications for a pupfish captive-breeding facility. 'We have a responsibility to keep these species alive as long as we can. So, what we try to do is share the fish as much as possible . . . so that there is hope if a species doesn't survive with us, because it's as easy as having a power cut over here,' says Angeles Villeda. The 'it' she is referring to is the loss of whole populations, and in some cases, entire species.

For several years, Lara Rivera and her colleagues at the university have been visiting the agricultural settlements near pupfish habitats to teach the children about pupfish. 'They love it, because most of the time they walk right by the habitat every day having no idea that there is a fish native to the area in their backyard,' says Lara Rivera. On printouts of simple line drawings of these species provided by the pupfish scientists, children practise colouring, filling in the water around the fish with green tangles of hornwort, and returning colour to the fishes' scales. No longer simply 'cachorritos', these fish are becoming unique in the minds of the children. Cultivating a sense of public pride is crucial to the long-term goal of conserving pupfish species. 'We tell them: "This is a species that a lot of people around the world care about and it lives here, right next to you! You should be proud of it,"' says Lara Rivera.

Much of what Lara Rivera, Angeles Villeda, and their colleagues do focuses on working with authorities to protect habitats and on bringing endangered species into captive breeding. However, the fates of pupfish lie not only in the hands of scientists, policy-makers, and agricultural companies, but also in the hands of the people living alongside the fish. As climate change continues to drive up temperatures, and desert springs and creeks continue to empty, they will play a part in determining whether species like the Catarina pupfish survive – whether lagoons remain lagoons or dry up and burn, like Ojo de Potosí. 'Hopefully . . .' says Lara Rivera before we say goodbye, 'when there is a problem in the future, when a spring is being drained for cattle or crops, the children will turn to their parents and say: "But what about the fish?"'

# *Epilogue*

I wrote *Lost Wonders* because I wanted to understand extinction. I hoped that by learning about recently extinct species from the people who studied and tried to save them, I'd gain a sense of the significance of these losses.

In the process of writing this book, it struck me just how unique an experience it is to be there right at the end of the story of a species, often millions of years after it began. The people I spoke with knew these species better than anyone, and their capacity to articulate the profundity of what they had witnessed frequently astounded me. Still, for all this, I'm left with the sense that extinction is something that is simply beyond comprehension. When a million years of evolution is extinguished right in front of you, what words suffice to describe this moment? Between Rich Switzer propping up the last po'ouli in a rolled-up towel, and later discovering it unresponsive – the world, in a moment, profoundly changed. In a cement kiln in Malaysia, during two nights on Christmas Island, in a fish tank in the basement of UC Berkeley, unfathomably long stories ended, leaving Earth just a little bit different, a little less rich.

Very few people have been in the presence of extinction. But it is my hope that this book has brought you a little closer to bearing witness and that the more these stories are shared the more we will learn from them. That way, the many other stories just like them currently unfolding around the world have a chance to end differently – or, perhaps, not at all.

The following is a record of the time each of the species in this book spent on Earth:

*Plectostoma sciaphilum*
Origin: tens of millions of years ago
Last seen: 2001

St Helena olive (*Nesiota elliptica*)
Origin: 12,000,000 BCE
Last seen: December 2003

*Partula labrusca*
Origin: 2,710,000 BCE
Last seen: 6 July 2002

Poʻouli (*Melamprosops phaeosoma*)
Origin: 5,800,000 – 5,700,000 BCE
Last seen: 26 November 2004

Bramble Cay melomys (*Melomys rubicola*)
Origin: unknown (possibly 900,000 BCE)
Last seen: 2009

Christmas Island pipistrelle (*Pipistrellus murrayi*)
Origin: unknown
Last seen: 26 August 2009 (23:29:38)

Christmas Island forest skink (*Emoia nativitatis*)
Origin: 18,000,000 – 9,000,000 BCE
Last seen: 31 May 2014

Alagoas foliage-gleaner (*Phylidor novaesi*)
Origin: millions of years ago
Last seen: September 2011

Cryptic treehunter (*Cichlocolaptes mazarbarnetti*)
Origin: millions of years ago
Last seen: 20 April 2007

Pinta Island tortoise (*Chelonoidis abindongii*)
Origin: 250,000 BCE
Last seen: 24 June 2012

Catarina pupfish (*Megupsilon aporus*)
Origin: 9,000,000 – 7,000,000 BCE
Last seen: 6 March 2015

# *Acknowledgements*

This book could not exist without the support, generosity, and expertise of many people, to whom I am eternally grateful. I could fill a whole other book with my thanks to them; but in lieu of a two-book deal, I will keep these thanks short.

First, a book always stands on the shoulders of other books. For *Lost Wonders*, this foundation was too broad to detail fully here, but the following books were particularly invaluable: for the chapter on *Partula labrusca*: *Snailing Around the South Seas* by Justin Gerlach; for the chapter on the po'ouli, *The Race to Save the World's Rarest Bird: The Discovery and Death of the Po'ouli* by Alvin Powell; for the chapter on the Christmas Island pipistrelle, *A Bat's End: The Christmas Island Pipistrelle and Extinction in Australia* by John Woinarski; and for the chapter on the Pinta Island tortoise, *Lonesome George: The Life and Loves of a Conservation Icon* by Henry Nichols, and *Galapagos Giant Tortoises* by James Gibbs, Linda Cayot, and Washington Tapia (Editors). I have tried my best to represent the stories of these species, and the scientists and conservationists who worked with them, accurately; however, if I have made any errors, I take full responsibility for them.

Almost fifty people – scientists, conservationists, hobbyists, and more – were interviewed for this book. For your expertise, openness, willingness to answer all of my questions, thank you. I have tried to include everyone here; I'm sorry if I have missed anyone.

Thank you to Craig Taylor-Hilton, for demystifying the complex and Sisyphean work of the IUCN, and for sending me the list of recent extinctions that started this all off.

Liew Thor-Seng – thank you for introducing me to the world of microsnails and limestone hills, and your support in writing this chapter, and for sharing with me the 3D model of *Plectostoma sciaphilum*; to Menno Schilthuizen, for illuminating what it is like to witness an extinction from space and so much more; to Junn Kitt Foon, for your infectious enthusiasm for snails and everything you shared with me about Plectostoma and the Crystal House, and Tekwyn Lim, for a wonderful chat about limestone hills and your work. Thanks also to Ruth Kim, Liz Price, Barbara van Benthem Jutting, Jon Ablett at NHM, and Toby Smith.

To Rebecca Cairns-Wicks and Vanessa Thomas-Williams for such a warm introduction to the world of St Helenian conservation, for sharing with me your memories of the olive and George Benjamin, my deepest thanks; and also to Quentin Cronk, for sharing your knowledge and recollections, and Phil Lambdon, for indulging my every question about St Helena's flora; Andrew Jackson – thank you for your openness and for telling me in such detail about your time spent with the last wild tree. And to Mike Fay and everyone at Kew, thank you for your generosity and for welcoming me so kindly, for tracking down a copy of the Project Popeye report – Mike, the chapter on the St Helena olive would not have been possible without the chats we had during the pandemic, and then in person. Also, thank you to Gayathri Anand, Laurence Carter, Shayla Ellick, Sebastian Kettley and Heather McLeod of Kew, Colin Fox, John Turner, and Ian Mathieson.

Justin Gerlach, thank you for the incredible chats about all things *Partula*, for everything you did for this chapter; Paul Pearce-Kelly, for finding time to speak about the Partula Propagation Group during the chaos of the pandemic, for the unforgettable visit to the Partula room at London Zoo. Thank you also to Christophe Brocherieux, Kirahu Howard, Laura Birch and team at ZSL, and Seth Cotterell at the California Academy of Sciences.

The chapter on the poʻouli would not have been possible without Tonnie Casey; I am unendingly grateful to you for bringing the poʻouli and its mysterious home so vividly to life for me, for all our conversations. To Jim Jacobi, too, thank you for sharing your knowledge about and experiences of the poʻouli and other Hawaiian birds, and Jim Groombridge, not only for speaking in depth about your experiences, but for showing me the ongoing legacy this species has with your students; to Kirsty Swinnerton for speaking about the capture of the last bird and everything that followed; and to Rich Switzer – thank you for speaking with such candour about your experience of caring for the last bird; to Hanna Mounce for speaking about the challenging work and resilience of the MFBRP.

John Woinarski – your help with the three chapters on Australian species was invaluable; thank you so much for your support, generosity and encouragement. Natalie Waller and Ian Gynther, thank you for sharing with me your experiences searching for the Bramble Cay melomys, and working on the cay. Lindy Lumsden, I am deeply grateful to you for speaking with me with such openness, for sharing with me your memories of the pipistrelle, for bringing it to life for me. David James, thank you for our fascinating conversation on the pipistrelle and skink, for being so candid. And Brendan Tiernan and Samantha Flakus, it was wonderful to speak with you both about lizard conservation on Christmas Island, and – Brendan – your memories of Gump. Thanks also to Roberto Portela Miguez and Patrick Cambell at NHM, Martin Schulz, Claire Ford, Alexia Jankowski, Lynnette Griffiths, the Ghost Net Collective, and Erub Arts.

Dante Buzzetti, it was wonderful to speak with you about the Alagoas foliage-gleaner and cryptic treehunter; but mostly, thank you for sharing with me your memories of Juan and your work together. Pedro Develey, for our wonderful chat about Serra do Urubu, SAVE Brazil, and bird conservation in Brazil; Phil Riris, for illuminating the distant past of the Atlantic Forest for me, and Marcos Raposo, for speaking with me about the cryptic treehunter. Thanks also to Stuart Butchart at Birdlife International and Luka Naka.

James Gibbs, it was wonderful to speak about your recollections of Lonesome George, and about the recent discoveries of giant tortoises on Vulcan Wolf. And George Dante Jr., what a pleasure it was to chat with you about taxidermy – thank you for sharing your fascinating world with me. Thanks also to Xavier Castro, Roderick Mickens, Jannatul Ahmed, the American Museum of Natural History, the Galápagos Conservancy and Galápagos National Park.

Chris Martin, thank you for such a fascinating introduction to pupfish, for sharing your memories of the last fish, and for your support and generosity with this chapter. John Brill – thank you for everything, for your deep well of knowledge and for sharing with me so much. Thank you also to Robert K. Lui, for sharing your memories too and a wonderfully warm chat, and Maria Elena Angeles Villeda and Ana Laura Lara Rivera for sharing with me your experiences of pupfish breeding and conservation, community work, and your hopes for the future. Thanks also to Priyadarsi Roy.

Many of the people mentioned here kindly read their relevant chapters for me before this book went to print. For that I'm also deeply grateful.

Beyond this book, thank you to everyone I spoke with for everything you did and continue to do for these species and others. Thank you also to the many other people not featured in the book who also played a part in attempts conserve these species. Learning about your work has given me hope that I did not think it was possible to have in these times.

## Acknowledgements

A few people in my life have been very supportive of this book and of my writing. First, Jane Lathan for seeing something in me before anyone else and encouraging me; I miss you. Paul and Kohko Hazelton, thank you for your constant cheerleading and support. Ruth Ozeki for our chat over dinner all that time ago, you gave me the confidence I needed to keep going. Misa Kim for your encouragement. Many thanks and love to my family. And special thanks to Yvonne for your infectious positivity and support!

I would also like to thank my publisher, Picador, for your patience while I researched and wrote this book. To my editor Andrea Henry, thank you for accompanying me on this journey, for your belief and encouragement, and for everything you've done to make *Lost Wonders* what it is. Thank you to Kieran Sangha, Nicholas Blake, Lindsay Nash, Stuart Wilson, and also the many other Picadores who've contributed without my knowledge (not least of all, my typesetter – I don't know who you are but thank you for all your hard work and patience . . . I see you!). Thanks also to Kris Doyle, who commissioned this book; your belief in this project remained a source of inspiration throughout. And Sam, my agent, thank you for finding a home for *Lost Wonders*, and everything you've done for it since.

Thank you also to Society of Authors. Your support for this book meant that it was possible to continue writing during the pandemic.

And finally, to Claire. None of this would have been remotely possible without you. You were there right when the idea landed in and, for a while, took over our lives. Thank you for every conversation we had about this book and the species in it, for reading every single word, for your unshakeable belief in the project and in me, and for your beautiful illustrations. I owe it all to genus hands and bears (and waffle!).

Thank you x

# Picture Acknowledgements

1.1 courtesy of Barbara van Benthem Jutting; 1.2 © The Trustees of the Natural History Museum, London; 1.3 courtesy of Thor-Seng Liew; 1.4 courtesy of Toby Smith / Rimba / Cambridge Conservation Initiative

2.1 courtesy of Andrew Jackson; 2.2 courtesy of Andrew Jackson; 2.3 courtesy of Quentin Cronk

3.2 © California Academy of Sciences. This image has been clarified with AI image enhancement software; 3.3 © London Zoo

4.1 and 4.2 courtesy of the Maui Forest Bird Recovery Project

5.1 courtesy of Ian Gynther; 5.2 © David Carter; 5.3 courtesy of Lynnette Griffiths

6.1 courtesy of Brendan Tiernan; 6.2 courtesy of Martin Schulz; 6.3 courtesy of Lindy Lumsden

7.2 courtesy of Parks Australia

8.2 courtesy of Dante Buzzetti

9.2 © Fotos593 / Shutterstock; 9.3 photo by Roderick Mickens / © AMNH

10.1 courtesy of Robert K. Liu; 10.2 courtesy of Priyadarsi D. Roy

# Notes and References

## 1 The Builders' Story

9 'shells measuring 5mm or less': Susan Milius, 'The fine art of hunting microsnails', *Science News* (accessed 20/04/23), https://www.sciencenews.org/article/fine-art-hunting-microsnails.

9 'the smallest, *Angustopila psammion*': Páll-Gergely, B., Jochum, A., Vermeulen, J. J., Anker, K., Hunyadi, A., Örstan, A., Szabó, Á., Dányi, L., & Schilthuizen, M., 'The world's tiniest land snails from Laos and Vietnam (Gastropoda, Pulmonata, Hypselostomatidae)', *Contributions to Zoology*, 91(1) (2022), pp. 62–78.

9 'Even if you were to put your nose to the ground': Tom Lathan, transcript of interview with Menno Schilthuizen (2023).

9 'After the Second World War . . . microsnail found nowhere else': Brenden Luyt, 'Michael Tweedie, Woutera van Benthem Jutting and the Mollusca of Malaya's limestone hills', *Archives of Natural History*, vol. 45.2 (2018), p. 247.

10 'They take this form': Zeinab Bakhshipouri, Husaini Omar, Zenoddin Yousof, Vahed Ghiasi, 'An Overview of Subsurface Karst Features Associated with Geological Studies in Malaysia', *Electronic Journal of Geotechnical Engineering*, 14, 2009.

10 'In 1947, Tweedie': W. S. S. van Benthem Jutting, 'The Malayan species of *Opisthostoma* (Gastropoda, Prosobranchia, Cyclophoridae), with a Catalogue of all the species hitherto described', *Bulletin of the Raffles Museum*, 24 (1952), p. 46.

10 'a special microsnail-collecting method': M. W. F. Tweedie, 'On certain Mollusca of the Malayan limestone hills', *Bulletin of the Raffles Museum*, 26 (1961), pp. 49–65.

11 'he turned to Woutera': Luyt, 'Michael Tweedie', p. 249.

11 'During the Nazi occupation of Holland': H. Engel, P. J. Van Der

Feen, 'The Life of Woutera S. S. van Benthem Jutting', *Beaufortia: Series of Miscellaneous Publications*, Zoological Museum – Amsterdam, no. 130, vol. 11 (1964), pp. 1–9.

13 'a delicate spiral sculpture': van Benthem Jutting, 'The Malayan species of *Opisthostoma*', pp. 45–7.

14 'tiny vampiric snails': Walter William Skeat, *Malay Magic: Being an Introduction to the Folklore and Popular Religion of the Malay Peninsula* (Malaysian Branch of the Royal Asiatic Society, 2005), p. 306.

14 'In 1952, van Benthem Jutting's findings were published': Jutting, 'The Malayan species of *Opisthostoma*', pp. 45–7.

15 'During the Carboniferous period . . . air above': Unnamed author, 'Carboniferous Period', *National Geographic* (accessed 03/04/24), https://www.nationalgeographic.com/science/article/carboniferous?loggedin=true&rnd=1717083026859; Ben M. Waggoner et al., 'The Carboniferous Period', Berkeley University of California, UC Museum of Palaeontology website (accessed 28/01/24), https://ucmp.berkeley.edu/carboniferous/carboniferous.php.

15 'a massive building project . . . all formed the bricks': I. Metcalfe, 'Palaeontology and age of the Panching limestone, Pahang', *Geology and Palaeontology of Southeast Asia*, vol. 21 (1980), pp. 11–17.

15 'tectonic movements . . . thriving rainforest': Reuben Clements, Navjot S. Sodhi, Menno Schilthuizen, Peter K. L. Lang, 'Limestone Karsts of Southeast Asia: Imperiled Arks of Biodiversity', *BioScience*, vol. 56, no. 9 (2006), p. 733.

16 'began to erode': W. S. S. van Benthem-Jutting, 'Non-marine Mollusca of the limestone hills in Malaya', *Proceedings of the Centenary and Bicentenary Congress of Biology* (1960), pp. 63–8; Lathan, interview with Junn Kitt Foon (2022).

16 'Erosion and dissolution . . . cave systems': M. R. Henderson, 'The Flora of the Limestone Hills of the Malay Peninsula', *Journal of the Malayan Branch of the Royal Asiatic Society*, vol. 17, no. 1 (October 1939), p. 13; Lathan, interview with Foon (2022).

16 'first species': Lathan, interview with Foon (2022).

16 'crowd together . . . three close neighbours': Thor-Seng Liew, Liz Price, Gopalasamy Reuben Clements, 'Using Google Earth to Improve the Management of Threatened Limestone Karst

Ecosystems in Peninsular Malaysia', *Tropical Conservation Science*, vol. 9:2 (2016), pp. 903–20.

16 'unidentified ancestor species': Lathan, interview with Thor-Seng Liew (2021); Lathan, interview with Schilthuizen (2023); Lathan, interview with Foon (2022).

16 'calcium ions': Natalie Hamilton (ed.), 'How Do Snails Get Their Shells? And More Questions From Our Readers', *Smithsonian Magazine* (accessed 03/04/24), https://www.smithsonianmag.com/smithsonian-institution/snails-get-shells-questions-readers-180978856/.

17 'the shape of the landscape . . . could not leave': Lathan, interview with Schilthuizen (2023); Lathan, interview with Liew (2021); Lathan, interview with Foon (2022).

17 'My personal favourite': Lathan, interview with Schilthuizen (2023).

18 'eighty-two species': Molluscabase, '*Plectostoma* H. Adams, 1865' (accessed 03/04/24), https://molluscabase.org/aphia.php?p=taxdetails&id=995232.

18 'telling stories': Lathan, interview with Foon (2022).

18 '*Atopos* slug': Thor-Seng Liew and Menno Schilthuizen, 'Association between shell morphology of micro-land snails (genus *Plectostoma*) and their predator's predatory behaviour', *PeerJ*, 2(1): e329 (2014).

19 'armour': Lathan, interview with Foon (2022).

19 'larvae of *Pteroptyx*': Liew and Schilthuizen, 'Association between shell morphology' (2014).

19 'unique patterns and shapes': Lathan, interview with Foon (2022).

19 'shell that coils on four axes': Thor-Seng Liew, Gopalasamy Reuben Clements, 'Whittenia, a new genus of land snails from Perak, Peninsular Malaysia (Gastropoda: Diplommatinidae)', *Raffles Bulletin of Zoology*, no. 35 (2020), pp. 143–8.

19 'Maybe it developed in this way': Lathan, interview with Foon (2022).

20 'mystery . . . succumb to infections': Lathan, interview with Liew (2021).

20 'vibrant array of wildlife . . . cliff faces and summit': Clements et al., 'Limestone Karsts of Southeast Asia: Imperiled Arks of Biodiversity' (2006), pp. 733–42; M. W. F. Tweedie, J. L.

Harrison, *Malayan Animal Life* (Singapore: Longman Group, 1954).

21 'natural reservoir . . . "waterfalls"': Lathan, interview with Foon (2022).

21 'soft clay called "moonmilk"': Dave Burnell, 'Moonmilk', National Speleological Society (accessed 03/04/24), https://caves.org/virtualcave/moonmilk/.

22 'millions of *Plectostoma sciaphilum*': Lathan, interview with Schilthuizen (2023).

22 'lichen and algae': Lathan, interview with Liew (2021).

22 'snails came together to mate . . . minuscule juveniles': Menno Schilthuizen et al., 'The ecology and demography of *Opisthostoma (Plectostoma) concinnum* s.l. (Gastropoda: Diplommatinidae) on limestone outcrops along the Kinabatangan River', in B. Maryati Mohamed et al., *Kinabatangan Scientific Expedition* (Universiti Malaysia Sabah, 2003), pp. 55–71.

23 'On Google Maps': Google Maps search for Bukit Pancing (accessed 03/04/24), https://www.google.co.uk/maps/place/Bukit+Pancing/@3.8922624,103.1348192,1557m/data=!3m1!1e3!4m10!1m2!2m1!1sbukit+panching!3m6!1sox31c8cocfd2515c71:0x184ef1fa719924f9!8m2!3d3.8913889!4d103.14!15sCg5idWtpdCBwYW5jaGluZ5IBBHBlYWvgAQA!16s%2Fg%2F12lrc972j?entry=ttu.

24 'These fires cause hazes': Kate Lamb, 'Indonesia's fires labelled a "crime against humanity" as 500,000 suffer', *The Guardian* (26/10/15) (accessed 28/01/24) https://www.theguardian.com/world/2015/oct/26/indonesias-fires-crime-against-humanity-hundreds-of-thousands-suffer.

24 'flames lap at the bases of the hills': Clements, Sodhi, Schilthuizen, Lang, 'Limestone Karsts of Southeast Asia: Imperiled Arks of Biodiversity' (2006), pp. 738–9.

24 'exposed so frequently': Lathan, interview with Liew (2021).

25 'Concrete is the second most used substance in the world': Jonathan Watts, 'Concrete: the most destructive material on Earth', *The Guardian* (25/01/2019) (accessed 28/01/24), https://www.theguardian.com/cities/2019/feb/25/concrete-the-most-destructive-material-on-earth.

25 'was created for the Pantheon in Rome': James Mitchell

Crow, 'The Concrete Conundrum', *Chemistry World* (March 2008), p. 63.

25 '30 billion tonnes each year': Editorial, 'Concrete needs to lose its colossal carbon footprint', *Nature* website, 28/09/21 (accessed 28/01/24), https://www.nature.com/articles/d41586-021-02612-5.

25 'enough to patio over every hill': Watts, 'Concrete: the most destructive material on Earth'.

25 'Cement, the magical binding agent': Editorial, 'Concrete needs to lose its colossal carbon footprint'.

26 'Quarrying began at Bukit Panching': Lathan, interview with Liew (2021).

26 '*Plectostoma sciaphilum* had gone extinct . . . "emotional impact"': Menno Schilthuizen, Reuben Clements, 'Tracking Land Snail Extinctions from Space', *Tentacle*, no. 16, pp. 8–9.

26 'I think the emotion comes': Lathan, interview with Schilthuizen (2023).

27 'declared extinct': Thor-Seng Liew, '*Plectostoma sciaphilum*', The IUCN Red List of Threatened Species 2015 (2015).

27 '$150 million': Toby Smith, 'Limestone Karst Ecosystems' (accessed 08/08/22), https://www.tobysmith.com/project/limestone-karst/.

28 'the company responsible': Lathan, interview with Liew (2021).

28 'Richard Owen had coined': D. F. N. Harrison, 'Sir Richard Owen (1804–1892): comparative anatomist and palaeontologist', *Journal of Medical Biography*, vol. 1 (1993), pp. 153–4.

29 'or tens of thousands of species': Fred Pearce, 'Global Extinction Rates: Why Do Estimates Vary So Wildly?', *Yale Environment 360* website, 17/08/15 (accessed 28/01/24), https://e360.yale.edu/features/global_extinction_rates_why_do_estimates_vary_so_wildly.

29 'weaponized the taxonomical process . . . existence of this snail': Jaap Vermeulen, Mohammad Effendi Marzuki, "*Charopa*' *lafargei* (Gastropoda, Pulmonata, Charopidae), a new, presumed narrowly endemic species from Peninsular Malaysia', *Bacteria* 78 (2014), pp. 31–4.

30 'biodiversity conservation agreement': Editorial, 'LafargeHolcim signs biodiversity agreement with Fauna & Flora International', *Global Cement* website, 12/07/17 (accessed 28/01/24), https://

www.globalcement.com/news/item/6325-lafargeholcim-signs-biodiversity-agreement-with-fauna-flora-international.

30 'of the 1,393 limestone hills in Malaysia': Thor-Seng Liew, Junn-Kitt Foon, Gopalasamy Reuben Clements, *Conservation of Limestone Ecosystems of Malaysia, Part I, Acknowledgements, Methodology, Overview of limestone outcrops in Malaysia, References, Detailed information on limestone outcrops of the states: Johor, Negeri Sembilan, Terengganu, Selangor, Perlis* (Institute for Tropical Biology and Conservation, Universiti Malaysia Sabah, 2021), p. 7.

30 'Ipoh, is surrounded': Natasha Zulaikha, 'The True Value of Limestone', *Macaranga* (04/01/22) (accessed 28/01/24), https://www.macaranga.org/the-true-value-of-limestone/.

31 'arks of biodiversity . . . on limestone hills': Clements, Sodhi, Schilthuizen, Lang, 'Limestone Karsts of Southeast Asia: Imperiled Arks of Biodiversity' (2006).

31 'A 2005 survey': K. Dittmar, Megan Porter et al., 'A Brief Survey of Invertebrates in Caves of Peninsular Malaysia', *Malayan Nature Journal*, vol. 57:2 (2005), pp. 221–33.

31 'contributed £88 million': Zulaikha, 'The True Value of Limestone', *Macaranga* (2022).

31 'Alongside biodiversity . . . Buddhist cave monastery': Zulaikha, 'The True Value of Limestone' (2022).

32 'Wessex Water': Wessex Water website (accessed 03/04/24), https://faq.wessexwater.co.uk/article/qed886/who-are-ytl.

32 'currently in the process of litigating': Zahratulhayat Mat Arif, 'Abbot of century-old Buddhist monastery expresses disappointment over court eviction', *New Straits Times* (12/09/23) (accessed 28/01/24), https://www.nst.com.my/news/nation/2023/09/954351/abbot-century-old-buddhist-monastery-expresses-disappointment-over-court.

32 'a petition': Change.org petition to save Sakyamuni Caves Monastery and Gunung Kanthan (accessed 03/04/24), https://www.change.org/p/his-royal-highness-sultan-nazrin-shah-save-sakyamuni-caves-monastery-save-gunung-kanthan-%E6%8B%AF%E6%95%91-%E9%87%8A%E8%BF%A6%E5%9C%A3%E6%B3%95%E5%B2%A9-%E4%B9%9F%E5%B0%B1%E6%98%AF%E5%9C%A8%E6%8B%AF%E6%95%91-%E6%8B%B1%E6%A1%A5%E5%B2%A9%E5%B1%B1.

## Notes and References

32 'economic value of bats': Zulaikha, 'The True Value of Limestone' (2022).
33 'up to six times more limestone:' Tan Cheng Li, 'A better place to quarry', *The Star* (13/07/15) (accessed 28/01/24), https://www.frim.gov.my/v1/fin/file/MY_899_20150713_N_Star_ST_pg4,5_394cb.pdf.
33 'during the Covid-19 pandemic . . . "the conservation of nature"': Zulaikha, 'The True Value of Limestone' (2022).
34 'setting a new production record': Ministry of Economy, Department of Statistic Malaysia Official Portal: Malaysia Cement Production (accessed 28/01/24), https://tradingeconomics.com/malaysia/cement-production.
34 'much that can be done . . . preventing future extinctions': Lathan, interview with Liew (2021).
35 'seven-part report': Liew, Foon, Clements, *Conservation of Limestone Ecosystems of Malaysia* (2021).
35 'a kind of handbook . . . "this is a way forward"': Lathan, interview with Liew (2021).
35 'collaborating with YTL Cement': Thor-Seng Liew et al., 'Cementing Relationships for Sustainability: A Collaboration Between Industry and Scientists to Improve Understanding of Land Snail Diversity and Ecology in Malaysia', *Tentacle*, No. 31 (2021), pp. 32–5.
38 'a 3D model': Thor-Seng Liew et al., 'A cybertaxonomic revision of the micro-landsnail genus Plectostoma Adam (Mollusca, Caenogastropoda, Diplommatinidae), from Peninsula Malaysia, Sumatra and Indochina', *Zookeys*, 393 (2014).
39 'Junn Kitt Foon told me a story': Lathan, interview with Foon (2022).

## 2 Paradise, Lost

45 'This Victorian glasshouse': Katie Avis-Riordan, 'Secrets of the Temperate House', *Royal Botanic Gardens, Kew* (20/06/2019) (accessed 27/05/24), https://www.kew.org/read-and-watch/secrets-kew-temperate-house.
45 'Travel the world in this glittering cathedral': 'Temperate House', *Royal Botanic Gardens, Kew* (accessed 2022), https://www.kew.org/kew-gardens/whats-in-the-gardens/temperate-house.

46 'mysterious misnomer': Lathan, interview with Mike Fay (2021).
46 'dark-brown branches . . . produced seeds': Q. C. B. Cronk, *The Endemic Flora of St Helena* (Oswestry: Anthony Nelson, 2000), p. 63; Andrew Jackson, 'Project Popeye – Saving the St Helena Olive. Preliminary report to the World Wide Fund for Nature', WWF Project no. 162/89 (unpublished report: Kew, 1991), p. 21.
47 'the design of those seeds': Roger Cousens, Calvin Dytham, Richard Law, *Dispersal in Plants: A Population Perspective* (Oxford, New York: Oxford University Press, 2008).
47 'One vast rock, perpendicular on every side': Bower et al., *The Modern Part of the Universal History, by the Authors of the Ancient Part*, 16 vols, folio, London, 1759, Volume 4, Chapter 6, Book XVIII, p. 399.
47 'Once rumoured to be the tip': G. C. Lawrence, The *St Helena "Wirebird"*, vol. 3, no. 99 (Jamestown, St Helena: J. A. Sim, 1963), p. 295.
47 'Further away from anywhere': Julia Blackburn, *The Emperor's Last Island: A Journey to St Helena* (New York: Vintage, 1991), p. 5.
48 'We know lots of reasons': Lathan, interview with Fay (2023).
48 'Without a single flap of its wings': Rhys Aneurin, 'Six Things You Didn't Know About Albatrosses', *Royal Society for the Protection of Birds* (accessed 2022), https://community.rspb.org.uk/getinvolved/wales/b/wales-blog/posts/six-things-you-didn-t-know-about-albatrosses.
48 'They spend most of their lives': G. Sachs et al., 'Flying at No Mechanical Energy Cost: Disclosing the Secret of Wandering Albatrosses', *PLOS ONE*, 7(9): e41449 (2012).
48 'unintended passenger': Lathan, transcript of interview with Fay (2021).
49 'St Helena emerged': J. Richardson, F. Weitz, M. Fay et al., 'Rapid and recent origin of species richness in the Cape flora of South Africa', *Nature*, 412 (2001), pp. 181–3.
49 'The message from this': Lathan, interview with Fay (2021).
50 'If anyone said to me . . . how would it turn out?': Lathan, interview with Phil Lambdon (2021).
51 'long spines and brightly fluoresces': Amy-Jayne Dutton, 'Spiky Yellow Woodlouse Pseudolaureola atlantica: Ecology and Habitat', *Saint Helena National Trust* (2017).

51 'sole occupant of the genus *Nesiota*': James E. Richardson et al., 'Genetics of the Endemic Species of Rhamnaceae from St Helena, the Tristan Da Cunha group and New Amsterdam: Implications for their Conservation' (London: Royal Botanic Gardens, Kew, 2003), p. 3; Phil Lambdon, *Flowering Plants & Ferns of St Helena* (Newbury, UK: Pisces Publications, 2012), p. 231.

51 'Today, over 500 endemic species': Natasha Stevens, 'Conserving St Helena's Endemic Invertebrates', *St Helena National Trust* (accessed 2022), http://www.trust.org.sh/conserving-st-helenas-endemic-invertebrates/

51 '45 vascular plant species': Lambdon, *Flowering Plants & Ferns of St Helena* (2012), p. 14.

51 'That means that 30 per cent': T. Churchyard et al., 'The UK's wildlife overseas: A stocktake of nature in our Overseas Territories' (Sandy, UK: RSPB, 2014), p. 57.

51 'it represents a tiny fraction': Lathan, interview with Lambdon (2021).

52 'many believed that the Garden of Eden': A. R. Azzam, *The Other Exile: The Remarkable Story of Fernão Lopes, the Island of St Helena and the Meaning of Human Solitude* (London: Icon Books, 2017), pp. 269–70.

52 'Remote islands were often seen': Richard H. Grove, *Green Imperialism: Colonial expansion, tropical island Edens and the origins of environmentalism, 1600–1860* (Cambridge, UK: Cambridge University Press, 1995), p. 32.

52 'In 1502, St Helena was discovered': Blackburn, *The Emperor's Last Island*, p. 14.

52 'Columbus had in letters described': Samuel Augustus Mitchell (Jr.), *An Accompaniment to Mitchell's Map of the World* (Philadelphia: Hinman and Dutton, 1888), p. 263.

52 'God revealed to him a small island': A. H. Schulenburg, 'The discovery of St Helena: the search continues', *Wirebird: The Journal of the Friends of St Helena*, no. 24 (Spring 2002), pp. 13–19.

53 'found what was believed to be the cross': Robin M. Jensen, *The Cross: History, Art, And Controversy* (Cambridge, MA, and London: Harvard University Press, 2017), p. 57.

53 'the island's first resident': Azzam, *The Other Exile*, pp. 188–98.

53 'Over time he cultivated . . . Lopes's goats': Blackburn, *The Emperor's Last Island*, pp. 11–18.

53 'The story of Lopes': Hugh Clifford, *Heroes in Exile* (London: John Murray, 1906), p. 17.

54 'comparisons with biblical stories': Azzam, *The Other Exile*, p. 270.

54 'Concurrently, however, the ecosystem': Jeremy Hance, 'Meet the Millennium Forest: A unique tropical island re-forestation project', *Mongabay* (02/11/22) (accessed 15/01/24), https://news.mongabay.com/2022/11/meet-the-millennium-forest-a-unique-tropical-island-reforestation-project/.

54 'In 1515': Neil MacGregor, *A History of the World in 100 Objects* (London: Penguin, 2012), pp. 412–16.

54 'It was the English . . . St Helena's identity and purpose': Stephen Royle, *The Company's Island: St Helena, Company Colonies and the Colonial Endeavour* (London: Bloomsbury Academic, 2008), p. 7.

55 'referred to by the EEIC as "factories"': Blackburn, *The Emperor's Last Island*, pp. 27–8.

55 'All told, tens of thousands': Lucina Caesar, 'St Helena, slavery and the abolition of the Trans-Atlantic slave trade', Museum of St Helena (2012), pp. 4–13.

55 'The horrors endured': Andrew Pearson, *Distant Freedom: St Helena and the Abolition of the Slave Trade, 1840–1872* (Liverpool: Liverpool University Press, 2016), p. 255; 'Slavery on St Helena', St Helena Island website (accessed 01/07/22), https://sainthelenaisland.info/slavery.htm.

55 'rocks, its trees, its soil . . . tan hides': Blackburn, *The Emperor's Last Island*, pp. 28–31.

56 'new plants invaded ecosystems': Lathan, interview with Lambdon (2021).

56 'the sea around St Helena reportedly turned black': Lathan, transcript of interview with Quentin Cronk (2023).

56 'Even that valley . . . become a paradise': Joseph Banks, *Journal of the Right Hon. Sir Joseph Banks, Bart., K.B., P.R.S.: During Captain Cook's First Voyage in H.M.S. Endeavour in 1768–71 to Terra Del Fuego, Otahite, New Zealand, Australia, the Dutch East Indies, Etc* (New York: Macmillan, 1896), p. 447.

56 'The Governor of St Helena': A. Beatson, *Tracts Relative to the Island of St. Helena; Written During a Residence of Five Years* (London: W. Bulmer, 1816), p. 192.

57 'I carefully sowed': Helen M. McKay, 'William John Burchell, Botanist', *The Journal of South African Botany*, vol. 7 (Kirstenbosch, South Africa: The National Botanic Gardens of South Africa, 1941), p. 5.
57 'Burchell also created': Lesley Shapland, 'Heartbroken on St Helena: the naturalist William John Burchell – Part One', Untold Lives, British Library website (23/01/20) (accessed 16/02/22), https://blogs.bl.uk/untoldlives/2020/01/heartbroken-on-st-helena-the-naturalist-william-john-burchell-part-one.html.
57 'plants (inevitably) escaped': Lambdon, *Flowering Plants*, p. 7.
57 'with 11,000 saplings planted': A. Beatson, *Tracts Relative to the Island of St. Helena; Written During a Residence of Five Years* (London: W. Bulmer, 1816), p. 193.
58 'the wood is dark-coloured': Ibid., p. 316.
58 'a plant craze': Donal P. McCracken, *Napoleon's Garden Island: Lost and old gardens of St Helena, South Atlantic Ocean* (London: Royal Botanic Gardens, Kew, 2022).
58 'a convalescent home': David Elliston Allen, quoted in McCracken, *Napoleon's Garden Island*, p. 12.
58 'By 1820, up to 6,000 . . . in the wild': McCracken, *Napoleon's Garden Island*, pp. 11–12.
59 'the everlasting daisy': Ibid., p. 150; Timothy L. Collins, Jeremy J. Bruhl, Alexander N. Schmidt-Lebuhn, Ian R. H. Telford, Rose L. Andrew, 'Tracing the origins of hybrids through history: monstrous cultivars and Napoléon Bonaparte's exiled paper daisies (Asteraceae; Gnaphalieae)', *Botanical Journal of the Linnean Society*, vol. 197: 2 (2021), pp. 277–89.
59 'the EEIC spent': Philip Gosse, *St Helena: 1502–1938* (London: Cassell, 1938), p. 301.
59 'control of St Helena was transferred': Arthur James Richens Trendell (ed.), *The Colonial Yearbook, 1890–92* (London: S. Low, Marston, Searle, & Rivington, 1890), p. 512.
60 'Having recently abolished slavery': Haroon Siddique, 'St Helena urged to return remains of 325 formerly enslaved people to Africa', *The Guardian*, 27/03/24 (accessed 01/04/24), https://www.theguardian.com/world/2024/mar/27/st-helena-urged-to-return-remains-of-325-formerly-enslaved-people-to-africa.
60 'In 2008, during the construction of St Helena Airport': Ibid.
60 'peaceful and respectful final resting place': Liberated African

Advisory Committee, 'The Trans-Atlantic Slave Memorial, St Helena: Master Plan, August 2020', pp. 7–28.

60 'The invention of the steamship': McCracken, *Napoleon's Garden Island*, p. xx.

60 'Meanwhile, the Suez Canal': Gosse, *St Helena: 1502–1938*, pp. 329–30.

60 'post being sent by the British': 'Post Office Statistics' on The Postal Museum website (accessed 28/01/24), https://www.postalmuseum.org/collections/statistics/.

61 'an evergreen perennial': '*Phormium tenax*' on The Royal Horticultural Society website (accessed 18/01/24), https://www.rhs.org.uk/plants/12791/phormium-tenax/details.

61 'Crucially, this hardy plant': Ken Fern, *Plants for a Future: Edible and Useful Plants for a Healthier World* (Clanfield, Hampshire: Permanent Publications, 2011), p. 188.

61 'In 1875, the first shipment': 'Colonial Reports – Annual: St Helena, report for 1928' (His Majesty's Stationery Office, 1929), p. 14.

61 'handsome indigenous plant': John Charles Melliss, *St Helena: A Physical, Historical and Topographical Description of the Island, including its geology, fauna, flora and meteorology* (London: L. Reeve, 1875), p. 256.

61 'During the two world wars': Stephen Thorning, 'Needs During First World War Resulted in Revival of Flax Growing', *Wellington Advertiser* (accessed 28/01/24), https://www.wellingtonadvertiser.com/needs-during-first-world-war-resulted-in-revival-of-flax-growing/.

61 'At its peak': Saint Helena Island website, 'The Flax Industry: *Phormium Tenax*, economic lifeline or ecological disaster?' (accessed 28/01/24), http://sainthelenaisland.info/flax.htm.

62 'All of this culminated': Matthew Engel, 'Last boat to St Helena', *The Financial Times*, 29/01/16.

62 'Wherever you went': Charles Frater, Bob Johnson, Esdon Frost, 'Island of St Helena', 1962 film (accessed 28/01/24), https://www.youtube.com/watch?v=BO5xxrWLowg.

62 'In the annual review': No author, bound document: *St Helena: Report for the Years 1964 and 1965* (Her Majesty's Stationery Office, 1967), p. 45.

62 'the flax didn't stop': D. Bormpoudakis, R. Fish, D. Leo, N. Smith,

'Cultural Ecosystem Services in St Helena: Final Report for the South Atlantic Overseas Territories Natural Capital Assessment' (South Atlantic Environmental Research Institute, 2019), p. 33.

62 'It was just the slightest interruption': Lathan, interview with Cronk (2023).
63 'Born in 1935': Basil George, 'One of St Helena's Great Sons', in Angela Wigglesworth, 'George Benjamin: A Tribute', *St Helena Connection*, no. 13 (2012), p. 9.
63 'harboured a fascination . . . presumed extinct': Lathan, interview with Cronk (2023).
64 'hadn't been seen since 1890': Phil Lambdon, *Flowering Plants*, p. 292.
64 'We got on extremely well': Lathan, interview with Cronk (2023).
64 'The rediscovery of the ebony . . . bare branches': Ibid.
65 'Project Popeye': Jackson, 'Project Popeye – Saving the St Helena Olive'.
65 'The first challenge was getting pollen . . . unfamiliar hoverfly appeared': Lathan, interview with Andrew Jackson (2024).
66 'another species endemic': P. H. van Doesburg (Sr.) and P. H. van Doesburg (Jr.) 1977, 'La faune terrestre de l'Île de Sainte-Hélène. Troisième partie' 13. Fam. Syrphidae', *Annales du Musée royal de l'Afrique centrale*, Ser. 8vo (Zool.) 215 (1976), pp. 63–74.
66 'I think the St Helena olive . . . artists' paintbrushes': Lathan, interview with Jackson (2024).
67 'self-incompatibility': Jackson, 'Project Popeye – Saving the St Helena Olive', pp. 23–4.
68 'transported to Kew for micropropagation . . . to see in the New Year': Lathan, interview with Fay (2021).
69 'A team of mycologists': Lathan, interview with Fay (2021).
69 'the last wild tree worsened . . . attacking the tree': Jackson, 'Project Popeye – Saving the St Helena Olive'.
70 'local radio station announced': Radio St Helena News segment, broadcast 14/10/1994 (accessed 10/06/2023), https://sainthelenaisland.info/rshnews19941014deathofsthelenaolive.mp3.
70 'I remember standing': Lathan, interview with Jackson (2024).
70 'The only hope now': Lathan, interview with Rebecca Cairns-Wicks and Vanessa Thomas-Williams (2021).

| | |
|---|---|
| 71 | 'They all went in exactly the same way': Ibid. |
| 71 | 'A 'cri de coeur'': Lathan, interview with Cronk (2023). |
| 72 | 'It was a really sad, slow demise': Lathan, interview with Cairns-Wicks and Thomas-Williams (2021). |
| 72 | 'Continuity': Ibid. |
| 73 | 'the identification of a successor': Jackson, 'Project Popeye – Saving the St Helena Olive', p. 40. |
| 73 | 'slipped between the gaps': Lathan, interview with Cairns-Wicks and Thomas-Williams (2021). |
| 73 | 'It was like burying': Ibid. |
| 73 | 'I found the extinction': Lathan, interview with Cronk (2023). |
| 74 | 'I think it broke his heart': Lathan, interview with Fay (2021). |
| 74 | 'a phylogenetic study': J. E. Richardson, M. F. Fay, Q. C. B. Cronk, D. Bowman and M. W. Chase, 'A phylogenetic analysis of Rhamnaceae using *rbcL* and *trnL-F* plastid DNA sequences', *American Journal of Botany*, 87 (2000), pp. 1309–24. |
| 74 | 'molecular clock': J. Richardson, F. Weitz, M. Fay et al., 'Rapid and recent origin of species richness in the Cape flora of South Africa', *Nature*, 412 (2001), pp. 181–3. |
| 75 | 'If you take those two dates': Lathan, interview with Fay (2021). |
| 76 | 'For a tree not to have any genetic variability': Ibid. |
| 75 | 'I think there is a real question': Lathan, interview with Cairns-Wicks and Thomas-Williams (2021). |
| 75 | 'Some people have suggested': Lathan, interview with Lambdon (2021). |
| 76 | 'A short walk from the Temperate House': visit to Jodrell Laboratory, Herbarium and Economic Botany Collection, at Kew, 4 November 2023, arranged and led by Fay. |
| 78 | 'a successor to Benjamin': Cairns-Wicks speaking, Lathan, interview with Cairns-Wicks and Thomas-Williams (2021). |
| 78 | 'His last words to me': Lathan, interview with Cairns-Wicks and Thomas-Williams (2021). |
| 79 | 'appointed MBE': Saint Helena Government website (accessed 10/05/23), https://www.sainthelena.gov.sh/2022/news/new-year-honours-award-2022/. |
| 79 | 'extinction is forever': Lathan, interview with Cairns-Wicks and Thomas-Williams (2021). |

## 3 A Tale of Three Snails

83 'he was ravenous': Lathan, interview with Justin Gerlach (2021).

83 'Marie-Isabelle had brought a snail': Justin Gerlach, *Snailing Round the South Seas: The Partula Story* (Cambridge, UK: Phelsuma Press, 2014), p. 189.

83 'white five-petalled flowers': Royal Botanic Gardens, Kew, '*Sclerotheca raiateensis* (Baill.) Pillon & J.Florence', Plants of the World Online (accessed 19/06/23), https://powo.science.kew.org/taxon/urn:lsid:ipni.org:names:77190175-1.

83 'These are Tiare Apetahi': Edward Dodd, *Polynesia's Sacred Isle* (New York: Dodd, Mead, 1976), p. 22.

83 'open[ing] simultaneously': Arthur W. Whistler, quoting Teuira Henry in 'Annotated List of Tahitian Plant Names', *Allertonia*, vol. 14 (2015), p. 17.

84 'said to be the embodiment': Tahiti Heritage, from the 'Collaborative Polynesian Encyclopedia' (accessed 29/03/2022), https://www.tahitiheritage.pf/legende-tiare-apetahi/.

84 'Stomach rumbling': Lathan, transcript of interview with Gerlach (2021).

84 '*Partula labrusca*, named after': Justin Gerlach, *Snailing Round the South Seas*, p. 63.

84 '*Partula* was named in 1821': Molluscabase website, 'Partula A. Férussac, 1821' (accessed 18/02/24), https://molluscabase.org/aphia.php?p=taxdetails&id=861628.

84 'Roman goddess of childbirth': Robin McKie, 'Precious escargot: the mission to return tiny snails to Pacific islands', *The Guardian* (28/09/19) (accessed 13/02/22), https://www.theguardian.com/environment/2019/sep/28/return-of-native-tiny-partula-snail-key-south-pacific-wildlife.

84 'once seventy-seven *Partula* species': Marwell Wildlife website (accessed 05/01/24), https://www.marwell.org.uk/conservation/action/species/polynesian-tree-snails/.

85 'These shells are': Justin Gerlach, *Icons of Evolution: Pacific Island tree-snails, family Partulidae* (Cambridge, UK: Phelsuma Press, 2016).

85 'endemic not only to Raiatea': Lathan, interview with Gerlach (2021).

85 'like they were platinum': Justin Gerlach, *Snailing Round the South Seas*, p. 189.
86 'Had he visited Raiatea thirty years earlier': Ibid., p. 245.
86 'During his stay . . . "find living survivors"': Ibid., pp. 191–6.
86 'his was the last sighting': Lathan, interview with Gerlach (2021).
86 'it all began in 1967': Sanjida O'Connell, 'Shell Shock', *The Independent* (June 2004) (accessed 18/02/22), https://www.independent.co.uk/news/science/shell-shock-733133.html.
87 '20cm in length . . . 142 rows of teeth': Datasheet on *Achatina fulica* (giant African land snail) in the 'Invasive Species Compendium', CABI (accessed 2020), https://www.cabi.org/isc/datasheet/2640; C. Y. L. Schotman, 'Data sheet on the giant African snail Achatina fulica Bowdich (Mollusca: Achatinidae)', RLAC-PROVEG, no. 19, pp. 16–21.
87 'They can eat everything': Zachary T. Sampson, 'Look out, Pasco: Here come giant African land snails', Phys.org, 30 June 2022.
87 'A single snail can consume': David Pimentel, 'Agricultural Invasions', in *Encyclopedia of Biodiversity (Second Edition)* (New York: Elsevier, 2013), pp. 75–84.
88 'French Polynesia's land area': CIA World Factbook, 'French Polynesia' (accessed 18/01/24), https://www.cia.gov/the-world-factbook/static/df837e2bb4cd62d299642042328bb369/FP-summary.pdf.
88 'Its 130 islands are divided': Francis James West, Sophie Foster, 'French Polynesia', *Encyclopedia Britannica* (31/05/24) (accessed 03/07/24), https://www.britannica.com/place/French-Polynesia.
88 'Clinging to the undersides': E. Dharmaraju, M. Laird, 'Transport and the spread of crop pests in tropical Polynesia', in M. Laird (ed.), *Commerce and the spread of pests and disease vectors* (Amsterdam: Praeger, 1984), pp. 257–72.
88 'in one orange grove': BBC Radio, 'SOS Snail', featuring Trevor Coote, Paul Pearce-Kelly, Rodrigo Navarro; BBC World Service (first broadcast: 02.32 Sunday 22 October 2017; accessed 2020).
89 'two wheelbarrows full': Elizabeth Murray, 'The Sinister Snail', *Endeavour*, New Series, vol. 17, no. 2 (1993), p. 81.
89 'send cars skidding': Matt Simon, 'Absurd Creature of the Week: Foot-Long, Sex-Crazed Snails That Pierce Tires and Devour Houses', *Wired* (2014: accessed 2022), https://www.wired.com/

2014/01/absurd-creature-of-the-week-foot-giant-african-land-snail/.

89 'turning "into projectiles"': 'Achatina fulica, Giant African Land snail', London Zoo website (accessed 16/01/24), https://www.londonzoo.org/whats-here/animals/giant-african-land-snail.

89 'disease risk': Joseph E. Alicata, 'Parasitic Infections of Man and Animals in Hawaii', *Hawaii Agricultural Experiment Station Technical Bulletin*, No. 61 (Honolulu: University of Hawaii, 1964), p. 33.

89 'Something had to be done': Daniel A. Vallero, *Learning from Environmental Mistakes, Mishaps and Misdeeds* (Oxford: Butterworth Heinemann, 2005), p. 279.

89 '80 per cent of their lives': K. T. Clifford et al., 'Slime-trail tracking in the predatory snail, Euglandina rosea', *Behavioural Neuroscience*, vol. 117, no. 5 (2003), pp. 1086–95.

89 'The key to this success': Justin Gerlach, 'The Ecology of the Carnivorous Snail: Euglandina Rosea', DPhil Thesis, Cambridge University (1994), p. 103.

89 'In the slime trails . . . to find a meal': Clifford et al., 'Slime-trail tracking'; Shui-Chen Chiu, Ken-Ching Chou, 'Observations on the Biology of the Carnivorous Snail Euglandina Rosea Ferussac', *Bulletin Institute of Zoology: Academia Sinica*, vol. 1 (1962).

90 'the guided missiles': Dave Clarke quoted in McKie, 'Precious escargot', *The Guardian*.

90 'In East Africa': Colin Everard, *Desert Locust Plagues: Controlling the Ancient Scourge* (London: I. B. Tauris, 2019), p. 190.

90 'The prickly pear cactus': No author, 'A successful example of biological control and its explanation', Australian National University, Research School of Biology, 2017 (accessed 2021), https://biology.anu.edu.au/news-events/news/successful-example-biological-control-and-its-explanation.

90 'The same moth': Nicole L. Elmer, 'Moth Threatens Prickly Pear Cactus', The University of Texas at Austin Biodiversity Centre blog (2020), p. 1.

90 'On 19 December 1974': J. P. Pointier, C. Blanc, 'Achatina fulica en Polynesie Française', *Bulletin de la Société des Études Océaniennes*, numéro 228, 1984 *Ana'ite* (accessed 15/01/24), https://bibnum.upf.pf/items/show/820.

| | |
|---|---|
| 91 | 'largely ignored their intended targets': Elizabeth Murray, 'The Sinister Snail', p. 78. |
| 91 | 'Or, if the species of *Partula*': Gerlach, *Snailing Round the South Seas*, p. 204. |
| 91 | 'Could anything else have been done': M. L. Fischer et al., 'O Caramuji Gigante Africano Achatina fulica no Brasil', *Champagnat Editora* (2010), p. 269. |
| 92 | '*Partula* were lucky': Lathan, interview with Gerlach (2021). |
| 92 | 'coined the phrase': Sebastian V. Grevsmühl, 'Laboratory Metaphors in Antarctic History: From Nature to Space', in Julia Herzberg, Christian Kehrt, Franziska Torma (eds.), *Ice and Snow in the Cold War* (New York: Berghahn Books, 2019), p. 5. |
| 92 | 'all the phenomena': Charlotte Bigg, David Aubin, Philipp Felsch (2009), 'Introduction: The Laboratory of Nature – Science in the Mountains', *Science in Context*, 22, p. 317. |
| 92 | 'instead of reproducing nature': Grevsmühl, 'Laboratory Metaphors', p. 6. |
| 92 | 'Geographic isolation': Murray, 'The Sinister Snail', p. 78. |
| 93 | 'the thing that made them special': Ibid. |
| 93 | 'A key topic of debate': Stephen Jay Gould, *Eight Little Piggies: Reflections in Natural History* (London: Jonathan Cape, 1993), p. 27. |
| 93 | '*Partula* snails were the perfect subjects': Murray, 'The Sinister Snail', p. 78. |
| 93 | 'extensive monographs': Gerlach, *Icons of Evolution*, p. 5. |
| 94 | 'Through this research, Crampton': Henry E. Crampton, 'Physiological problems of the geographical distribution of Partula in Polynesia, with demonstration of specimens', *Scientific Proceedings for the Society for Experimental Biology and Medicine*, 1908 meeting, pp. 57–8. |
| 95 | 'In 1961, two young geneticists': Gerlach, *Snailing Round the South Seas*, p. 121. |
| 95 | 'being watched': Lathan, interview with Gerlach (2021). |
| 95 | 'Together, these four traipsed': Murray, 'The Sinister Snail', p. 81. |
| 96 | 'Peering into': Lathan, interview with Gerlach (2021). |
| 96 | 'Natural selection': Gerlach, *Snailing Round the South Seas*, p. 123. |
| 96 | 'We only wish we could': Murray, 'The Sinister Snail', p. 81. |
| 96 | 'We have had the depressing distinction': Ibid., p. 82. |
| 96 | 'devised a simple experiment': Ibid., p. 78. |
| 97 | 'Both had been ignored': Ibid., p. 81. |

## Notes and References

- 97 'project changed abruptly': Gerlach, *Snailing Round the South Seas*, p. 172.
- 97 'The species not held already': Ibid., p. 173.
- 97 'Soon, zoos around the world': Ibid., p. 177.
- 97 'cardboard tubes': D. Clarke, 'EAZA Best Practice Guidelines for Polynesian tree snails (*Partula* spp)', EAZA (2019).
- 98 'The ark that had saved these species': Gerlach, *Snailing Round the South Seas*, pp. 164–75.
- 99 'a sixty-one-page document . . . plastic glove': Clarke, 'EAZA Best Practice Guidelines'.
- 99 'Grass pellets 300g': Ibid., p. 25.
- 100 'They can be surprisingly speedy': Ibid., p. 21.
- 100 'The snails knew': Lathan, interview with Gerlach (2021).
- 100 'whenever she had a day off': Gerlach, *Snailing Round the South Seas*, p. 219.
- 100 'It was extraordinary': Lathan, interview with Gerlach (2021).
- 101 'Sometimes, no matter': Lathan, interview with Paul Pearce-Kelly (2021).
- 101 'humanely euthanize': D. Clarke, 'EAZA Best Practice Guidelines', pp. 31–2.
- 101 'There have been remarkable . . . population numbering 296': Gerlach, *Snailing Round the South Seas*, pp. 177–9 and pp. 227–8.
- 102 'the most precise moment': Ibid., p. 225.
- 102 '1.5 million years BC': Lydia Pyne, *Endlings: Fables for the Anthropocene* (Minneapolis: University of Minnesota Press, 2022) (accessed 16/01/24), https://manifold.umn.edu/read/endlings/section/048eec69-61a8-483b-89ec-534785f37fdc.
- 102 'Gerlach delivered the last *Partula labrusca*': Lathan, interview with Gerlach (2024).
- 103 'Four years later, in 1996': Gerlach, *Snailing Round the South Seas*, p. 233.
- 103 'only five': Trevor Coote, Eric Loeve, 'From 61 species to five: endemic tree snails of the Society Islands fall prey to an ill-judged biological control programme', *Oryx*, vol. 37, no. 1 (2003), p. 91.
- 104 'Crampton's discovery': H. E. Crampton, C. M. Cooke, 'New species of *Partula* from South-eastern Polynesia', *Occasional Papers of the Bernice P. Bishop Museum*, vol. 21 (1953), pp. 135–59.
- 104 'Each day . . . cleaned': Clarke, 'EAZA Best Practice Guidelines'.

104 'We just don't know': Lathan, interview with Gerlach (2021).
104 '1979 study on the courtship dances': Gerlach, *Snailing Round the South Seas*, p. 134.
105 'I'd love to know more about this': Lathan, interview with Gerlach (2021).
105 'the ancestor of all *Partula*': Taehwan Lee, Jingchun Li et al., 'Evolutionary history of a vanishing radiation: isolation-dependent persistence and diversification in Pacific Island partulid tree snails', *BMC Evolutionary Biology* (September 2014), p. 2.
106 '*Homo sapiens* migrated eastwards': Michael H. Fisher, *Migration: A World History* (Oxford: Oxford University Press, 2014), p. 127.
106 'became the centre': Richard Leviton, *The Galaxy on Earth: a Traveler's Guide to the Planet's Visionary Geography* (Charlottesville, Virginia: Hampton Roads, 2002), p. 416.
106 'Soon, stones began migrating': Bernard Salvat, Tamara Maric, Tyler Goepfert, Anton Eisenhauer, 'The marae of Taputapuatea (Raiatea, Society Islands) in 2016: nature, age and origin of coral erected stones', *Journal de la Société des Océanistes*, no. 149 (2019), p. 282.
106 'pierced and then threaded together': Hei Upo'o, Auckland museum collection (ornament 12517), https://www.aucklandmuseum.com/collection/object/am_humanhistory-object-557923.
106 'Some necklaces': Anna Laura Jones, 'Women, Art and the Crafting of Ethnicity in Contemporary French Polynesia', *Pacific Studies*, vol. 15, no. 4 (1992), p. 138.
107 'A 2007 study found': T. Lee et al., 'Prehistoric inter-archipelago trading of Polynesian tree snails leaves a conservation legacy', *Proc Biol Sci* 2007 (November 2022).
108 'the transit of Venus is': Royal Astronomical Society, 'Transits of Venus', RAS website (2012, accessed 16/01/24), https://ras.ac.uk/education-and-careers/for-everyone/125.
108 'In 1691 . . . island of Tahiti': D. A. Teets, 'Transits of Venus and the Astronomical Unit', *Mathematics Magazine*, 76(5) (2003), pp. 337–8.
108 'Arriving in Tahiti': Gerlach, *Snailing Round the South Seas*, pp. 9–18.
108 'sealed secret orders': Lorraine Boissoneault, 'Captain Cook's 1768 Voyage to the South Pacific Included a Secret Mission',

*Smithsonian Magazine* (24/08/2018) (accessed 16/01/24), https://www.smithsonianmag.com/history/captain-cooks-1768-voyage-south-pacific-included-secret-mission-180970119/.

109 'I then hoisted': James Cook, ed. W. J. L. Wharton, *Captain Cook's Journal during his First Voyage Round the World made in H.M. Bark 'Endeavor' 1768–71, A Literal transcription of the Original MSS* (London: Elliott Stock, 1893) (Project Gutenberg; no page numbers).

109 'from the species *Partula faba*': Justin Gerlach, 'Captain Cook's bean snail: *Partula faba*' (2016, accessed 2019), https://islandbiodiversity.com/faba.htm.

109 'Polynesian traditions': no cited author, 'Polynesia, 1800–1900 A.D.', in *Heilbrunn Timeline of Art History*, New York: The Metropolitan Museum of Art (2004, accessed 2021), https://www.metmuseum.org/toah/ht/10/ocp.html.

110 'the *Partula* genus was named': Gerlach, *Snailing Round the South Seas*, p. 63.

110 'also responsible for weaving death': S. Breemer, J. H. Waszink, 'Fata Scribunda', *Mnemosyne*, Third Series, vol. 13:4 (1947), pp. 263–4.

110 'The *Partula* room at London Zoo is tucked away': visit to the *Partula* Room at ZSL London Zoo, led by Paul Pearce-Kelly (2022).

111 'What we've found': Lathan, interview with Gerlach (2021).

112 'devastating': Lathan, interview with Pearce-Kelly (2021).

112 'When you get down to one': Lathan, interview with Gerlach (2021).

112 'he'd worked feverishly': ibid.; Gerlach, *Icons of Evolution*, p. 224.

112 'But despite adjusting': Lathan, interview with Gerlach (2021).

113 'This extinction': T. Coote, *Partula faba*. The IUCN Red List of Threatened Species 2009: e.T16288A5597344 (accessed 16/01/24), https://dx.doi.org/10.2305/IUCN.UK.2009-2.RLTS.T16288A5597344.en.

113 'Another species, *Achatinella apexfulva*': Christie Wilcox, 'Lonely George the tree snail dies, and a species goes extinct', *National Geographic* (2019), https://www.nationalgeographic.com/animals/article/george-the-lonely-snail-dies-in-hawaii-extinction.

113 'a looming cabinet': visit to the *Partula* Room.

114 'Given the numbers': Lathan, interview with Gerlach (2021).

114 'Polynesian artisans ... "easily exploited"': Anna Laura Jones, 'Women, Art, and the Crafting of Ethnicity in French Polynesia', *Pacific Studies*, Vol. 15:4 (1992), pp. 137–54.

114 'Since the introduction ... shut down': T. Coote, E. Loeve, 'From 61 species to five', pp. 91–6.

115 'Taputapuatea, faces out': UNESCO World Heritage list (accessed 2021), https://whc.unesco.org/en/list/1529/.

115 'human and animal sacrifices': Douglas L. Oliver, *Ancient Tahitian Society* (Hawaii: University of Hawaii Press, 1974), p. 1213.

115 'Taputapuatea was a ruin': P. H. Buck (Te Rangi Haroa), *Vikings of the Sunrise* (Philadelphia: Lippincott, 1938), pp. 85–6.

115 'The bleak wind of oblivion': Ibid.

115 'Taputapuatea had been restored': Anne Salmond, *Aphrodite's Island: The European Discovery of Tahiti* (Berkeley: University of California Press, 2011), p. 29.

116 'A local cultural association': UNESCO World Heritage List.

116 'Upon death, a person's soul': Ake Hultkrantz, *The North American Indian Orpheus Tradition: A Contribution to Comparative Religion* (Stockholm: Caslon Press, 1957), p. 197.

116 'During the autopsy': Gerlach, *Icons of Evolution*, p. 276.

116 'A thin sliver of the last snail's foot': Lathan, interview with Gerlach (2021).

116 'The Frozen Ark – a "biobank"': Costa Bento, Mafalda, 'Frozen Ark Project', Cardiff University School of Biosciences website (accessed 17/01/24), https://www.cardiff.ac.uk/biosciences/research/projects/frozen-ark-project.

117 'Had the Clarkes and Murrays not': The Frozen Ark website (accessed 07/01/24), https://www.frozenark.org/our-history.

117 'safeguard against extinction': Jim Al-Khalili (presenter), 'Ann Clarke on the Frozen Ark', *The Life Scientific*, BBC Radio 4 (first broadcast: 21.30 Tuesday 2 May 2017; accessed 2019).

117 'a carnivorous flatworm': Global Invasive Species Database (2024) (accessed 17/01/2024), http://www.iucngisd.org/gisd/100_worst.php.

117 'and climate change means': Lathan, interview with Gerlach (2021).

117 'However, in 1994': D. Clarke, 'EAZA Best Practice Guidelines for Polynesian tree snails (*Partula* spp)', EAZA (2019), p. 12.

118 'still present, living as a cannibal': Murray, 'The Sinister Snail', p. 83.
118 'It was feared': Gerlach, *Snailing Round the South Seas*, pp. 217–18.
118 'In the Palm House': Paul Pearce-Kelly, Georgina M. Mace, Dave Clarke, 'The release of captive bred snails (*Partula taeniata*) into a semi-natural environment', *Biodiversity and Conservation*, no. 4, pp. 645–63.
118 'most intensively monitored snails ever': Gerlach, *Snailing Round the South Seas*, p. 219.
119 'Concerns that the snails': Pearce-Kelly et al., 'The release of captive bred snails', pp. 645–63.
119 'A 20 × 20m . . . electric fence': Gerlach, *Snailing Round the South Seas*, p. 221.
118 'three hundred *Partula* snails': Clarke, 'EAZA Best Practice Guidelines', p. 12.
119 'Trevor Coote was': Lathan, interview with Pearce-Kelly (2021).
119 'While performing fieldwork': Lathan, interview with Gerlach (2024).
119 'The dry and dusty trunks': Clarke, 'EAZA Best Practice Guidelines', p. 13.
119 'While the introduced flatworm': Clarke, 'EAZA Best Practice Guidelines', p. 14.
119 'Reintroductions started in 2015': Lathan, interview with Gerlach (2021).
120 'Coote oversaw': Paul Pearce-Kelly, 'Remembering Trevor Coote', ZSL website (accessed 08/01/24), https://www.zsl.org/news-and-events/feature/remembering-dr-trevor-coote.
120 'In addition to': Paul Pearce-Kelly, 'In Memoriam: Trevor Coote (1953–2021)', *Tentacle*, no. 29 (March 2021), p. 2.
120 'largest ever reintroduction': Lathan, interview with Gerlach (2024).
120 '21,000 snails': P. Barkham, 'Polynesian snails release is biggest ever of "extinct-in-the-wild" species', *The Guardian* (27/04/23) (accessed 17/01/24), https://www.theguardian.com/environment/2023/apr/27/polynesian-partula-snails-extinct-in-the-wild-species.
120 'conservationists are now waiting': Lathan, interview with Gerlach (2024).
120 'Our snails move slowly': Christophe Brocherieux, Robert

H. Cowie, 'Areho, natural and cultural heritage: report of a seminar on land snail conservation, August 2019, Tahiti', *Tentacle*, no. 28 (March 2020), pp. 36–9.

121 'Back in 2017 . . . "Welcome home, guys.": BBC Radio, 'SOS Snail'.

## 4 The Little Masked Bird

125 'where the rain rarely stops . . . The air is alive with their calls': Lathan, interview with Tonnie Casey (2023).

126 'that creeps up and down': T. K. Pratt, S. G. Fancy, and C. J. Ralph, 'Maui Nukupuu (*Hemignathus affinis*)', version 1.0, in A. F. Poole and F. B. Gill (eds), *Birds of the World* (Ithaca, NY: Cornell Lab of Ornithology, 2020) (accessed 26/01/24), https://doi.org/10.2173/bow.nukupu1.01.

126 'land has remained pristine': Alvin Powell, *The Race to Save the World's Rarest Bird* (Mechanicsburg, PA: Stackpole Books, 2008), p. 41.

126 'In amongst the hubbub . . . while the mother keeps the chick warm': Cameron Kepler, Thane Pratt et al., 'Nesting behavior of the Poo-uli', *Wilson Bulletin*, 108(4) (1996), pp. 620–38.

128 'jeans he has tried to make waterproof': Powell, *The Race to Save the World's Rarest Bird*, p. 36.

128 'They are university students . . . until now': Lathan, interview with Casey (2023); Powell, *The Race to Save the World's Rarest Bird*, pp. 37–55.

128 'They didn't really know what humans were': Lathan, interview with Casey (2023).

129 'just materialized': Powell, *The Race to Save the World's Rarest Bird*, p. 44.

129 'little yellow marshmallow . . . formal species identification': Lathan, interview with Casey (2023).

131 'That was a really challenging trip . . . like a popsicle': Lathan, interview with Jim Jacobi (2023).

131 'Casey then hand-delivered': Lathan, interview with Casey (2023).

131 'scientific name *Melamprosops phaeosoma*': Tonnie L. C. Casey, James D. Jacobi, 'A new genus and species of bird from the island of Maui, Hawaii (Passeriformes: Drepanididae)', *Occasional Papers of the Bernice P. Bishop Museum*, 24 (12) (1974), pp. 215–26.

132  'We felt that this was important': Lathan, interview with Casey (2023).
132  'often onomatopoeic': Lathan, interview with Jacobi (2023).
132  'Pukui chose a name': Lathan, interview with Casey (2023).
132  'One of the oldest': Robert C. Fleischer et al., 'Phylogenetic Placement of the Poʻouli, Melamprosops Phaeosoma, based on Mitochondrial DNA Sequence and Osteological Characters', *Studies in Avian Biology*, 22 (2002), pp. 98–103.
132  '5.8–7.2 million years ago': H. R. Lerner et al., 'Multilocus resolution of phylogeny and timescale in the extant adaptive radiation of Hawaiian honeycreepers', *Current Biology*, 21(21) (8 November 2011), pp. 1838–44.
132  'lost over the Pacific': Powell, *The Race to Save the World's Rarest Bird*, p. 63.
132  'geological processes': G. P. L. Walker, 'Review article: geology and volcanology of the Hawaiian Islands', *Pacific Science*, 44 (4) (1990), pp. 315–47.
133  'At long last . . . no amphibians, lizards': Powell, *The Race to Save the World's Rarest Bird*, p. 58.
133  'the only native land mammal': A. L. Russell et al., 'Two Tickets to Paradise: Multiple Dispersal Events in the Founding of Hoary Bat Populations in Hawaiʻi', *PLOS ONE*, 10:6 (2015).
134  'Some developed special tongues . . . insects and sap': Douglas H. Pratt, *The Hawaiian Honeycreepers* (Oxford: Oxford University Press, 2005).
134  'the lost flock of forest birds': Powell, *The Race to Save the World's Rarest Bird*, pp. 37–55.
134  'where one ancestor species': Rosemary Gillespie, 'Oceanic Islands: Models of Diversity', *Encyclopedia of Biodiversity* (Amsterdam: Elsevier Inc., 2007), p. 1.
134  'a broad range of ecological niches': Pratt, *The Hawaiian Honeycreepers*, p. 3.
134  'Māui is said to have painted': Cultural Significance: Maui Forest Bird Recovery Project website, 'Woven through Hawaiʻi's history' (accessed 22/02/24), https://www.mauiforestbirds.org/cultural-significance/.
135  'when Dean Amadon examined': Lathan, interview with Casey (2023).
135  'All Hawaiian honeycreepers smell': T. K. Pratt, C. B. Kepler,

T. L. Casey (2020), 'Poo-uli (*Melamprosops phaeosoma*)', version 1.0, in Poole and Gill (eds), *Birds of the World*; Douglas H. Pratt, 'Is the Poo-uli a Hawaiian Honeycreeper?', *The Condor*, vol. 94, no. 1 (Berkeley: University of California Press, 1992), pp. 172–80.

135 'DNA analysis': Fleischer et al., 'Phylogenetic Placement', pp. 98–103.

136 'Polynesians arrived': Sumner La Croix, *Hawai'i: Eight Hundred Years of Political and Economic Change* (Chicago: University of Chicago Press, 2019).

136 'tucked in the hulls . . . shredding native vegetation': Pratt, *The Hawaiian Honeycreepers*, pp. 23–32.

137 'a modern-day crescendo': William Allen, 'Restoring Hawaii's Dry Forests: Research on Kona slope shows promise for native ecosystem recovery', *BioScience*, vol. 50, no. 12 (2000), pp. 1037–41.

137 'a dreadful seed': Samuel Kamakau, *Na Mo'olelo a ka Po'e Kahiko* ('Tales and Traditions of the People of Old'), from a section omitted from the English translation and quoted in Anna Della Subin, 'How to kill a god: the myth of Captain Cook shows how the heroes of empire will fall', *The Guardian* (18/01/2022) (accessed 22/01/24), https://www.theguardian.com/news/2022/jan/18/how-to-kill-a-god-captain-cook-myth-shows-how-heroes-of-empire-will-fall.

138 'In 1826, Hawaiians': Powell, *The Race to Save the World's Rarest Bird*, pp. 81–2.

138 'singing in the ear': D. L. Van Dine, 'Mosquitoes in Hawaii', *Hawaii Agricultural Experiment Station: Bulletin no. 6* (Hawaiian Gazette Company, Ltd., 1904), p. 7, https://www.ctahr.hawaii.edu/oc/freepubs/pdf/B-6.pdf.

138 'far more dangerous things': Cardé T. Ring and Vincent H. Resh (eds)., *A World of Insects: The Harvard University Press Reader* (Cambridge, MA: Harvard University Press, 2012), p. 179.

139 'migratory birds . . . on the legs': R. E. Warner, 'The Role of Introduced Diseases in the Extinction of the Endemic Hawaiian Avifauna', *The Condor*, 70 (1968), pp. 101–20.

139 'The 'ōhi'a blossoms': H. W. Henshaw, 'Complete List of the Birds of the Hawaiian possessions, with notes on their habits',

appearing in *Hawaiian Almanac and Annual For 1902* (Thos. G. Thrum: Honolulu, 1901), p. 59.

140 'Extinction followed extinction': Powell, *The Race to Save the World's Rarest Bird*, pp. 67–8.

140 'In 1939, the American painter': Tony Perrottet, 'O'Keeffe's Hawaii', *The New York Times* (30/11/2012) (accessed 03/07/22), https://www.nytimes.com/2012/12/02/travel/georgia-okeeffes-hawaii.html.

140 'It is really a beautiful world': Jennifer Saville (ed.), 'Off in the Far Away: Georgia O'Keeffe's Letters Home from Hawai'i', *The Hawaiian Journal of History*, vol. 46 (2012), p. 101.

140 'The country is very paintable': Ibid., p. 117.

140 'So many of the flowers': Ibid., p. 95.

140 'it is really a wonderful flower': Ibid., p. 117.

140 'many things are so beautiful': Ibid., p. 100.

141 'Below 800m elevation': Pratt, *The Hawaiian Honeycreepers*, p. 24.

141 'different than': Saville, 'Off in the Far Away', p. 102.

142 '*Melamprosops*, with a very localized range': Casey, Jacobi, 'A new genus and species of bird from the island of Maui, Hawaii', pp. 215–26.

142 'The po'ouli was added': US Fish and Wildlife Service website profile for the po'ouli (accessed 27/01/24), https://www.fws.gov/species/poo-uli-melamprosops-phaeosoma.

142 'long period of silence . . . 80–99 per cent': Powell, *The Race to Save the World's Rarest Bird*, pp. 106–17.

143 'Despite the exclosure': Ibid., p. 148.

143 'a 1994 search turned up five': Ibid., pp. 155–6.

143 'We presume the po'ouli': Paul Baker, 'Status and Distribution of the Po'ouli in the Hanawi Natural Area Reserve between December 1995 and June 1997', *Studies in Avian Biology*, no. 22 (2001), pp. 144–50.

144 'mired in indecision and infighting': Lathan, interview with Casey (2023); Powell, *The Race to Save the World's Rarest Bird*, pp. 174–6.

144 'Actually, that photo . . . shaking in the video': Lathan, interview with Jim Groombridge (2023).

146 'The video he's referring to . . . "Unbelievable!"': 'Po'ouli Translocation Video News Release', Flycatcher Films Ltd., video shared by Groombridge.

147 'Groombridge and his team released . . . "What Next?"': Lathan, interview with Groombridge (2023).
148 'riding in a helicopter . . . "this was the last po'ouli"': Lathan, interview with Kirsty Swinnerton (2024).
148 'at the captive-breeding centre': Powell, *The Race to Save the World's Rarest Bird*, p. 11.
148 'And, to help the po'ouli': Lathan, interview with Swinnerton (2024).
149 'more curious than afraid': Powell, *The Race to Save the World's Rarest Bird*, p. 10.
149 'Rich Switzer was just thirty . . . "It's mosquitoes"': Lathan, interview with Rich Switzer (2024).
150 'trips to capture . . . "went downhill"': Lathan, interview with Swinnerton (2024).
150 '12ft by 4ft aviary': Powell, *The Race to Save the World's Rarest Bird*, p. 15.
151 'It can all be over': Lathan, interview with Switzer (2024).
151 'On 15 November . . . continued to deteriorate': Powell, *The Race to Save the World's Rarest Bird*, pp. 19–24.
152 '"Typically, if you're caring" . . . He wipes his eyes': Lathan, interview with Switzer (2024).
153 'It was very dark . . . Packaged up and frozen': Lathan, interview with Swinnerton (2024).
154 'systemic fungal infection': Lathan, interview with Switzer (2024).
154 'world's largest collection': Elizabeth Kolbert, *The Sixth Extinction: An Unnatural History* (London: Bloomsbury, 2015), pp. 384–6.
154 'declared extinct': BirdLife International, 'Melamprosops phaeosoma', The IUCN Red List of Threatened Species 2019 (accessed 17/06/23), https://www.iucnredlist.org/species/22720863/153774712.
154 'had spent months': Lathan, interview with Casey (2023).
155 'One doesn't have to be a scientist': Powell, *The Race to Save the World's Rarest Bird*, pp. 217–18.
155 'Perhaps the most important lesson': Eric VanderWerf et al., 'Decision analysis to guide recovery of the po'ouli, a critically endangered Hawaiian honeycreeper', *Biological Conservation*, 129 (2006), pp. 383–92.
156 'Fossilized remains': Powell, *The Race to Save the World's Rarest Bird*, p. 79 and p. 236; J. M. Scott et al., 'Forest bird communities

of the Hawaiian islands: their dynamics, ecology, and conservation', *Studies in Avian Biology*, No. 9 (Lawrence, Kansas: Allen Press, 1986), p. 183.

157 'Hawaii is a fascinating place': Lathan, interview with Jacobi (2023).

157 'the extinction capital': Benji Jones, 'Welcome to the extinction capital of the world', Vox website (accessed 03/13/24), https://www.vox.com/down-to-earth/2023/12/14/23990382/extinction-capital-hawaii-endangered-species-act.

157 'a third of all species': Center for Biological Diversity, '9 Species From Hawai'i Lost to Extinction' (29/09/21) (accessed 27/01/24), https://biologicaldiversity.org/w/news/press-releases/9-species-from-hawai'i-lost-to-extinction-2021-09-29/.

157 'It's not just about': Lathan, interview with Jacobi (2023).

157 'In the years . . . "I miss the birds"': Lathan, interview with Casey (2023).

157 'in a lecture room': Jim Groombridge lecture on the poʻouli to Kent University undergraduate students, audio recording (2023).

158 'I think that if there's one uplifting': Lathan, interview with Groombridge (2023).

159 'Satellite images of Maui': 'Devastation in Maui', NASA The Earth Observatory website (accessed 27/01/24), https://earthobservatory.nasa.gov/images/151688/devastation-in-maui.

160 'CCTV footage': 'MBCC fire – security cam footage', video upload by the Hawaii Department of Land and Natural Resources (accessed 27/01/24), https://vimeo.com/853861911.

160 'The fire could have destroyed . . . disbanding the project altogether': Lathan, interview with Hanna Mounce (2024).

160 'the total kiwikiu population': The Maui Forest Bird Working Group, *Kiwikiu Reintroduction Plan: August 2019*, 2018, p. 11.

160 'In October 2019, fourteen kiwikiu': C. C. Warren et al., 'Kiwikiu Translocation Report 2019', Internal Report, 2020, p. 11.

160 'The disease landscape . . . this time around': Lathan, interview with Mounce (2024).

163 'In 2022, the US Department of the Interior': US Department of the Interior, 'Department of the Interior Releases Multiagency Strategy for Preventing Imminent Extinction of Hawai'i Forest Birds' (15/12/22) (accessed 27/01/24), https://www.doi.gov/

pressreleases/department-interior-releases-multiagency-strategy-preventing-imminent-extinction.
163 'A lot of people ask me': Lathan, interview with Mounce (2024).
163 'We have brought eight': email from Mounce (2024).

## 5 Driftwood Stowaway

167 'I follow Roberto': visit to Natural History Museum collections, led by Roberto Potela Miguez (2023).
167 'the museum's vast collection': Natural History Museum website (accessed 02/06/24), https://www.nhm.ac.uk/our-science/services/collections/zoology.html.
167 'headlines': Michael Slezak, 'Revealed: first mammal species wiped out by human-induced climate change', *The Guardian* (14 June 2016) (accessed 17/03/2022), https://www.theguardian.com/environment/2016/jun/14/first-case-emerges-of-mammal-species-wiped-out-by-human-induced-climate-change.
168 'the first recorded': Ian Gynther, Natalie Waller, Luke K.-P. Leung, 'Confirmation of the extinction of the Bramble Cay melomys *Melomys rubicola* on Bramble Cay, Torres Strait: results and conclusions from a comprehensive survey in August–September 2014', June 2016, University of Queensland, Australia, p. i.
168 'Bramble Cay melomys holotype': Natural History Museum Data Portal, Object: 1846.8.26.7, https://data.nhm.ac.uk/dataset/56e711e6-c847-4f99-915a-6894bb5c5dea/resource/05ff2255-c38a-40c9-b657-4ccb55ab2feb/record/3596538.
168 'You can see where': Miguez, during Natural History Museum visit (2023).
168 '274 small islands': Western Australian Museum website (accessed 17/03/2022), https://museum.wa.gov.au/explore/online-exhibitions/1968-torres-strait-islander-track-laying-world-record/acknowledgement-c-1.
168 'Despite being 227km . . . foraminifera': Joanna C. Ellison, 'Natural History of Bramble Cay, Torres Strait', Washington DC, National Museum of Natural History, Smithsonian Institution, 1998, Atoll Research Bulletin no. 455.
169 'tiny stars and spirals': Kristie Nobes, Sven Uthicke, 'Benthic Foraminifera of the Great Barrier Reef: A guide to species

potentially useful as Water Quality Indicators', Report to the Marine and Tropical Sciences Research Facility (Cairns: Reef and Rainforest Research Centre Limited, 2008).

169 'state of constant flux': Andrew Dennis and Daryn Storch, 'Conservation and Taxonomic Status of the Bramble Cay Melomys, *Melomys rubicola*', Endangered Species Program, Project no. 598, Progress Report for Environment Australia Endangered Species Program (1998), p. 7 and p. 14.

169 'Apart from a spotty covering': Ellison, 'Natural History of Bramble Cay'.

169 'wandering white butterflies': No author, 'Bramble Cay', *The Bulletin*, vol. 55: 2838 (04/07/1939), p. 21.

169 'and were distinguished': Slezak, 'Revealed: first mammal species wiped out by human-induced climate change'.

169 'In April 1845': Gynther et al., 'Confirmation of the extinction', p. 30.

170 'A month later': G. R. Fulton, 'Bramble Cay Melomys Melomys rubicola Thomas 1924: specimens in the Macleay Museum', *Proceedings of the Linnean Society of New South Wales*, 138 (2016), pp. 59–60.

170 'This is Michael Rogers Oldfield Thomas': Roberto Potela Miguez, 'Oldfield Thomas: In his own words', Natural Sciences Collections Association (28 March 2019) (accessed 01/12/23), https://natsca.blog/2019/03/28/oldfield-thomas-in-his-own-words/.

170 'he added the Bramble Cay melomys . . . feet white': O. Thomas, 'A subdivision of the genus Uromys', *Annals and the Magazine of Natural History*, 9 (1922), pp. 260–1.

171 'This lurch is the Fly River': Frank Jacobs, 'Who Bit My Border?', *New York Times* (13 March 2012) (accessed 01/12/23), https://archive.nytimes.com/opinionator.blogs.nytimes.com/2012/03/13/who-bit-my-border/.

171 'Every second': P. T. Harris, E. K. Baker, 'The nature of sediments forming the Torres Strait turbidity maximum', *Australian Journal of Earth Sciences*, 38:1, pp. 65–78, p. 67.

171 'is the northernmost tip': Ellison, 'Natural History of Bramble Cay', p. 1.

171 'Clumps of earth . . . earthen rafts': Gynther, Waller, Leung, 'Confirmation of the extinction', p. 18.

171 'It isn't known ... stowed away on board': Dennis and Storch, 'Conservation and Taxonomic Status,' p. 3.
172 'eleven plant species': Ellison, 'Natural History of Bramble Cay', p. 8.
172 'Several depressions': Ibid., p. 6.
172 'small winding tunnels': Ibid., p. 14.
172 'Studies of related *Melomys*': Tasmin Rymer, James Cook University, on Tree-Kangaroo and Mammal Group website: https://www.tree-kangaroo.net/mammals-wet-tropics/fawn-footed-melomys.
173 'no predators': Peter Latch, 'Recovery Plan for the Bramble Cay Melomys Melomys rubicola', Report to Department of the Environment, Water, Heritage and the Arts (Canberra: Environmental Protection Agency, Brisbane, 2008), p. 7.
173 'thousands of nesting birds': Dennis and Storch, 'Conservation and Taxonomic Status', p. 18.
173 'the slate-grey ... frigate bird': Ellison, 'Natural History of Bramble Cay', pp. 9–12.
173 'contact was sometimes inevitable': C. J. Limpus, C. J. Parmenter, C. H. S. Watts, 'Melomys rubicola, an endangered murid rodent endemic to the Great Barrier Reef of Queensland', *Australian Mammalogy*, 6 (1983), pp. 77–9.
173 'venturing into turtle and bird nests': Dennis, Storch, 'Conservation and Taxonomic Status', p. 3.
174 'significant hazard to these ships': Ellison, 'Natural History of Bramble Cay', p. 5 and p. 24.
174 'Groups of two flashes': Clem L. Lack, 'The Taming of the Great Barrier Reef', *Journal of the Royal Historical Society of Queensland*, vol. 6:1 (Brisbane, 1959), p. 154.
174 'specific sequence of flashes': Lighthouse Preservation Society website (accessed 02/09/23), https://lighthousepreservation.org/facts.
174 'omelettes, boiled and scrambled eggs': Ellison, 'Natural History of Bramble Cay', p. 4.
174 'The Erubam Le': Queensland government website, 'Erub (Darnley Island)' (accessed 02/09/23), https://www.qld.gov.au/firstnations/cultural-awareness-heritage-arts/community-histories/community-histories-e-i/community-histories-erub.

| | |
|---|---|
| 174 | 'With expert knowledge': Seamew, 'The Legend of Bramble Cay', *The Queenslander* (19 December 1921) (accessed 02/09/23), https://trove.nla.gov.au/newspaper/article/22617102. |
| 174 | 'In 1906, a boat transporting': Colonel Kenneth Mackay, *Across Papua: Being an account of a voyage round, and a march across, the territory of Papua, with the Royal Commission* (London: Witherby & Co, 1909), p. 172. |
| 174 | 'anecdotal evidence': Gynther, Waller, Leung, 'Confirmation of the extinction', p. 18. |
| 176 | '1998 by two conservationists . . . "preserve biodiversity wherever possible"': Dennis and Storch, 'Conservation and Taxonomic Status'. |
| 178 | 'I remember saying': Lathan, interview with Natalie Waller (2023). |
| 178 | 'population estimates': Gynther, Waller, Leung, 'Confirmation of the extinction', p. 3. |
| 179 | 'Desolate . . . People didn't care so much': Lathan, interview with Waller (2023). |
| 180 | 'on one condition': Lathan, interview with Waller (2023). |
| 181 | 'When I saw the cay': Lathan, interview with Ian Gynther (2024). |
| 181 | 'diminished dramatically': Gynther, Waller, Leung, 'Confirmation of the extinction', p. 13. |
| 181 | '"It hit me like a sledgehammer" . . . finally set off': Lathan, interview with Gynther (2024). |
| 182 | '"There wasn't much hope" . . . melomys was gone': Lathan, interview with Waller (2023). |
| 183 | 'I remember flying home': Lathan, interview with Gynther (2024). |
| 183 | 'The tide had been rolling in': Liz Minchin, 'Going Under', *The Sydney Morning Herald* (12 August 2006) (accessed 02/09/23), https://www.smh.com.au/national/going-under-20060812-gdo5uo.html. |
| 184 | 'A confluence': D. Green et al., 'An assessment of climate change impacts and adaptation for the Torres Strait Islands, Australia', *Climatic Change*, 102 (2010), pp. 405–33. |
| 184 | 'Reports have emerged of seawater': Minchin, 'Going Under'. |
| 184 | 'the plight of Torres Strait Islanders': Livia Albeck-Ripka, 'Their Islands Are Being Eroded. So Are Their Human Rights, They Say', *The New York Times* (12 May 2019) (accessed 02/09/23), |

https://www.nytimes.com/2019/05/12/world/australia/climate-change-torres-strait-islands.html.

184 'the average sea-level rise': Dennis and Storch, 'Conservation and Taxonomic Status', p. 26.

185 '6–8mm': John Rainbird, from the Torres Strait Regional Authority, 'Adapting to sea-level rise in the Torres Strait', Case Study for CoastAdapt (Gold Coast: National Climate Change Adaptation Research Facility, 2016), p. 8.

185 'found in favour . . . "right to life"': Daniel Billy and others v Australia (Torres Strait Islanders Petition), case summary for lawsuit, Decision (23/09/22) and Complaint (13/05/19) (accessed 28/01/24), http://climatecasechart.com/non-us-case/petition-of-torres-strait-islanders-to-the-united-nations-human-rights-committee-alleging-violations-stemming-from-australias-inaction-on-climate-change/.

185 'signs of seawater inundation . . . "the root cause of the loss"': Gynther, Waller, Leung, 'Confirmation of the extinction'.

186 '"natural" extinction': Dennis and Storch, 'Conservation and Taxonomic Status', p. 27.

186 'IUCN declared the extinction': J. Woinarski, A. A. Burbidge, 'Bramble Cay melomys: *Melomys rubicola*', The IUCN Red List of Threatened Species website (2016) (accessed 02/09/23), https://dx.doi.org/10.2305/IUCN.UK.2016-2.RLTS.T13132A195439637.en.

186 'Here's the sad part': segment from *The Late Show with Stephen Colbert*, https://www.youtube.com/watch?v=m8MjSWihBhw&t=109s.

186 'the dice was so loaded': Lathan, interview with John Woinarski (2021).

187 'an alternative explanation': J. Woinarski, 'A very preventable mammal extinction', *Nature* (2016), vol. 535, p. 493.

187 'recovery plan': Latch, 'Recovery Plan for the Bramble Cay Melomys'.

187 'It was just one of those flukes': Lathan, interview with Woinarski (2021).

187 'It's just tragic, really': Lathan, interview with Waller (2023).

188 'never implemented': J. Woinarski, 'A very preventable mammal extinction'.

188 'a mackerel fisherman': Gynther, Waller, Leung, 'Confirmation of the extinction', pp. 17–18.
189 'There is a myth . . . six stone statues': Story told by Idagi to Rev. W. H MacFarlane in *Reports of the Cambridge Anthropological Expedition to Torres Straits, Vol. 1* (London: Cambridge University Press, 1935), p. 192.
190 'native title rights': National Native Title Tribunal website, 'Torres Strait Islanders' native title recognised through agreements' (08/12/2004) (accessed 02/09/23), https://www.nntt.gov.au/News-and-Publications/latest-news/Pages/Torres_Strait_Islanders_native_title_re.aspx.
190 'We create things from the sea': Brendan Mounter, 'Torres Strait artists give extinct native rodent new life while flagging first climate change loss' (18 August 2021) (accessed 02/09/23), https://www.abc.net.au/news/2021-08-19/bramble-cay-melomys-revived-through-art/100386670.

## 6 Bird of the Evening

193 'Every October to November': Red Crab Migration page on Parks Australia website, including clip from BBC Earth ('What are 100 Million Crabs Doing Here?', *The Trials of Life*, David Attenborough, BBC Earth, 1990) (accessed 18/07/23), https://parksaustralia.gov.au/christmas/discover/highlights/red-crab-migration/.
195 'exploration lines': John Woinarski, *A Bat's End: The Christmas Island Pipistrelle and Extinction in Australia* (Clayton South, VIC, Australia: CSIRO Publishing, 2018), pp. 58–9.
195 'A Chinese businessman': Hwee Hoon Lee, 'Ong Sam Leong', National Library Board, Singapore website (accessed 18/07/23), https://www.nlb.gov.sg/main/article-detail?cmsuuid=eb023b12-d44e-4c0f-b948-0042e083a6b8.
195 'The labourers': 'The coolies of Kasumasu speak – 1910' and 'The early coolie houses on Christmas Island', Christmas Island Archives website (accessed 18/07/23), https://christmasislandarchives.com/coolies-of-kasumasu-speak/ and https://christmasislandarchives.com/coolie-houses/.
195 'Portuguese and Dutch navigators': Jan Tent, 'The Ghosts of

Christmas (Island) Past: An Examination of its Early Charting and Naming', *Terrae Incognitae*, 48:2 (2016), pp. 160–82.

196 'The 1887 discovery . . . bulldozers arrived': Woinarski, *A Bat's End*, pp. 16–22.

196 'The pipistrelle measured': Ibid., p. 96.

196 'widespread distribution': C. R. Tidemann, 'A study of the status, habitat requirements and management of the two species of bats on Christmas Island (Indian Ocean)', Report to Australian National Parks and Wildlife Service, Canberra (1985).

196 'Like all small bats': Woinarski, *A Bat's End*, p. 97.

197 'the pipistrelle foraged': Martin Schulz and Lindy Lumsden, 'National Recovery Plan for the Christmas Island Pipistrelle *Pipistrellus murrayi*', prepared for the Department of the Environment and Heritage (2004) (accessed 18/01/24), https://www.dcceew.gov.au/environment/biodiversity/threatened/recovery-plans/national-recovery-plan-christmas-island-pipistrelle-pipistrellus-murrayi).

197 'The earliest population estimate': Tidemann, 'A study of the status'.

197 'Early scientific records': C. A. Gibson-Hill, 'A note on the mammals of Christmas Island', *Bulletin of the Raffles Museum*, 18 (1947), p. 166.

197 'Tiny bats used to fly in': M. Neale, *We were the Christmas Islanders – Reminiscences and Recollections of the People of an Isolated Island – The Australian Territory of Christmas Island, Indian Ocean* (Chapman, A.C.T.: Bruce Neale, 1988), p. 25.

197 'Bird of the evening': 'Vespertilio', Merriam-Webster.com Dictionary (accessed 18/01/24), https://www.merriam-webster.com/dictionary/vespertilio.

197 'It is from this source' Woinarski, *A Bat's End*, p. 93.

197 'True to this ancient epithet': Alan Sieradzki, Heimo Mikkola, 'Bats in Folklore and Culture: A Review of Historical Perceptions around the World', *Bats – Disease-prone but Beneficial* (IntechOpen, 2022).

198 'a widespread genus': 'Pipistrellus pipistrellus', *Bats Life* website (accessed 28/01/24), https://batslife.eu/item/pipistrellus-pipistrellus/).

198 'The most striking factor': J. J. Lister, 'On the Natural History of

Christmas Island', *Proceedings of Zoological Society* (London, 1889), p. 530.

198 'To date, over 250': D. James, N. Milly, 'A biodiversity inventory database for Christmas Island National Park' (Canberra: Department of Environment and Heritage, 2006).

198 'There are also sixteen': Woinarski, *A Bat's End*, p. 40.

198 'some other-worldly creatures': T. Namiotko et al., 'On the origin and evolution of a new anchialine stygobitic *Microceratina* species (Crustacea, Ostracoda) from Christmas Island (Indian Ocean)', *Journal of Micropalaeontology*, 23 (2004), pp. 49–59; W. F. Humphreys et al., 'On the origin of *Danielopolina baltanasi* sp. n. (Ostracoda, Thaumatocypridoidea) from three anchialine caves on Christmas Island, a seamount in the Indian Ocean', *Crustaceana*, 82, pp. 1177–203, 2009. doi:10.1163/156854009X423157

198 'Alongside these endemic species': D. J. James et al., 'Endemic species of Christmas Island, Indian Ocean', *Records of the Western Australian Museum* 055-114 (2019) DOI: 10.18195/issn.0312-3162.34(2).2019.055-114.

198 'the Galápagos of the Indian Ocean': Christmas Island National Park on Parks Australia website, 'Nature' (accessed 28/01/24), https://parksaustralia.gov.au/christmas/discover/nature/.

199 'They could be found': D. J. James, K. Retallick, 'Christmas Island Biodiversity Monitoring Programme: research into the conservation status and threats of the Christmas Island Pipistrelle (Pipistrellus murrayi), 2004–2006' (Canberra: Department of Environment Water Heritage and the Arts, 2007).

199 'strange texture somewhere between': Woinarski, *A Bat's End*, p. 97.

199 'semi-hibernation': John Altringham, *Bats: From Evolution to Conservation* (Oxford: Oxford University Press, 2011), pp. 97–111.

199 'the bats usually roosted . . . chosen foraging sites': L. Lumsden et al., 'Investigation of the threats to the Christmas Island Pipistrelle' (Melbourne: Arthur Rylah Institute for Environmental Research, 2007).

200 'By contracting the larynx': Alain Van Ryckegham, 'How do bats echolocate and how are they adapted to this activity?', *Scientific American* (21 December 1998) (accessed 29/01/24), https://www.scientificamerican.com/article/how-do-bats-echolocate-an/.

200 'Wherever flying insects gathered': Tidemann, 'A study of the status'.

200 'a German alchemist': K. Ashley, D. Cordell, D. Mavinic, 'A brief history of phosphorus: From the philosopher's stone to nutrient recovery and reuse', *Chemosphere*, 84 (2011), pp. 737–46.

201 'light-bearing': etymology of phosphorus, Merriam Webster dictionary (accessed 19/01/24), https://www.merriam-webster.com/dictionary/phosphorus.

201 'The chemical element phosphorus': Phosphorus page, periodic table, Royal Society of Chemistry (accessed 19/01/24), https://www.rsc.org/periodic-table/element/15/phosphorus.

201 'The very backbone of DNA': Phosphate Backbone page, updated 24 January 2024, National Human Genome Research Institute (accessed 19/01/24), https://www.genome.gov/genetics-glossary/Phosphate-Backbone.

201 'In human bodies, phosphorus': 'Phosphorus', in *The Nutritional Source* (Harvard T. H. Chan School of Public Health) (accessed 19/01/24), https://www.hsph.harvard.edu/nutritionsource/phosphorus/.

201 'the safety match': Alan Dronsfield and Pete Ellis, 'The medicinal history of phosphorus', Royal Society of Chemistry (1 September 2020) (accessed 19/01/24), https://edu.rsc.org/feature/the-medicinal-history-of-phosphorus/2020257.article.

201 'in the lives of plants': Hassan Etesami, 'Enhanced Phosphorus Fertilizer Use Efficiency with Microorganisms', in R. S. Meena (ed.), *Nutrient Dynamics for Sustainable Crop Production* (Singapore: Springer Nature Singapore, 2020), p. 217.

202 'It was John Murray': Woinarski, *A Bat's End*, p. 16.

202 'Murray contracted Charles Andrews': Ibid., pp. 51–2.

202 'His 1900 monograph': David James and Ian McAllan, 'The birds of Christmas Island, Indian Ocean: A review', *Australian Field Ornithology*, 31 (2014), p. 9

202 'There was also something else': C. W. Andrews (ed.), *A Monograph of Christmas Island (Indian Ocean) – Physical Features and Geology, with Descriptions of the Fauna and Flora by Numerous Contributors* (London: British Museum Trustees, 1900), pp. 20–1.

203 'Tamarinds': Woinarski, *A Bat's End*, p. 67; Sturgis B. Rand, 'The Romance of Christmas Island: True Story of a Recently-Discovered "Treasure Island"', *McClure's Magazine*, vol. 18, no. 1 (November 1901) (New York & London: S. S. McClure,

1902) pp. 64–71 (accessed 26/07/23), https://repository.library.brown.edu/studio/item/bdr:533042/PDF/.
203 'Murray noted': Woinarski, *A Bat's End*, p. 67.
203 'In a few years': Andrews, *A Monograph*, p. 20.
203 'That year, a ship': Jane Pickering and Christopher A. Norris, 'New Evidence Concerning the Extinction of the Endemic Murid *Rattus Macleari* from Christmas Island, Indian Ocean', *Australian Mammalogy*, 19 (1996), pp. 19–25.
203 'Inadvertently, it also dropped off': P. T. Green, 'Mammal extinction by introduced infectious disease on Christmas Island (Indian Ocean): the historical context', *Australian Zoologist*, 37 (2014), pp. 1–14.
203 'in a dying condition . . . swarmed over the whole island': C. W. Andrews, 'On the fauna of Christmas Island', *Proceedings of the Zoological Society of London*, 79 (1909), pp. 101–3.
204 'A third endemic species': Pickering and Norris, 'New Evidence Concerning,' pp. 19–25.
204 'All of this Andrews had': Andrews, 'On the fauna', pp. 101–3.
204 'After annexation': Woinarski, *A Bat's End*, p. 16.
204 'It has not hitherto': Murray, in Andrews, *A Monograph*, p. x.
205 'Between 1900 and 1904 . . . This included the pipistrelle': Woinarski, *A Bat's End*, pp. 16–19.
205 'had named *Pipistrellus murrayi*': Andrews, *A Monograph*, p. 26.
205 'with large, bare, mined-out': J. B. Nelson, D. Powell, 'The breeding ecology of Abbott's booby *Sula abbotti*', *Emu – Austral Ornithology*, 86:1 (1986), pp. 33–46.
205 'Poaching': House of Representatives Standing Committee on Environment and Conservation (1974) Conservation of Endangered Species on Christmas Island. House of Representatives Standing Committee on Environment and Conservation, Canberra.
206 'first ever ecological study': Tidemann, 'A study of the status'.
206 'there seems little reason': C. R. Tidemann, 'Survey of the terrestrial mammals on Christmas Island (Indian Ocean)', unpublished 1989 report to Australian National Parks and Wildlife Service, Canberra, p. 10.
206 'One of the things . . . emerge from their roosts': Lathan, interview with Lindy Lumsden (2023.)
207 'in 1994, Lumsden found catching': L. Lumsden, K. Cherry,

'Report on a preliminary investigation of the Christmas Island pipistrelle, Pipistrellus murrayi, in June-July 1994' (Melbourne: Arthur Rylah Institute for Environmental Research, 1997).

207 'They were starting to grow over': Lathan, interview with Lumsden (2023).

208 'Around the size of a football': Emily Osterloff, 'Coconut crabs: the bird-eating behemoths thriving on isolated tropical islands', Natural History Museum website (accessed 28/01/24), https://www.nhm.ac.uk/discover/coconut-crabs-bird-eating-giants-on-tropical-islands.html.

208 '"They'd just demolish anything"' . . . both crabs and rain': Lathan, interview with Lumsden (2023).

209 'Between 8 June and 19 July': Lumsden and Cherry, 'Report on a preliminary'.

209 'The data just didn't fit': Lathan, interview with Lumsden (2023).

209 'Over six weeks': Schulz and Lumsden, 'National Recovery Plan for the Christmas Island Pipistrelle'; L. Lumsden, J. Silins, M. Schulz, 'Population dynamics and ecology of the Christmas Island Pipistrelle Pipistrellus murrayi on Christmas Island', (Melbourne: Arthur Rylah Institute for Environmental Research, 1999).

210 'I was really hoping': Lathan, interview with Lumsden (2023).

210 'Between 1994 and 1998': Lumsden, Silins, Schulz, 'Population dynamics and ecology of the Christmas Island Pipistrelle'.

210 'From the late 1970s': House of Representatives Standing Committee.

211 'The highest-quality reserves': W. W. Sweetland, 'Commission of Inquiry into the viability of the Christmas Island Phosphate Industry' (Canberra: Australian Government Publishing Service, 1980), (accessed 26/07/23), https://parlinfo.aph.gov.au/parlInfo/search/display/display.w3p;query=Id%3A%22publications%2Ftabledpapers%2FHPP032016007272%22.

211 'a variety of solutions': Woinarski, *A Bat's End*, pp. 31–5.

211 'The MV *Tampa*': Ben Doherty, 'The Tampa affair, 20 years on: the ship that capsized Australia's refugee policy', *The Guardian* (21 August 2021) (accessed 28/01/24), https://www.theguardian.com/australia-news/2021/aug/22/the-tampa-affair-20-years-on-the-ship-that-capsized-australias-refugee-policy.

212 '$400 million immigration detention centre': D. Marr, 'The

Indian Ocean solution: Christmas Island', *The Monthly* (September 2009) (accessed 26/07/23), https://www.themonthly.com.au/monthly-essays-david-marr-indian-solution-christmas-island-1940#mt; M. Kile, 'Christmas Island's other industry', *Quadrant Online* (2013) (accessed 26/07/23), https://quadrant.org.au/opinion/qed/2013/10/christmas-islands-industry/.

212 'Invoking "the national interest"': F. Pearce, 'Australia's asylum plans clash with bird on the brink', *New Scientist*, 29 (June 2002) (accessed 26/07/23), https://www.newscientist.com/article/dn2460-australias-asylum-plans-clash-with-bird-on-the-brink/.

212 'choosing dead trees for their roosts': Lumsden et al., 'Investigation of the threats to the Christmas Island Pipistrelle'.

213 'A strong gust of wind': L. Corbett, F. Crome, G. Richards, 'Fauna survey of mine lease applications and National Park reference areas, Christmas Island, August 2002' (Darwin: report by EWL Sciences Pty Ltd for Phosphate Resources Limited, 2003), p. 79.

213 'apology money': Woinarski, *A Bat's End*, p. 154.

213 'one of the most intensely': Schulz and Lumsden, 'National Recovery Plan for the Christmas Island Pipistrelle'.

213 'James and his colleague': James, Retallick, 'Christmas Island Biodiversity,' p. 8.

214 'the signature of something else': Schulz and Lumsden, 'National Recovery Plan for the Christmas Island Pipistrelle'.

214 'fundamental law of biological inheritance': H. Darras et al., 'Obligate chimerism in male yellow crazy ants', *Science*, 380 (2023), pp. 55–8.

214 'The exact purpose of this chimeric trait': E. Callaway, 'Crazy ants' strange genomes are a biological first', *Nature* (2023) (26/07/23), https://www.nature.com/articles/d41586-023-01002-3; Dino Grandoni, 'Scientists discover bizarre type of sex in this "crazy" ant', *The Washington Post* (2023) (accessed 26/07/23) https://www.washingtonpost.com/climate-environment/2023/04/06/yellow-crazy-ant-sex/.

214 'Having spread widely': '*Anoplolepis gracilipes*', IUCN Invasive Species Specialist Group website (accessed 28/01/24), http://www.iucngisd.org/gisd/species.php?sc=110.

214 'nankeen kestrel': M. Lewis, 'Australian kestrels Falco cenchroides feeding on bats', *Australian Bird Watcher*, 12 (1987), pp. 126–7;

Lumsden, Silins, Schulz, 'Population dynamics and ecology of the Christmas Island Pipistrelle'.

214 'Cats, adept climbers': Ricardo Rocha, 'Look what the cat dragged in: Felis silvestris catus as predators of insular bats and instance of predation on the endangered Pipistrellus maderensis', *Barbastella*, 8 (2015), pp. 18–21; T. S. Doherty et al., 'Invasive predators and global biodiversity loss', *Proceedings of the National Academy of Sciences of the United States of America*, 113 (2016), pp. 11261–5; J. Scrimgeour, A. Beath, M. Swanney, 'Cat predation of short-tailed bats (Mystacina tuberculata rhyocobia) in Rangataua Forest, Mount Ruapehu, Central North Island, New Zealand', *New Zealand Journal of Zoology*, 39:3 (2012), pp. 257–60, doi: 10.1080/03014223.2011.649770

215 'black rat, whose extinction scorecard': Doherty et al., 'Invasive predators', pp. 11261–5.

215 'It's for their erratic movements': I. H. Haines, J. B. Haines, J. M. Cherrett, 'The impact and control of the crazy ant Anoplolepis longipes (Jerd.) in the Seychelles', in D. F. Williams (ed.), *Exotic Ants: Biology, Impact and Control of Introduced Species* (Boulder, Colorado: Westview Press, 1994), pp. 206–18.

215 'gigantic ant-megapolises': Dennis J. O'Dowd et al., 'Status, Impact and Recommendations for Research and Management of Exotic Invasive Ants in Christmas Island National Park: Report to Environment Australia', (1999), pp. 2–7; K. L. Abbott, 'Supercolonies of the invasive yellow crazy ant, Anoplolepis gracilipes, on an oceanic island: forager activity patterns, density and biomass', *Insectes Sociaux*, 52 (2005), pp. 266–73.

215 'like a mustard gas attack': 'Christmas Island – Island Life' 3/6 – *Go Wild*, documentary film (accessed 28/01/24), https://www.youtube.com/watch?v=5v8ZCy1_YmY.

215 'invasional meltdown': D. J. O'Dowd, P. T. Green, P. S. Lake, 'Invasional "meltdown" on an oceanic island', *Ecology Letters*, 6 (2003), pp. 812–17.

216 'Now, all the detritus': Woinarski, *A Bat's End*, p. 78.

216 'Another devastating side effect': Ibid., p. 76.

216 'farmed their scale insects . . . timing seems to line up': Ibid., pp. 125–6.

217 'The yellow crazy ant was such a big issue': Lathan, interview with Lumsden (2023).

217 'a 2002 eradication programme': K. L. Abbott, P. T. Green, 'Collapse of an ant-scale mutualism in a rainforest on Christmas Island' *Oikos*, 116 (2007), pp. 1238–46; P. T. Green et al., 'Recruitment dynamics in a rainforest seedling community: context-independent impact of a keystone consumer', *Oecologia*, 156 (2008), pp. 373–85.

217 'Christmas Island's feral cats': C. R. Tidemann, H. D. Yorkston, A. J. Russack, 'The diet of cats, Felis catus, on Christmas Island, Indian Ocean', *Wildlife Research*, vol. 21 (1994), pp. 279–86.

217 'Likewise, despite the suspicious timing . . . considerations too': Martin Schulz, Lindy Lumsden, 'Diet of the Nankeen Kestrel *Falco cenchroides* on Christmas Island', *Australian Field Ornithology*, Vol. 26 (2009), pp. 28–32; Schulz and Lumsden, 'National Recovery Plan'; James, Retallick, 'Christmas Island Biodiversity'; Woinarski, *A Bat's End*.

218 'But Lumsden and others': Lathan, interview with Lumsden (2023).

218 'The common wolf snake': L. A. Smith, 'Lycodon aulicus capucinus a colubrid snake introduced to Christmas Island, Indian Ocean', *Records of the Western Australian Museum*, 14 (1988), pp. 251–2.

218 'Wolf snakes weren't thought': Lathan, interview with Lumsden (2023).

218 'A decade later': H. Rumpff, 'Distribution, population structure and ecological behaviour of the introduced South-East Asian Wolf Snake Lycodon aulicus capucinus on Christmas Island, Indian Ocean', report to Australian National Parks and Wildlife Service, Christmas Island (1992).

218 'By 1998, wolf snakes': Lumsden, Silins, Schulz, 'Population dynamics and ecology of the Christmas Island Pipistrelle'; H. G. Cogger, R. A. Sadlier, 'The Terrestrial Reptiles of Christmas Island: A Reappraisal of Their Status' (Sydney: The Australian Museum, 1999).

218 'damning piece of evidence': James, Retallick, 'Christmas Island Biodiversity'.

218 'proved impossible for Lumsden . . . four pipistrelles': Lathan, interview with Lumsden (2023).

218 'There is an extremely high risk': Lindy Lumsden and Martin Schulz, 'Captive breeding and future in-situ management of the

Christmas Island Pipistrelle *Pipistrellus murrayi*: a report to the Director of National Parks' (Melbourne: Arthur Rylah Institute for Environmental Research, 2009), p. 7.

219 'To predict what was going to happen': Lathan, interview with Lumsden (2023).

219 'Lumsden suggested building': Lumsden and Schulz, 'Captive breeding'.

219 'The Australian Environment Department's response': Woinarski, *A Bat's End*, p. 168.

220 'There were really good people': Lathan, interview with Lumsden (2023).

220 'Lumsden and the society's president': L. Lumsden, 'The extinction of the Christmas Island Pipistrelle', *The Australasian Bat Society Newsletter*, no. 33 (November 2009), pp. 22–6.

220 'I pleaded with him': Lathan, interview with Lumsden (2023).

220 'On 8 August 2009': Susan Campbell, 'The last of its kind?', *The Australasian Bat Society Newsletter*, no. 33 (November 2009) pp. 29–33.

221 'In January 2009': Lumsden, 'The extinction', pp. 22–6.

221 'the Australasian Bat Society': Lathan, interview with Lumsden (2023).

221 'We picked up a call . . . only one pipistrelle left': Lathan, interview with Lumsden (2023).

221 'it had to be captured . . . never looking back': Lumsden, 'The extinction,' pp. 22–6; Lathan, interview with Lumsden (2023).

223 'At 23:29': email from Lindy Lumsden (2024).

223 'You know you asked me': Lathan, interview with Lumsden (2023).

223 'The last call of the pipistrelle': Zoos Victoria YouTube (accessed 28/01/24), https://www.youtube.com/watch?v=HERL71LLHWQ.

224 'press release', Peter Garrett, 'Media Release: Bat Mission Sadly Fails on Christmas Island', *The Australasian Bat Society Newsletter*, no. 33 (November 2009), p. 27.

224 'It was a day of lots of people': Lathan, interview with Lumsden (2023).

224 'In 2006, the Environment Department': Woinarski, *A Bat's End*, p. 148.

225 'an authority on every kind': Eric Pace, 'Dr. Karl F. Koopman,

77, An Authority on Bats, Is Dead', *The New York Times* (30 September 1997) (accessed 28/01/24), https://www.nytimes.com/1997/09/30/nyregion/dr-karl-f-koopman-77-an-authority-on-bats-is-dead.html.
225 'In 1973, Koopman declared': Woinarski, *A Bat's End*, p. 100.
225 'a Latin word': (accessed 28/01/24), https://www.oxfordlearnersdictionaries.com/definition/english/tenuous.
225 'Australian bodies . . . four bats to one': Woinarski, *A Bat's End*, p. 100.
225 'In 2023, Koopman's classification . . . in a manner, at least – saved': Lathan, interview with Lumsden (2023).
226 'full licence and authority': 'The Christmas Island Agreement Act 1958–73', quoted in Woinarski, *A Bat's End*, p. 21.
226 'major management and policy failings': Woinarski, *A Bat's End*, p. 177–8.
226 'ecotourism': Ibid., p. 33.
227 'Christmas Island provides': Parks Australia statement, 2002, quoted in Ibid., p. 176.
227 'When you do this sort of work': Lathan, interview with Lumsden (2023).
228 'Australian Mammal of the Year': Imma Perfetto, 'The golden-tipped bat is the 2023 Australian Mammal of the Year!', Cosmos Magazine (24/08/23) (accessed 18/01/24), https://cosmosmagazine.com/nature/amoty/the-golden-tipped-bat-is-the-2023-australian-mammal-of-the-year/.
228 'Later, she shares some footage': email from Lindy Lumsden (2023).

## 7 Run, Forest, Run!

231 'A few months after': Lathan, interview with Samantha Flakus (2023).
231 'tens of thousands': Lathan, interview with Brendan Tiernan (2023).
232 'What was all this for?': M. J. Smith et al., 'An Oceanic Island Reptile Community Under Threat: the Decline of Reptiles on Christmas Island, Indian Ocean', *Herpetological Conservation and Biology*, 7 (2012), pp. 206–18.

232 'An intricate mosaic': G. A. Boulenger, 'Reptiles', *Proceedings of the Zoological Society London*, 55 (1887), pp. 515–17.

232 'At 20cm long': H. G. Cogger, R. R. Sadlier, E. E. Cameron, 'The terrestrial reptiles of Australia's Island territories' (Canberra: Australian National Parks and Wildlife Service, 1983).

232 'the majority of its time': J. Woinarski, H. Cogger, 'Australian endangered species: Christmas Island Forest Skink', *The Conversation* (2013) (accessed 28/07/23), https://theconversation.com/australian-endangered-species-christmas-island-forest-skink-18053.

232 'Any break . . . coastal fringes': Cogger, Sadlier, Cameron, 'The terrestrial reptiles'.

232 'Yet introduced threats': M. J. Smith et al., 'An Oceanic Island Reptile Community Under Threat'.

232 'In response, a plan was put into action': Lathan, interview with Flakus (2023).

233 'On certain days': Lathan, interview with Tiernan (2023).

233 'By August 2010': M. J. Smith et al., 'An Oceanic Island Reptile Community Under Threat'.

233 'caught using a unique method': Lathan, interview with Tiernan (2023).

233 'Look around at': C. Booth, T. Low, 'GONE: Australian animals extinct since the 1960s' (Invasive Species Council Inc., 2023), p. 34.

233 'one of which immediately escaped . . . another one at all': Lathan, interview with Tiernan (2023).

234 'Shortly after the forest skinks': Lathan, interview with John Woinarski (2023).

234 'Workers returned': Lathan, interview with Tiernan (2023).

234 '130cm x 65cm x 65cm tank': Kieron McLeonard, 'Gump faces extinction', Australian Broadcasting Corporation radio broadcast (09/10/13).

235 '9 and 18 million years ago': Paul M. Oliver et al., 'Insular biogeographic origins and high phylogenetic distinctiveness for a recently depleted lizard fauna from Christmas Island, Australia', *Biology Letters*, vol. 14 (2018), p. 2.

235 'Once an undersea volcano': Ken Grimes, 'Karst features of Christmas Island (Indian Ocean)', *Helictite*, 37 (2001), pp. 41–58.

235 'drowned coral atoll': J. R. Ali and J. C. Aitchison, 'Time of

re-emergence of Christmas Island and its biogeographical significance', *Palaeogeography, Palaeoclimatology, Palaeoecology*, 537 (2020), pp. 1–5.

235 'calcium secretors': Foraminifera, Discovering Fossils and Geological Time page on British Geological Survey website (accessed 28/07/23), https://www.bgs.ac.uk/discovering-geology/fossils-and-geological-time/foraminifera/.

235 'Five million years ago': Ali and Aitchison, 'Time of re-emergence of Christmas Island'.

235 'Gump's ancestor could make the journey': Jason Ali, Jonathan Aitchison and Shai Meiri, 'Redrawing Wallace's Line based on the fauna of Christmas Island, eastern Indian Ocean', *Biological Journal of the Linnean Society* (2020), pp. 1–13.

235 '630,000 years ago': Ali et al., 'Redrawing Wallace's Line'; Richard Bintanja et al., 'Modelled atmospheric temperatures and global sea levels over the past million years', *Nature*, vol. 437 (01/09/2015), pp. 125–8.

236 'the seafloor plummets': Marine Waters website (Government of Western Australia), 'Indian Ocean Territories: The remote location of Christmas Island and the Cocos (Keeling) Islands is the key to their unique and spectacular marine biodiversity' (accessed 28/07/23), https://marinewaters.fish.wa.gov.au/explore/indian-ocean-territories/.

236 'riding westward currents': Ali et al., 'Redrawing Wallace's Line'.

236 'The Wallace Line': Charles Hirschman, 'The Wallace Line', The Henry M. Jackson School of International Studies, Center for Southeast Asia and its Diasporas, University of Washington (accessed 28/07/23), https://jsis.washington.edu/csead/resources/educators/where-in-southeast-asia/the-wallace-line/.

236 'It was discovered in 1859': John van Wyhe, 'Wallace's Help: The Many People Who Aided A. R. Wallace in the Malay Archipelago', *Journal of the Malaysian Branch of the Royal Asiatic Society*, 91 (314) (2018), pp. 41–68.

236 'Charles Darwin is often described': Jane R. Camerini, 'Evolution, Biogeography, and Maps: An Early History of Wallace's Line', *Isis*, 84 (1993), pp. 700–27.

236 'keeping apart': A. R. Wallace, 'On the Physical Geography of the Malay Archipelago', *The Journal of the Royal Geographical Society of London*, vol. 33 (1863), pp. 217–34.

237 'Christmas Island, 1,100km ... smorgasbord': Jason Ali, Jonathan Aitchison and Shai Meiri, 'Redrawing Wallace's Line based on the fauna of Christmas Island, eastern Indian Ocean', *Biological Journal of the Linnean Society* (2020), pp. 1–13, doi: 10.1093/biolinnean/blaa018/5802023.

237 'Benign': Lathan, interview with Woinarski (2023).

237 'broadest palate': Cogger, Sadlier and Cameron, 'The terrestrial reptiles'.

238 'more sociable side': Anaël Gorge, 'Christmas Island Biodiversity Monitoring Programme: Survey of two endemic declining reptiles: the blue tailed Skink and the forest Skink, for the period June to August 2005'.

238 'After mating': Cogger, Sadlier and Cameron, 'The terrestrial reptiles'.

238 'In 1887, a strange new skink': Boulenger, 'Reptiles'.

239 'a very different kind of reptile': G. A. Boulenger, 'Catalogue of the Snakes of the British Museum (Natural History), vol. I' (1893).

239 'one of the most formidable': A. Günther, *The Reptiles of British India* (London: Taylor & Francis, 1864).

239 'Curving sharply back': K. Jackson, T. Fritts, 'Dentitional specialisations for durophagy in the Common Wolf snake, Lycodon aulicus capucinus', *Amphibia-Reptilia*, 25(3) (2004), pp. 247–54.

240 'In Boulenger's drawing': Boulenger, 'Catalogue of the Snakes'.

240 'the Christmas Island blind snake': G. A. Boulenger, 'Reptiles', *Proceedings of the Zoological Society London*, 55 (1887), pp. 515–17.

240 'preserved remains of a wolf snake': Thomas H. Fritts, 'The common wolf snake, Lycodon aulicus capucinus, a recent colonist of Christmas Island in the Indian Ocean', *Wildlife Research*, vol. 20 (1993), pp. 261–6.

240 'Should it become established': L. Smith, 'Lycodon aulicus capucinus a colubrid snake introduced to Christmas Island, Indian Ocean', *Records of the Western Australian Museum*, vol. 14 (1988), pp. 251–2.

240 'wiped out multiple ... on the island': Thomas H. Fritts, Dawn Leasman-Tanner, 'The Brown Tree Snake on Guam: How the Arrival of One Invasive Species Damaged the Ecology, Commerce, Electrical Systems and Human Health on Guam:

A Comprehensive Information Source' (US Fish and Wildlife Service, 2001).

242 'potent an ecological threat': Fritts, 'The common wolf snake', pp. 261–6.

242 'Flowerpot blind snakes': Shreya Dasgupta, 'The snake that decapitates its prey', *Mongabay*, (27/07/2015) (accessed 28/07/23), https://news.mongabay.com/2015/07/dasgupta-the-snake-that-decapitates-its-prey/.

242 'Bowring's writhing skinks . . . into a new frontier': Jon-Paul Emery et al., 'The lost lizards of Christmas Island: A retrospective assessment of factors driving the collapse of a native reptile community', *Conservation Science and Practice*, vol. 3 no. 2 (2021).

242 'clamp its jaws': K. Jackson, T. Fritts, 'Dentitional specialisations for durophagy in the Common Wolf snake'.

243 'the vicinity . . . led the charge': Emery et al., 'The lost lizards of Christmas Island'.

243 'fewer forest skinks than usual . . . were everywhere': David J. James, 'Christmas Island Biodiversity Monitoring Programme: Summary Report, December 2003 to April 2006', for Department of Finance & Administration and Department of the Environment & Water Resources, for Parks Australia North, unpublished, March 2007, pp. 23–34.

243 'In 2005, James sent': Lathan, interview with David James (2023).

243 'a reptile monitoring survey': Threatened Species Scientific Committee, *Emoia nativitatis* (forest skink) Draft Conservation Advice, March 2013, p. 1.

244 'Other than Gump': Lathan, interview with Flakus (2023).

244 'Sluggish': Lathan, interview with Woinarski (2023).

245 'Life had been completely transformed . . . three days per week': Lathan, interview with Flakus (2023); Lathan, interview with Tiernan (2023); Nicole Gill, 'Every lizard counts: A day at Christmas Island's Lizard Lodge', *The Monthly* (February 2016) (accessed 28/07/23), https://www.themonthly.com.au/issue/2016/february/1454245200/nicole-gill/every-lizard-counts#mtr.

246 'I remember her being': Lathan, interview with Tiernan (2023).

247 'Once a month, keepers': Lathan, interview with Flakus (2023).

247 'Comfortable, if lonely': Lathan, interview with Woinarski (2023).

247 'When staff arrived': Lathan, interview with Flakus (2023); Lathan, interview with Tiernan (2023).
247 'It was extremely sad': Lathan, interview with Tiernan (2023).
247 'the moment she learned of Gump's death': Lathan, interview with Flakus (2023).
248 'A 2017 paper': John C. Z. Woinarski et al., 'The contribution of policy, law, management, research, and advocacy failings to the recent extinctions of three Australian vertebrate species', *Conservation Biology*, vol. 31 no. 1 (2016), pp. 13–23.
248 'Coroners': O. Quick, 'Coronial Investigations and Inquests. In Regulating Patient Safety: The End of Professional Dominance?', in *Cambridge Bioethics and Law* (Cambridge: Cambridge University Press, 2017), pp. 128–43.
248 'the language of ecology ... "These are sorry tales"': Woinarski et al., 'The contribution of policy'.
249 'trying to balance despair and hope': Lathan, interview with Woinarski (2023).
249 'Our primary interest ... positive legacy': Woinarski et al., 'The contribution of policy'.
250 'I'd been shown specimens': visit to Natural History Museum collections, led by Roberto Potela Miguez (2023).
250 'is what biologists call': A. S. Hitchcock, 'The Type Concept in Systematic Botany', *American Journal of Botany*, vol. 8, no. 5 (1921), pp. 251–5.
251 'At one point we had nearly two thousand': Lathan, interview with Flakus (2023).
252 'new species of bacterium': Loren Smith, 'Christmas Island reptile-killer identified', The University of Sydney, News (18/03/21) (accessed 28/07/23), https://www.sydney.edu.au/news-opinion/news/2021/03/18/christmas-island-reptile-killer-identified.html.
252 'Pretty much anywhere on Christmas Island': Lathan, interview with Flakus (2023).
252 'Today, a change of address': No author, 'Saving the Blue-Tailed Skink', on Taronga Conservation Society Australia website (accessed 28/07/23), https://taronga.org.au/media-release/2019-09-12/saving-blue-tailed-skink.
253 'There are lizards everywhere': Lathan, interview with Tiernan (2023).

## 8 THE GRITADOR

253 'It would have been successful': Lathan, interview with Flakus (2023).

257 'whose 11,500-year-old skull': James MacDonald, 'Brazil's Museu Nacional Was More Than Just a Museum', *JSTOR Daily* (2018) (accessed 17/10/23), https://daily.jstor.org/brazils-museu-nacional-was-more-than-just-a-museum/.

257 'Freshly exhumed': Lathan, interview with Dante Buzzetti (2024).

257 '19cm and 22cm': J. Mazar Barnett and D. R. C. Buzzetti, 'A new species of Cichlocolaptes Reichenbach 1853 (Furnariidae), the "gritador-do-nordeste", an undescribed trace of the fading bird life of northeastern Brazil', *Revista Brasileira de Ornitologia*, 22:2 (2014), pp. 75–94.

258 'cryptic species': D. Bickford et al., 'Cryptic species as a window on diversity and conservation', *Trends in Ecology & Evolution*, 22:3 (2007), pp. 148–55.

258 'Barnett is no stranger': Luciano N. Naka, 'In Memoriam: Juan Mazar Barnett, 1972–2012', *The Condor*, 115-3 (2013), pp. 688–92.

259 'Barnett and Buzzetti scrutinize': Lathan, interview with Buzzetti (2024).

259 'Using a dial calliper': Barnett and Buzzetti, 'A new species of Cichlocolaptes'.

259 'In 1979, the type . . . "uüarrr, uüarrr"': D. M. Teixeira and L. P. Gonzaga, 'Um novo Furnariidae do nordeste do Brasil: *Philydor novaesi* sp. nov. Boletim do Museu Paraense Emílio Goeldi', *Série Zoologia*, 124 (1983), pp. 1–22.

260 'On 12 October 2002, the pair': Lathan, interview with Buzzetti (2024).

260 '"rattle" . . . Furnariidae family of birds': Barnett and Buzzetti, 'A new species of Cichlocolaptes'.

261 'It's also the densest': World Wildlife Fund website, page on animals living in and facts about the Amazon (accessed 13/08/23), https://www.worldwildlife.org/stories/what-animals-live-in-the-amazon-and-8-other-amazon-facts.

261 'erroneous belief': Katarina Zimmer, 'Why the Amazon doesn't really produce 20% of the world's oxygen', *National Geographic* (28 August 2019).

261 'Running south': Milton Cezar Ribeiro et al., 'The Brazilian Atlantic Forest: How much is left, and how is the remaining forest distributed? Implications for conservation', *Biological Conservation*, vol. 142, no. 6 (2009), pp. 1141–53; 'Places We Protect: The Atlantic Forest', The Nature Conservancy website (accessed 13/08/23), https://www.nature.org/en-us/get-involved/how-to-help/places-we-protect/atlantic-forest/.

262 'a global biodiversity hotspot': Raf de Lima et al., 'The erosion of biodiversity and biomass in the Atlantic Forest biodiversity hotspot', *Nature Communications* (2020).

262 'Plant diversity is so high': 'The Atlantic Forest', The Nature Conservancy website.

262 '2,000 species of epiphytes': F. N. Ramos, S. R. Mortara et al., 'Atlantic Epiphytes: a data set of vascular and non-vascular epiphyte plants and lichens from the Atlantic Forest', *Ecology*, 100:2 (2019).

262 'Species diversity in the Atlantic Forest': L. Patricia, C. Morellato and Celio F. B. Haddad, 'Introduction: The Brazilian Atlantic Forest', *Biotropica*, vol. 32, no. 4b (2000), pp. 786–92.

262 'the second largest rainforest': Francis Dov Por, *Sooretama: the Atlantic rain forest of Brazil* (University of Texas, 1992).

262 'tail end of the last glacial': Leonora Costa, 'The historical bridge between the Amazon and the Atlantic Forest of Brazil: A study of molecular phylogeography with small mammals', *Journal of Biogeography*, no. 30 (2003). pp. 71–86.

263 'at some unknown point': Elizabeth P. Derryberry et al., 'Lineage Diversification and Morphological Evolution in a Large-scale Continental Radiation: The Neotropical Ovenbirds and Woodcreepers (Aves: Furnariidae)', *Evolution*, vol. 65:10 (The Society for the Study of Evolution, 2011), pp. 2973–86; Martin Irestedt et al., 'Convergent evolution, habitat shifts and variable diversification rates in the ovenbird-woodcreeper family (Furnariidae)', *BMC Evolutionary Biology*, vol. 9: 268 (BioMed Central, 2009).

263 'It is a word with Celtic roots': Merriam Webster online dictionary (accessed 13/08/23), https://www.merriam-webster.com/dictionary/gleaner.

263 'flit between orchids': C. J. Sharpe et al., 'Alagoas Foliage-gleaner (*Philydor novaesi*), version 2.0', in Keeney (ed.), *Birds of the World*,

Cornell Lab of Ornithology (08/04/22) (accessed 13/08/23), https://doi.org/10.2173/bow.alfgle1.02.

263 'a clump of bromeliad leaves': G. M. Kirwan et al., 'Cryptic Treehunter (*Cichlocolaptes mazarbarnetti*), version 2.0', in Keeney (ed.), *Birds of the World*, Cornell Lab of Ornithology (25/03/22) (accessed 13/08/23), https://doi.org/10.2173/bow.crytre1.02.

263 'went about their busy work': Sharpe et al., 'Alagoas Foliage-gleaner'; Kirwan et al., 'Cryptic Treehunter'.

264 'small ashy-grey bird': T. S. Schulenberg and G. M. Kirwan, 'Alagoas Antwren (*Myrmotherula snowi*)', in Keeney (ed.), *Birds of the World*.

264 'cinereous antshrike': C. J. Sharpe et al., 'Alagoas Foliage-gleaner (Philydor novaesi), version 2.0', in Keeney (ed.), *Birds of the World*, Cornell Lab of Ornithology (08/04/22) (accessed 13/08/23), https://doi.org/10.2173/bow.alfgle1.02.

264 'there were also giants': Ed Yong, 'How Climate Change Unleashed Humans Upon South America's Megabeasts', *The Atlantic* (17 June 2016) (accessed 13/08/23), https://www.theatlantic.com/science/archive/2016/06/how-climate-change-unleashed-humans-upon-south-americas-megabeasts/487502/.

265 'things began to change': Ibid.; Luciano Prates, S. Ivan Perez, 'Late Pleistocene South American megafaunal extinctions associated with rise of Fishtail points and human population', Nature Communications 12, 2175 (2021).

265 'Estimates for the date of human colonization': Simon Romero, 'Discoveries Challenge Beliefs on Humans' Arrival in the Americas', *The New York Times* (27/03/14) (accessed 13/08/23), https://www.nytimes.com/2014/03/28/world/americas/discoveries-challenge-beliefs-on-humans-arrival-in-the-americas.html.

265 'Humans can't help but change': Lathan, interview with Phil Riris (2023).

266 'Charcoal and the remains of stone tools': Ibid.; Romero, 'Discoveries Challenge Beliefs on Humans'.

266 'Then there is pollen . . . "indigenous groups in southern Brazil"': Lathan, interview with Riris (2023).

266 'the tree has played': Maurício Sedrez dos Reis, Ana Ladio, Nivaldo Peroni, 'Landscapes with Araucaria in South America: evidence for a cultural dimension', *Ecology and Society*, vol. 19, no. 2 (2014), p. 1

266 'traditional dietary, spiritual, symbolic': Ann Garibaldi and Nancy J. Turner, 'Cultural Keystone Species: Implications for Ecological Conservation and Restoration', *Ecology and Society*, vol. 9, no. 3 (2004), p. 1.

266 'the remains of these seeds': Lathan, interview with Risis (2023).

266 'On 23 April 1500, a Portuguese': A. C. Metcalf, *Go-betweens and the Colonization of Brazil: 1500–1600* (University of Texas Press, 2005), p. 17.

267 'When Cabral and his crew': Chris Allan, 'Brazilwood: A Brief History', University of Minnesota, Bell Library website (accessed 13/08/23), https://web.archive.org/web/20200727034659/https://www.lib.umn.edu/bell/tradeproducts/brazilwood#n3.

268 'Soon, fire cleared the way . . . ripe for exploitation': Alexandro Solorzano, Lucas Brasil, Rogério Oliveira, 'The Atlantic Forest Ecological History: From Pre-colonial Times to the Anthropocene', chapter 2 in Marcia C. M. Marques and Carlos E. V. Grelle (eds.), *The Atlantic Forest: History, Biodiversity and Opportunities of the Mega-diverse Forest* (Cham: Springer, 2021), pp. 25–44.

269 'the Atlantic Forest shrank back': Ribeiro et al., 'The Brazilian Atlantic Forest'; Lathan, interview with Pedro Develey (2023).

269 'changes hands for the sake of a bird': Lathan, interview with Develey (2023); BirdLife International, 'Serra do Urubu Pernambuco State, Brazil: long-term conservation of a threatened Atlantic Forest' (2015) (accessed 28/09/23), https://datazone.birdlife.org/sowb/casestudy/serra-do-urubu-pernambuco-state-brazil:-long-term-conservation-of-a-threatened-atlantic-forest.

269 'We purchased' . . . fragment of just 400 hectares': Lathan, interview with Develey (2023).

270 'global conservation priorities': 'Places We Protect: The Atlantic Forest', The Nature Conservancy website.

270 'fifteen hereditary fiefdoms': Cameron J. G. Dodge, 'A Forgotten Century of Brazilwood: The Brazilwood Trade from the Mid-Sixteenth to Mid-Seventeenth Century', *e-Journal of Portuguese History*, Brown Digital Repository, Brown University Library (2018) (accessed 13/08/23), https://doi.org/10.7301/Z0VH5MBT.

270 'indicators of good forest': Lathan, interview with Develey (2023).

271 'forest of hope': Bennett Hennessey, 'Into Serra do Urubu– Brazil's Forest of Hope', American Bird Conservancy

website (16/02/17) (accessed 13/07/23), https://abcbirds.org/into-serra-do-urubu-brazils-forest-of-hope/.

271 'When you have a really small population . . . we arrived too late': Lathan, interview with Develey (2023).

273 'Barnett, having long struggled': Naka, 'In Memoriam: Juan Mazar Barnett'.

273 'Juan was a great friend': Lathan, interview with Buzzetti (2024).

273 'the pride of a generation': Naka, 'In Memoriam: Juan Mazar Barnett'.

273 'Buzzetti struggled . . . in 2014': Lathan, interview with Buzzetti (2024).

273 'the second author dedicates': Barnett and Buzzetti, 'A new species of Cichlocolaptes', pp. 75–94.

275 'This name was Juan's idea': Lathan, interview with Buzzetti (2024).

275 'On the evening of Sunday 2 September': Mariana Lenharo, Meghie Rodrigues, 'Can a National Museum Rebuild its Collection Without Colonialism?', *The New York Times* (09/11/22) (accessed 13/08/23), https://www.nytimes.com/2022/11/09/magazine/brazil-national-museum-indigenous.html.

275 'In the auditorium': Evan Nicole Brown, 'Air Conditioning Caused the Fire that Claimed Brazil's National Museum', *Atlas Obscura* (09/04/19) (accessed 13/08/23), https://www.atlasobscura.com/articles/cause-of-brazil-museum-fire.

275 'the city's residents looked on': Alejandro Chacoff, 'Brazil lost more than the past in the National Museum fire', *The New Yorker* (16/09/18) (accessed 13/08/23), https://www.newyorker.com/news/dispatch/brazil-lost-more-than-the-past-in-the-national-museum-fire.

275 'A lobotomy of the Brazilian memory': Dom Phillips, 'Brazil museum fire: "incalculable" loss as 200-year-old Rio institution gutted', *The Guardian* (03/09/18) (accessed 13/08/23), https://www.theguardian.com/world/2018/sep/03/fire-engulfs-brazil-national-museum-rio.

275 'including "Luzia"': Brown, 'Air Conditioning Caused the Fire'.

275 'It was a loss so significant': Constance Witham, 'Heritage on Fire', British Council (March 2019) (accessed 13/08/23), https://www.britishcouncil.org/research-insight/heritage-on-fire.

276 'artefacts belonging to the indigenous peoples': Manuela Andreoni, Ernesto Londoño, 'Loss of Indigenous Works in Brazil Museum Fire Felt "Like a New Genocide"', *The New York Times* (13/09/18) (accessed 13/08/23), https://www.nytimes.com/2018/09/13/world/americas/brazil-museum-fire-indigenous; James Gorman, 'The Brazil Museum Fire: What Was Lost', *The New York Times* (04/09/18) (accessed 13/08/23), https://www.nytimes.com/2018/09/04/science/brazil-museum-fire.html.

276 'The collection of flora and fauna': Peter Moon, 'A Brazilian mourns what was lost in the National Museum fire', *Mongabay* (06/09/18) (accessed 13/08/23), https://news.mongabay.com/2018/09/a-brazilian-mourns-what-was-lost-in-the-national-museum-fire/.

276 'So it was that, in November 2018': Stuart H. M. Butchart, Stephen Lowe, Rob W. Martin et al., 'Which bird species have gone extinct? A novel quantitative classification approach', *Biological Conservation*, 227 (2018), pp. 9–18.

276 'thousands of items': Meilan Solly, 'Around 2,000 Artifacts Have Been Saved From the Ruins of Brazil's National Museum Fire', *Smithsonian Magazine* (15/02/19) (accessed 13/08/23), https://www.smithsonianmag.com/smart-news/around-2000-artifacts-have-been-saved-ruins-brazils-national-museum-fire-180971510/.

276 'In 2019, the IUCN declared': 'Alagoas Foliage-gleaner: *Philydor novaesi*', IUCN Red List (2019) (accessed 13/08/23), https://www.iucnredlist.org/species/22702869/156126928; Cryptic Treehunter: *Cichlocolaptes mazarbarnetti*', IUCN Red List (2019) (accessed 13/08/23), https://www.iucnredlist.org/species/103671170/155880473.

276 'I can tell you that': email from confidential source (2023).

277 'I'm here to take away your doubts': Lathan, interview with Marcos Raposo (2024).

278 'our requests for permission': Barnett and Buzzetti, 'A new species of Cichlocolaptes', p. 88.

278 'transcript of a 2016 discussion': 'Proposal (714) to South American Classification Committee: Recognize *Cichlocolaptes mazarbarnetti* as a valid species', South American Classification Committee of the American Ornithological Society (2016)

(accessed 21/03/24), https://www.museum.lsu.edu/~Remsen/SACCprop714.htm.

## 9 Ghost of the Galápagos

283 'On 1 December 1971': Henry Nicholls, *Lonesome George: The Life and Loves of a Conservation Icon* (London: Macmillan, 2006), pp. 1–8.

284 'Over the past 122 years': B. R. Scheffers et al., 'The World's Rediscovered Species: Back from the Brink?', *PLOS ONE*, Public Library of Science (2011).

284 'the Kandyan dwarf toad': L. J. Mendis Wickramasinghe et al., 'Back from the dead: The world's rarest toad Adnomus kandianus rediscovered in Sri Lanka', *Zootaxa* (2012), pp. 63–8.

285 'the tiny bird': N. J. Collar, 'Extinction by assumption; or, the Romeo Error on Cebu', *Oryx*, vol. 32, no. 4 (2008), pp. 239–44.

285 'The short period the bird was considered extinct': Ibid.

285 'American comedian Johnny Carson': Linda Cayot, *The Lonesome George Story: Where Do We Go From Here?* (Galápagos Conservancy, 2014). p. 6.

285 'The real Lonesome George': Paul Chambers, *A Sheltered Life: The Unexpected History of the Giant Tortoise* (Oxford: Oxford University Press, 2004), p. 246.

286 'His fans included': Jorge Mancero, 'My Life Story – By Lonesome George' (2017) (31/05/22), https://www.galapagosislands.com/blog/lonesome-george-history/.

286 'wrote to him instead': Cayot, *The Lonesome George Story*, pp. 18–9.

286 'lacked the natural curiosity': Lathan, interview with James Gibbs (2022).

286 'When female tortoises': Cayot, *The Lonesome George Story*, p. 16; Nicholls, *Lonesome George*, p. 21.

286 'a distant dusky ridge': Herman Melville, *Putnam's Monthly Magazine of American Literature, Science, and Art, Volume 3* (New York: G. P. Putnam, 1854), p. 319.

286 'estimated, around 1910': 'Lonesome George', Galápagos Conservancy website (accessed 31/05/22), https://www.galapagos.org/about_galapagos/lonesome-george/.

286 'come together to share waterholes': Jack Frazier, 'Galápagos tortoises: Protagonists in the spectacle of life on Earth', in James

P. Gibbs, Linda J. Cayot, Washington A. Tapia, *Galápagos Giant Tortoises* (London: Academic Press, Elsevier, 2021), p. 32.

286 'Male tortoises fight over females': Erika Kubisch and Nora R. Ibarguengoytia, 'Reproduction', in Gibbs, Cayot, Tapia, *Galápagos Giant Tortoises*, p. 160.

287 'tied by his ankle to a cactus': Dr. Ole Hamann, quoted in Cayot, *The Lonesome George Story*, p. 10.

287 'laid on the floor of a dinghy': Nicholls, *Lonesome George*, p. 10.

287 'bananas became a favourite': Dr. Craig McFarland, quoted in Cayot, *The Lonesome George Story*, p. 7.

288 'in 1923, an American naturalist': William Beebe, *Galápagos: World's End* (New York: Dover Publications, 1988); Nicholls, *Lonesome George*, p. 35.

288 'Ever since its discovery in 1535 ... sea around the Galápagos Islands': William Howarth, 'Earth Islands: Darwin and Melville in the Galápagos', *The Iowa Review*, vol. 30, no. 3 (2000), p. 98.

288 'Marine iguanas': Luis Vinueza and Judith Denkinger (eds.), *The Galápagos Marine Reserve: A Dynamic Social-Ecological System* (New York: Springer International Publishing, 2014), p. 101.

288 'The wings of flightless cormorants': John C. Kricher, *Galápagos: A Natural History* (Princeton: Princeton University Press, 2006), p. 104.

288 'Red-lipped batfish': Randy Moore, *Galápagos: An Encyclopedia of Geography, History and Culture* (Santa Barbara: ABC-CLIO, 2021), p. 186.

289 'These huge, lumbering reptiles': James Gibbs, Harrison Goldspiel, 'Growth' and 'Age and Longevity', in Gibbs, Cayot, Tapia, *Galápagos Giant Tortoises*, p. 241 and p. 251.

289 'It was after these goliaths': Moore, *Galápagos*, p. 42.

289 'to my fancy': Charles Darwin, *The Voyage of the Beagle* (New York: Random House, 2001; originally published 1835), p. 629.

289 'One early theory': Nicholls, *Lonesome George*, pp. 37–8.

289 'In 1837, Albert Günther': Albert C. L. G. Günther, 'The Gigantic Land-Tortoises (Living and Extinct) in the Collection of the British Museum' (London: British Museum, 1887).

290 'experiments in 1891 using sound waves': Nicholls, *Lonesome George*, p. 44.

## Notes and References

290   'floated to the archipelago . . . one-knot current': Nicholls, *Lonesome George*, p. 34–6.

290   'I could see the throat vibrate': Beebe, *Galápagos: World's End*, p. 228.

291   'In 1999, Adalgisa Caccone': Adalgisa Caccone, James P. Gibbs, Valerio Ketmaier et al., 'Origin and evolutionary relationships of giant Galápagos tortoises', *Proceedings of the National Academy of Sciences*, vol. 96, no. 23 (1999), pp. 13223–8.

291   'more recent genetic analyses': Adalgisa Caccone, 'Evolution and phylogenetics', in Gibbs, Cayot, Tapia, *Galápagos Giant Tortoises*, p. 122.

292   'Giant tortoise fossils': Caccone, Gibbs, Ketmaier et al., 'Origin and evolutionary relationships', pp. 13223–8).

292   'specialist Linda Cayot': Linda Cayot in 'When it Rains', *Galápagos News* (Spring–Summer 2016), p. 6.

292   'The El Niño Southern Oscillation': Michael J. McPhaden, Angus Santoso, Wenju Cai, *El Niño Southern Oscillation in a Changing Climate* (Hoboken, NJ & Washington, DC: John Wiley & Sons and the American Geophysical Union, 2021), p. 3.

292   'he floated, bounced against rocks': Linda Cayot in 'When it Rains'.

292   'fairly regularly during El Niño': Nicholls, *Lonesome George*, pp. 59–60.

293   'Some time between 2 and 3 million years ago': Adalgisa Caccone et al., 'Phylogeography and history of giant Galápagos tortoises', *Evolution*, 56:10 (Oct 2002), pp. 2052–66.

293   'carrying a clutch of fertilized eggs': B. D. Palmer, L. J. Guillette Jr., 'Histology and functional morphology of the female reproductive tract of the tortoise *Gopherus polyphemus*', *American Journal of Anatomy*, vol. 183, no. 3 (1988), pp. 200–11.

293   'Around 250,000 years ago . . . directly to Pinta Island': Caccone, 'Evolution and phylogenetics', in Gibbs, Cayot, Tapia, *Galápagos Giant Tortoises*, p. 123.

293   'usually barren-looking island': Marylee Stephenson, *The Galapagos Islands: Exploring, Enjoying & Understanding Darwin's Enchanted Islands* (Seattle: Mountaineers Books, 2005), p. 33.

294   'ecological engineers': Elizabeth A. Hunter, Stephen Blake, Linda J. Cayot, James P. Gibbs, 'Role in ecosystems', in Gibbs, Cayot, Tapia, *Galápagos Giant Tortoises*, p. 299.

294 'paths appeared, criss-crossing ... water as they grew': Stephen Blake, Patricia Isabella Tapia, Kamran Safi and Diego Ellis-Soto, 'Diet, behaviour and activity patterns', in Gibbs, Cayot, Tapia, *Galápagos Giant Tortoises*, pp. 216–17.

295 'As tortoises moved about': Diego Ellis-Soto et al., 'Plant species dispersed by Galápagos tortoises surf the wave of habitat suitability under anthropogenic climate change', *PLOS ONE* (2017); Hunter et al., 'Role in ecosystems', pp. 299–312.

295 'This detour': F. L. Rose, F. W. Judd, 'The biology and status of Berlandier's tortoise (Gopherus berlandieri)', in R. B. Bury, *North American Tortoises: Conservation and Ecology*, Fish and Wildlife Service Research Report 12 (1982), pp. 57–70.

295 'Shorter and thinner ... deep conversation with each other': Hunter et al., 'Role in ecosystems', pp. 299–312.

297 'If a clutch of eggs': Erika Kubisch and Nora R. Ibarguengoytia, 'Reproduction', in Gibbs, Cayot, Tapia, *Galápagos Giant Tortoises*, p. 160.

297 'Pinta Island soil is unusually rich': Nicholls, *Lonesome George*, p. 129.

297 'their own clumsiness': Cayot, *The Lonesome George Story*, p. 9.

298 'The clumsy hind legs': James P. Gibbs, Harrison Goldspiel, 'Population biology', Gibbs, Cayot, Tapia, *Galápagos Giant Tortoises*, p. 256.

298 'Galápagos hawks': Kubisch and Ibarguengoytia, 'Reproduction', p. 170.

298 'only weapon was': Ibid., p. 168.

298 'Above them, Galápagos hawks': Nicholls, *Lonesome George*, p. 152.

298 'once a Pinta Island tortoise': Kubisch and Ibarguengoytia, 'Reproduction', p. 170.

298 'Prehistoric giant tortoises': Jack Frazier, 'The rise of large and giant tortoises', in Gibbs, Cayot, Tapia, *Galápagos Giant Tortoises*, pp. 25–31.

299 'managed to hide': Anders G. J. Rhodin et al., 'Turtles and Tortoises of the World During the Rise and Global Spread of Humanity: First Checklist and Review of Extinct Pleistocene and Holocene Chelonians', *Turtle Extinctions Working Group* (2015), p. 17.

299 'It was completely by accident': Frazier, 'The Galapagos: Island

| | home of giant tortoises', in Gibbs, Cayot, Tapia, *Galápagos Giant Tortoises*, p. 30. |
|---|---|
| 299 | 'Berlanga took a leaf': Moore, *Galápagos: An Encyclopedia*, p. 119. |
| 299 | 'tortoises so great': Jack Frazier, 'The Galápagos: Island home of giant tortoises,' in Gibbs, Cayot, Tapia, *Galápagos Giant Tortoises*, p. 3. |
| 299 | 'In 1798': Nicholls, *Lonesome George*, p. 113. |
| 299 | 'At the time, whale oil': Eric Jay Dolin, *Leviathan: The History of Whaling in America* (New York & London: W. W. Norton & Company, Inc., 2007), pp. 12, 23, and 86. |
| 300 | 'invaluable resource for whalers': Cayot, *The Lonesome George Story*, p. 8. |
| 300 | 'Gold Rush': Frazier, 'The Galapagos: Island home of giant tortoises', p. 7. |
| 300 | 'By 1870, an estimated 200,000': Mark B. Bush et al., 'Human-induced ecological cascades: Extinction, restoration, and rewilding in the Galápagos highlands', *PNAS*, Vol. 119: 24 (2022), pp. 1–2. |
| 301 | 'Darwin also developed a taste': Darwin, *The Voyage of the Beagle*. |
| 302 | 'It's estimated that tens of thousands': Cayot, *The Lonesome George Story*, p. 8. |
| 302 | 'first to make a written account': Nicholls, *Lonesome George*, pp. 113–14. |
| 302 | 'Numerous stories': Ibid., pp. 133–5. |
| 303 | 'a very fat male': Ibid., p. 135. |
| 303 | '¡Muerte al Solitario Jorge!' . . . their prime target': Ibid., p. 91. |
| 304 | 'Fucosylated glycosaminoglycan': H. Li et al., 'Low-molecular-weight fucosylated glycosaminoglycan and its oligosaccharides from sea cucumber as novel anticoagulants: A review', *Carbohydrate Polymers*, vol. 251 (2021), p. 1; R. Ru et al., 'Cancer Cell Inhibiting Sea Cucumber (Holothuria leucospilota) Protein as a Novel Anti-Cancer Drug', *Nutrients*, vol. 14, 4 (2022), p. 786. |
| 304 | 'Traditional Chinese Medicine': R. Pangestuti, Z. Arifin, 'Medicinal and health benefit effects of functional sea cucumbers', *Journal of Traditional and Complementary Medicine*, vol. 8, no. 3 (2018), pp. 341–51. |
| 304 | 'Fishermen in the Galápagos Islands': Nicholls, *Lonesome George*, p. 93. |
| 305 | 'Most shocking of all': Linda Cayot, Ed Lewis, 'Recent Increase in |

Killing of Giant Tortoises on Isabela Island', *Noticias de Galápagos*, no. 54 (1994), pp. 2–5.

305 'bodies were hacked apart': Nicholls, *Lonesome George*, p. 95.
305 'a huge fire broke out on Isabela': Cruz Maruez, Jacinto G. Gordillo, Arnaldo Tupiza, 'The fire of 1994 and herpetofauna of southern Isabela', *Noticias de Galápagos*, no. 54 (1994), pp. 8–10.
305 'three-month quota': Nicholls, *Lonesome George*, pp. 96–7.
305 'Ominous chants': Ibid., p. 93.
306 'a similar protest': Cayot, *The Lonesome George Story*, p. 18.
306 'In 1993, Sveva Grigioni': Ibid., p. 15.
307 'the pepineros pressed in': Nicholls, *Lonesome George*, p. 97.
307 'We are prepared to take': original statement and list of demands sent by Veliz by email (1995) (accessed 2022), http://paleonet.org/archive/1995/msg01501.html.
307 'On public radio . . . the protesters had left': Nicholls, *Lonesome George*, pp. 100–1; Heidi M. Snell, 'Conservation Gets Personal', *Noticias de Galápagos*, no. 56 (1996), p. 16.
308 'multiplied to over 40,000': Cayot, *The Lonesome George Story*, pp. 10–11.
308 'For George or any of his potential offspring': Chambers, *A Sheltered Life*, p. 246.
309 '100,000 feral goats': Nicholls, *Lonesome George*, pp. 156–7.
309 'With the goats gone': Lathan, interview with Gibbs (2022).
309 'Judas goats': Daniel Stone, 'On the Galápagos: The Betrayal of Judas Goats', *National Geographic* (2014) (accessed 13/08/22), https://www.nationalgeographic.com/culture/article/on-the-Galápagos-the-betrayal-of-judas-goats; Nicholls, *Lonesome George*, pp. 156–7.
309 'In 2010, thirty-nine': Abigail Rowley, 'Restoration of Pinta Island through the Repatriation of Giant Tortoises,' *Testudo*, vol. 7, no. 4 (2012) p. 50.
309 'the same friendly, familiar face': Cayot, *The Lonesome George Story*, p. 23; Lathan, interview with Gibbs (2022).
310 '"What's wrong?" . . . "How are you?"': Alejandra Martins, 'With the death of the world's rarest creature, ranger loses his best friend, Lonesome George', *Mongabay* (2012) (13/08/22), https://news.mongabay.com/2012/06/with-the-death-of-the-worlds-rarest-creature-ranger-loses-his-best-friend-lonesome-george/.

## Notes and References 425

310 'find him sleeping . . . embraced': Cayot, *The Lonesome George Story*, pp. 15–16.
310 'When Lonesome George died': Ibid., p. 3.
310 'I feel like I've lost': Fausto Llerena in Martins, 'With the death of the world's rarest creature'.
310 'I grieved': Adalgisa Caccone in Henry Nicholls, 'How scientists hope to raise Lonesome George from the dead', *The Guardian* (2012) (accessed 13/08/22), https://www.theguardian.com/environment/blog/2012/nov/22/scientists-lonesome-george.
311 'The news of George's death': Cayot, *The Lonesome George Story*, p. 25.
311 'Today we have witnessed extinction': Ibid., p. 33.
311 'I'd love to tell you a nice story': Lathan, interview with Gibbs (2022).
312 'San Diego Zoo's Frozen Zoo': Cayot, *The Lonesome George Story*, p. 25.
312 'Gibbs would accompany the tortoise': Lathan, interview with Gibbs (2022).
313 'On 11 March 2013, the frozen remains': Lathan, interview with George Dante Jr. (2022).
313 'Taxidermy occupies': Rachel Poliquin, *The Breathless Zoo: Taxidermy and the Cultures of Longing* (Pennsylvania: Penn State Press, 2012), p. 12.
313 'public museums sprang up': Elissavet Ntoulia, 'The bird of the public museum', Wellcome collection (2017) (accessed 13/08/22), https://wellcomecollection.org/articles/W_okHhEAADUAbHiJ.
314 'Carl Akeley, an American taxidermist': AMNH staff, 'The Man Who Made Habitat Dioramas', American Museum of Natural History (2016) (accessed 13/08/22), https://www.amnh.org/explore/news-blogs/news-posts/carl-akeley-dioramas.
314 'Delia Akeley': Lauren Davis, 'The Woman Who Shot Elephants for America's Natural History Museums' (2013) (accessed 13/08/22), https://gizmodo.com/the-woman-who-shot-elephants-for-americas-natural-hist-1466170378; Delia Akeley, *Jungle Portraits* (New York: Macmillan, 1930).
314 'Dante Jr. first saw . . . first to see him': Lathan, interview with Dante Jr. (2022).
315 'A documentary': AMNH, 'Preserving Lonesome George'

(accessed 13/08/22), https://www.amnh.org/explore/preserving-lonesome-george.
315 'transported back': Lathan, interview with Dante Jr. (2022).
316 'In 2008, Gibbs and CDRS researchers . . . "save the species"': Lathan, interview with Gibbs (2022).
317 'recent genetic analysis': 'BREAKING: Expedition to Wolf Locates Tortoise with Pinta Genes', Galápagos Conservancy website (31/01/20) (accessed 13/08/22), https://www.galapagos.org/newsroom/breaking-expedition-to-wolf-locates-tortoise-with-pinta-genes/.
317 '"This story is far from over" . . . lives on': Lathan, interview with Gibbs (2022).

## 10 THE EYE OF POTOSÍ

321 'Rain rarely falls': Priyadarsi D. Roy et al., 'Subsurface fire and subsidence at Valle del Potosí (Nuevo León, Mexico): Preliminary observations', *Boletín de la Sociedad Geológica Mexicana*, vol. 66: 3 (2014), pp. 553–7.
321 'On some mornings . . . paddling in the shallows': Rodrigo Flores, 'La tierra que arde', Zócalo (03/06/17) (accessed 20/12/23) https://www.zocalo.com.mx/la-tierra-que-arde/; Fernando Seriñá Garza YouTube documentary, 'Rarísmo Incendio Subterráneo en Nuevo León, Mexico' (accessed 10/12/23), https://www.youtube.com/watch?v=w-wDE1J8uuo; Lathan, interview with Maria Elena Angeles Villeda (2023).
322 'zombie fires': Matt Simon, 'How to Kill a Zombie Fire', *Wired* (24/03/21) (accessed 10/12/23), https://www.wired.com/story/how-to-kill-a-zombie-fire/.
323 'here and only here': R. R. Miller, V. Walters, 'A new genus of cyprinodontid fish from Nuevo Leon, Mexico', *Contributions in Science*, no. 233 (1972).
323 'succumbing to extinction in March 2015': Lathan, interview with Chris Martin (2023).
323 'most recent in this book': 'Catarina Pupfish: *Megupsilon aporus*', The IUCN Red List of Threatened Species 2019 (2019) (accessed 10/12/23), https://www.iucnredlist.org/species/13013/511283.
323 'Pumpkin seed': Kevin C. Brown, *Devils Hole Pupfish: The*

*Unexpected Survival of an Endangered Species in the Modern American West* (Reno, Las Vegas: University of Nevada Press, 2021), p. 28.

323 'pupfish are no more than 5cm': Anthony A. Echelle, Alice E. Echelle, 'Cyprinodontidae: pupfishes', in Melvin L. Warren Jr, Brooks M. Burr (eds.), *Freshwater fishes of North America: Characidae to Poeciliidae, Volume 2* (Baltimore: JHU Press, 2020), p. 621.

323 'They play just like puppies': Carl L. Hubbs, 'Devil's Hole Pupfish May Be Doomed', *The Los Angeles Times* (6 February 2006), p. 15.

324 'As a group, what makes pupfish unusual': Lathan, interview with Martin (2023).

324 'Julimes pupfish': 'World of difference: hot fish, one-horned rhinos and green soya – find out the good news behind each', WWF Action Magazine (October 2011), p. 6.

324 'hotter than hot tub water': Lathan, interview with Martin (2023).

324 'Atlantic pupfish': Frank G. Nordlie, 'Osmotic regulation in the Sheephead Minnow *Cyprinodon variegatus* Lacepede', *Journal of Fish Biology*, vol. 26, no. 2 (February 1985), p. 1.

324 'So, they have the widest temperature': Lathan, interview with Martin (2023).

325 'ecological loneliness': Christopher J. Norment, *Relicts of a Beautiful Sea: Survival, Extinction, and Conservation in a Desert World* (Chapel Hill, North Carolina: University of North Carolina Press, 2003), p. 15.

325 'didn't have cephalic sensory canal pores': R. R. Miller, V. Walters, 'A new genus of cyprinodontid fish from Nuevo Leon, Mexico', *Contributions in Science*, no. 233 (1972); Echelle, Echelle, 'Cyprinodontidae: pupfishes', in Warren Jr, Burr (eds.), *Freshwater fishes of North America*, p. 621.

325 'These pores are believed': Sylvia Sáez, Roberto Jaramillo, Luis Vargas-Chacoff, 'Gross morphology of the cephalic sensory canal pores in Patagonian toothfish *Dissostichus eleginoides* Smitt, 1898 from southern Chile (Perciformes: Nototheniidae)', *Latin American Journal of Aquatic Research*, vol. 48, no. 5 (2020).

325 'It was just absolutely fascinating': Lathan, interview with Martin (2023).

325 'How pupfish dispersed': Norment, *Relicts of a Beautiful Sea*, p. 123.

325 'A pupfish egg can survive': Lathan, interview with Martin (2023).
325 '*Cyprinodon simus*': Christopher H. Martin, Peter C. Wainwright, 'On the Measurement of Ecological Novelty: Scale-Eating Pupfish Are Separated by 168 my from Other Scale-Eating Fishes', *PLOS ONE*, vol. 8: 8, p. 6.
325 '*Cyprinodon desquamator*': 'Cyprinodon desquamator: Scale-eating pupfish', Seriously Fish website (accessed 10/12/23), https://www.seriouslyfish.com/species/cyprinodon-desquamator/.
325 'Even the flow of water': Christopher H. Martin, Peter C. Wainwright, 'Trophic Novelty is Linked to Exceptional Rates of Morphological Diversification in Two Adaptive Radiations of *Cyprinodon* Pupfish', *Evolution*, vol. 65: 8 (The Society for the Study of Evolution, 2011), p. 2199.
326 '7–9 million years ago': Echelle, Echelle, 'Cyprinodontidae: pupfishes', in Warren Jr, Burr (eds.), *Freshwater fishes of North America*, p. 620.
326 'is a relict': Anthony A. Echelle et al., 'Historical Biogeography of the New-World Pupfish Genus *Cyprinodon* (Teleostei: Cyprinodontidae)', *Copeia*, vol. 2 (2005), p. 332.
326 'The species became isolated': A. Valdés González et al., 'The extinction of the Catarina pupfish Megupsilon aporus and the implications for the conservation of freshwater fish in Mexico', *Oryx*, 54:2 (2020), p. 156.
326 '*Cyprinodon*, which became fifty-five species': Echelle, Echelle, 'Cyprinodontidae: pupfishes', in Warren Jr, Burr (eds.), *Freshwater fishes of North America*, p. 611.
326 'not really changing at all': Lathan, interview with Martin (2023).
326 'During the Pluvial period': Miller, Walters, 'A new genus of cyprinodontid fish from Nuevo Leon, Mexico', p. 6.
326 'In 1948, a Mexican biologist': Salvador Contreras-Balderas, Maria de Lourdes Lozano-Vilano, 'Extinction of most Sandia and Potosí valleys (Nuevo León, Mexico) endemic pupfishes, crayfishes and snails', *Ichthyological Exploration of Freshwaters*, vol. 7: 1 (1996), pp. 33–40.
326 'two American ichthyologists': Miller, Walters, 'A new genus of cyprinodontid fish from Nuevo Leon, Mexico'.
327 'mat of hornwort': A. Valdés González et al., 'The extinction of the Catarina pupfish', pp. 155–6.
327 'circle the females': Robert K. Liu, Anthony A. Echelle,

'Behaviour of the Catarina Pupfish (Cyprinodontidae: *Megupsilon aporus*), a Severely Imperiled Species', *The Southwestern Naturalist*, Vol. 58: 1 (2013), p. 3.

327 'silt would rise up': Lathan, interview with Robert K. Liu (2023).
328 'Robert K. Liu was a student . . . S-shaped whole': Lathan, interview with Liu (2023); Liu, Echelle, 'Behaviour of the Catarina Pupfish'.
329 'S-shaped waves of desire': J. Norment, *Relicts of a Beautiful Sea*, p. 30.
329 'Bodies interlocked . . . "that distinguished them"': Lathan, interview with Liu (2023); Liu, Echelle, 'Behaviour of the Catarina Pupfish'.
330 'Not occurring in any': R. R. Miller, V. Walters, 'A new genus of cyprinodontid fish from Nuevo León, Mexico', *Contributions in Science*, no. 233 (1972).
330 '*Aporus*, from the Latin': Ibid.
331 'In the early morning of 6 July 1972 . . . "the spring had been modified"': Lathan, interview with Liu (2023).
331 'Technically, this process': Contreras-Balderas, Lozano-Vilano, 'Extinction of most Sandia and Potosí valleys', p. 36.
331 'Now, there was a pumphouse . . . there is conflict': Lathan, interview with Liu (2023).
331 'depend on the same desert waters': Norment, *Relicts of a Beautiful Sea*, p. 12.
332 'The first recorded contribution . . . swarmed the shallows': González et al., 'The extinction of the Catarina pupfish', p. 156; Contreras-Balderas, Lozano-Vilano, 'Extinction of most Sandia and Potosí valleys', pp. 33–40.
333 'the lake swelled': Lathan, interview with Angeles Villeda (2023).
333 'all but disappeared . . . gone from the lagoon by 1976': González et al., 'The extinction of the Catarina pupfish', p. 156; Contreras-Balderas, Lozano-Vilano, 'Extinction of most Sandia and Potosí valleys', pp. 33–40.
333 'alfalfa, and potatoes . . . crisps': Lathan, interview with Martin (2023).
333 'Fruit orchards': Flores, 'La tierra que arde'.
333 'was paid in water . . . dregs of habitat': Contreras-Balderas, Lozano-Vilano, 'Extinction of most Sandia and Potosí valleys', pp. 36–8.

334 'Rescued from a shrinking lagoon . . . water extraction continued': Lathan, interview with Angeles Villeda (2023).

336 'In 1992, Charles J. Yancey': Barrett L. Christie et al., 'The End of the Dallas Aquarium/Children's Aquarium At Fair Park?', *Drum and Croaker*, vol. 52 (February 2021), pp. 99–122.

336 'And by 1994': A. Valdés González et al., 'The extinction of the Catarina pupfish'.

336 'in 1998 a fire spread': Lathan, interview with Angeles Villeda (2023).

337 'sensitive to environmental changes': A. Valdés González et al., 'The extinction of the Catarina pupfish'.

337 'The females were able to breed': Lathan, interview with Angeles Villeda (2023).

337 'Over the next two decades': Lathan, interview with Martin (2023).

337 'It didn't look like anything was wrong': Lathan, interview with Angeles Villeda (2023).

338 'A slew of *Mycobacteria*': A. Valdés González et al., 'The extinction of the Catarina pupfish'.

338 'Charles Yancey had successfully': Christie et al., 'The End of the Dallas Aquarium', pp. 99–122.

338 'Nobody had had as much success': Lathan, interview with Liu (2023).

338 'The population dwindled': A. Valdés González et al., 'The extinction of the Catarina pupfish'.

338 'It was lucky that Chris Martin . . . Martin's tanks came alive': Lathan, interview with Martin (2023).

341 'S-shaped waves of desire': J. Norment, *Relicts of a Beautiful Sea*, p. 30.

341 'hundreds of eggs . . . the day that the Catarina pupfish went extinct': Lathan, interview with Martin (2023).

342 'de-extinction is a branch': Beth Shapiro, *How to Clone a Mammoth: The Science of De-Extinction* (Princeton, NJ: Princeton University Press, 2016), p. ix.

342 'Talk at the conference turned . . . "I was just like "Shit!"'': Lathan, interview with Martin (2023).

343 'Gently scooping . . . just completely gone now': Lathan, interview with Martin (2023).

| | |
|---|---|
| 344 | 'When I was ten or twelve . . . basement fish room': Lathan, interview with John Brill (2023). |
| 345 | 'Cuatro Ciénegas killifish': 'Cuatrocienegas Killifish: *Lucania interioris*', The IUCN Red List (accessed 24/01/24), https://www.iucnredlist.org/species/12395/506299. |
| 345 | 'They were wild fish': Lathan, interview with Brill (2023). |
| 345 | 'The late Al Morales': Lathan, interview with Martin (2023). |
| 346 | 'In 1992, Morales set up': Lathan, interview with Brill (2023). |
| 346 | 'There were probably ten': Lathan, interview with Martin (2023). |
| 346 | 'biannual e-newsletter': email communication with Martin (2023). |
| 347 | 'Right now, in northern Mexico': Lathan, interview with Martin (2023). |
| 347 | 'On the Caribbean island of San Salvador': Martin, Peter C. Wainwright, 'On the Measurement of Ecological Novelty'. |
| 347 | 'We know that they'll almost certainly': Lathan, interview with Martin (2023). |
| 347 | 'the endemic Maya pupfish': Julian M. Humphries, Robert Rush Miller, 'A Remarkable Species Flock of Pupfishes, Genus Cyprinodon, from Yucatán, México', *Copeia*, No. 1 (American Society of Ichthyologists and Herpetologists, 1981), p. 52. |
| 347 | 'Farmed Nile tilapia . . . I hope this makes it into your book': Lathan, interview with Martin (2023). |
| 348 | 'We're seeing the effects of climate change': Lathan, interview with Ana Laura Lara Rivera (2023). |
| 348 | 'By the time we realize': Lathan, interview with Angeles Villeda (2023). |
| 349 | 'Crop or cattle producers': Lathan, interview with Lara Rivera (2023). |
| 349 | 'We have a responsibility': Lathan, interview with Angeles Villeda (2023). |
| 349 | 'They love it . . . But what about the fish?': Lathan, interview with Lara Rivera (2023). |

## About the Author

Tom Lathan is an author and journalist living in Kent. His writing has appeared in publications including *The Guardian*, the *Financial Times*, and *The Times Literary Supplement*. He volunteers with Kent Wildlife Trust, working as part of their initiative to introduce wild bison to the UK at West Blean and Thornden Woods Nature Reserve. *Lost Wonders* is Lathan's first book and was the recipient of a Society of Authors award.

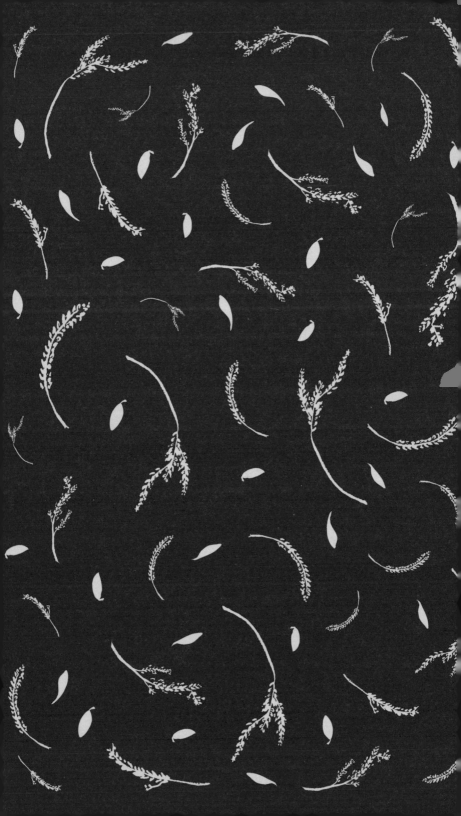